SCHAUM'S OUTLINE OF

Theory and Problems of

GEOMETRY

includes plane, analytic, and transformational geometries

Third Edition

BARNETT RICH, Ph.D.

Former Chairman, Department of Mathematics
Brooklyn Technical High School, New York City

REVISED BY

PHILIP A. SCHMIDT, Ph.D.

Associate Provost for Academic Services
and Professor of Mathematics
Berea College

Schaum's Outline Series

McGRAW-HILL

New York   San Francisco   Washington, D.C.   Auckland   Bogotá   Caracas   Lisbon
London   Madrid   Mexico City   Milan   Montreal   New Delhi
San Juan   Singapore   Sydney   Tokyo   Toronto

BARNETT RICH held a doctor of philosophy degree (Ph.D.) from Columbia University and a doctor of jurisprudence (J.D.) from New York University. He began his professional career at Townsend Harris Hall High School of New York City and was one of the prominent organizers of the High School of Music and Art where he served as the Administrative Assistant. Later he taught at CUNY and Columbia University and held the post of Chairman of Mathematics at Brooklyn Technical High School for 14 years. Among his many achievements are the 6 degrees that he earned and the 23 books that he wrote, among them Schaum's Outlines of *Elementary Algebra*, *Modern Elementary Algebra*, and *Review of Elementary Algebra*.

PHILIP A. SCHMIDT has a B.S. from Brooklyn College (with a major in mathematics), an M.A. in mathematics and a Ph.D. in mathematics education from Syracuse University. He was an Associate Professor at Berea College until 1985, served as Professor of Secondary Education and Dean of the School of Education at SUNY at New Paltz, and currently is Associate Provost for Academic Services and Professor of Mathematics at Berea College. He is the author of numerous articles and books.

Sponsoring Editor: Barbara Gilson
Production Supervisor: Tina Cameron
Editing Liaison: Maureen B. Walker
Project Supervision: Keyword Publishing Services Ltd.

Library of Congress Cataloging-in-Publication Data

Rich, Barnett, 1906–
    Schaum's outline of theory and problems of geometry: includes plane, analytic, and transformational geometries / Barnett Rich; revised by Philip A. Schmidt.—
3rd ed.
    p. cm.—(Schaum's outline series)
Includes index.
ISBN 0-07-052766-0
1. Geometry. I. Title: Theory and problems of geometry. II. Schmidt, Philip A. III. Title. IV. Series.
QA445.R53 1999
516—dc21   99-054044

McGraw-Hill
A Division of The McGraw-Hill Companies

PREFACE TO THE THIRD EDITION

Barnett Rich's original *Plane Geometry* has been reprinted some twenty-five times since it was first published in 1968. The challenge in revising such a text is to update the material as necessary while retaining the prose and pedagogy that are responsible for its success. In the case of *Plane Geometry*, the challenge was particularly great. Dr. Rich's command of geometry and its pedagogy was enormous. Conversations I have had with past students and colleagues of Dr. Rich substantiate that his ability to convey ideas in mathematics was unsurpassable.

In this revision I have attempted to maintain Barnett Rich's "spirit of explanation" while making the text suitable to the geometry that is currently being taught in schools and colleges. Terminology and notation have been changed to match the current texts and curriculum. I have continued to use the more common "congruent segments" and "measure of the angle" phrasing and have made textual changes to support that terminology. Outdated material has been deleted, and the supplementary problems have been modified. In addition, the calculator has been introduced in various topics, and an appendix devoted to the HP 38G graphing calculator has been added.

I owe thanks to several people for their assistance during this revision: Mrs. Barnett Rich, who has helped with support and friendship; my wife Jan Z. Schmidt and my son Reed Schmidt, who have been loving supporters in all my work; and finally, Dr. Barnett Rich, for providing me with such a rich text to revise, and for teaching geometry so meaningfully to so many people.

PHILIP A. SCHMIDT
Berea, Kentucky

PREFACE TO THE FIRST EDITION

The central purpose of this book is to provide maximum help for the student and maximum service for the teacher.

PROVIDING HELP FOR THE STUDENT:

This book has been designed to improve the learning of geometry far beyond that of the typical and traditional book in the subject. Students will find this text useful for these reasons:

(1) *Learning Each Rule, Formula, and Principle*

Each rule, formula, and principle is stated in simple language, is made to stand out in distinctive type, is kept together with those related to it, and is clearly illustrated by examples.

(2) *Learning Each Set of Solved Problems*

Each set of solved problems is used to clarify and apply the more important rules and principles. The character of each set is indicated by a title.

(3) *Learning Each Set of Supplementary Problems*

Each set of supplementary problems provides further application of rules and principles. A guide number for each set refers a student to the set of related solved problems. There are more than 2000 additional related supplementary problems. Answers for the supplementary problems have been placed in the back of the book.

(4) *Integrating the Learning of Plane Geometry*

The book integrates plane geometry with arithmetic, algebra, numerical trigonometry, analytic geometry, and simple logic. To carry out this integration:

(*a*) A separate chapter is devoted to analytic geometry.

(*b*) A separate chapter includes the complete proofs of the most important theorems together with the plan for each.

(*c*) A separate chapter fully explains 23 basic geometric constructions. Underlying geometric principles are provided for the constructions, as needed.

(*d*) Two separate chapters on methods of proof and improvement of reasoning present the simple and basic ideas of formal logic suitable for students at this stage.

(*e*) Throughout the book, algebra is emphasized as the major means of solving geometric problems through algebraic symbolism, algebraic equations, and algebraic proof.

(5) *Learning Geometry Through Self-study*

The method of presentation in the book makes it ideal as a means of self-study. For the able student, this book will enable him to accomplish the work of the standard course of study in much less time. For the less able, the presentation of numerous illustrations and solutions provides the help needed to remedy weaknesses and overcome difficulties and in this way keep up with the class and at the same time gain a measure of confidence and security.

(6) *Extending Plane Geometry into Solid Geometry*

A separate chapter is devoted to the extension of two-dimensional plane geometry into three-dimensional solid geometry. It is especially important in this day and age that the student understand how the basic ideas of space are outgrowths of principles learned in plane geometry.

PROVIDING SERVICE FOR THE TEACHER:

Teachers of geometry will find this text useful for these reasons:

(1) *Teaching Each Chapter*

Each chapter has a central unifying theme. Each chapter is divided into two to ten major subdivisions which support its central theme. In turn, these chapter subdivisions are arranged in graded sequence for greater teaching effectiveness.

(2) *Teaching Each Chapter Subdivision*

Each of the chapter subdivisions contains the problems and materials needed for a complete lesson developing the related principles.

(3) *Making Teaching More Effective Through Solved Problems*

Through proper use of the solved problems, students gain greater understanding of the way in which principles are applied in varied situations. By solving problems, mathematics is learned as it should be learned—by doing mathematics. To ensure effective learning, solutions should be reproduced on paper. Students should seek the why as well as the how of each step. Once a student sees how a principle is applied to a solved problem, he is then ready to extend the principle to a related supplementary problem. Geometry is not learned through the reading of a textbook and the memorizing of a set of formulas. Until an adequate variety of suitable problems has been solved, a student will gain little more than a vague impression of plane geometry.

(4) *Making Teaching More Effective Through Problem Assignment*

The preparation of homework assignments and class assignments of problems is facilitated because the supplementary problems in this book are related to the sets of solved problems. Greatest attention should be given to the underlying principle and the major steps in the solution of the solved problems. After this, the student can reproduce the solved problems and then proceed to do those supplementary problems which are related to the solved ones.

OTHERS WHO WILL FIND THIS TEXT ADVANTAGEOUS:

This book can be used profitably by others besides students and teachers. In this group we include: (1) the parents of geometry students who wish to help their children through the use of the book's self-study materials, or who may wish to refresh their own memory of geometry in order to properly help their children; (2) the supervisor who wishes to provide enrichment materials in geometry, or who seeks to improve teaching effectiveness in geometry; (3) the person who seeks to review geometry or to learn it through independent self-study.

BARNETT RICH
Brooklyn Technical High School
April, 1963

CONTENTS

CHAPTER 1   **Fundamentals of Algebra: Laws and Operations**   **1**

1.1 Relating Fundamentals of Arithmetic and Algebra
1.2 Commutative Law of Addition   **1.3** Commutative Law of
Multiplication   **1.4** Symbolizing the Fundamental Operations in
Algebra: Division by Zero   **1.5** Expressing Addition and
Subtraction Algebraically   **1.6** Expressing Multiplication and
Division Algebraically   **1.7** Expressing Algebraically Statements
Involving Two or More Operations   **1.8** Associative Laws of
Addition and Multiplication   **1.9** Order in Which Fundamental
Operations Are Performed   **1.10** Terms, Factors, and
Coefficients   **1.11** Repeated Multiplying of a Factor: Base,
Exponent, and Power

CHAPTER 2   **Fundamentals of Algebra: Equations and Formulas**   **20**

2.1 Variables and Equations   **2.2** Translating Verbal Problems
into Equations   **2.3** Solving Simple Equations Using Inverse
Operations   **2.4** Rules for Solving Equations   **2.5** Using
Division to Solve an Equation   **2.6** Using Multiplication to
Solve an Equation   **2.7** Using Subtraction to Solve an
Equation   **2.8** Using Addition to Solve an Equation   **2.9** Using
Two or More Operations to Solve an Equation   **2.10** Deriving
Formulas   **2.11** Transforming Formulas   **2.12** Finding the Value
of a Variable in a Formula   **2.13** Using a Calculator to Perform
the Fundamental Operations of Arithmetic   **2.14** Graphs
2.15 Midpoint of a Segment   **2.16** Distance Between Two Points
2.17 Slope of a Line

CHAPTER 3   **Lines, Angles, and Triangles**   **64**

3.1 Historical Background of Geometry   **3.2** Undefined Terms
of Geometry: Point, Line, and Plane   **3.3** Line Segments
3.4 Circles   **3.5** Angles   **3.6** Triangles   **3.7** Pairs of Angles

CHAPTER 4   **Methods of Proof**   **83**

4.1 Proof by Deductive Reasoning   **4.2** Postulates
(Assumptions)   **4.3** Basic Angle Theorems   **4.4** Determining
the Hypothesis and Conclusion   **4.5** Proving a Theorem

CHAPTER 5   **Congruent Triangles**   **100**

5.1 Congruent Triangles   **5.2** Isosceles and Equilateral Triangles

CHAPTER 6 **Parallel Lines, Distances, and Angle Sums** **114**
6.1 Parallel Lines **6.2** Distances **6.3** Sum of the Measures of
the Angles of a Triangle **6.4** Sum of the Measures of the
Angles of a Polygon **6.5** Two New Congruency Theorems

CHAPTER 7 **Parallelograms, Trapezoids, Medians, and Midpoints** **143**
7.1 Trapezoids **7.2** Parallelograms **7.3** Special Parallelograms:
Rectangle, Rhombus, Square **7.4** Three or More Parallels;
Medians and Midpoints

CHAPTER 8 **Circles** **161**
8.1 The Circle; Circle Relationships **8.2** Tangents
8.3 Measurement of Angles and Arcs in a Circle

CHAPTER 9 **Similarity** **189**
9.1 Ratios **9.2** Proportions **9.3** Proportional Segments
9.4 Similar Triangles **9.5** Extending a Basic Proportion
Principle **9.6** Proving Equal Products of Lengths of
Segments **9.7** Segments Intersecting Inside and Outside a
Circle **9.8** Mean Proportionals in a Right Triangle
9.9 Pythagorean Theorem **9.10** Special Right Triangles

CHAPTER 10 **Areas** **225**
10.1 Area of a Rectangle and of a Square **10.2** Area of a
Parallelogram **10.3** Area of a Triangle **10.4** Area of a
Trapezoid **10.5** Area of a Rhombus **10.6** Polygons of the
Same Size or Shape **10.7** Comparing Areas of Similar
Polygons **10.8** Areas in Analytic Geometry

CHAPTER 11 **Regular Polygons and the Circle** **242**
11.1 Regular Polygons **11.2** Relationships of Segments in
Regular Polygons of 3, 4, and 6 Sides **11.3** Area of a Regular
Polygon **11.4** Ratios of Segments and Areas of Regular
Polygons **11.5** Circumference and Area of a Circle
11.6 Length of an Arc; Area of a Sector and a Segment
11.7 Areas of Combination Figures

CHAPTER 12 **Locus** **259**
12.1 Determining a Locus **12.2** Locating Points by Means of
Intersecting Loci **12.3** Proving a Locus **12.4** Locus in
Analytic Geometry

CHAPTER 13   **Inequalities and Indirect Reasoning**   272
13.1 Inequalities   **13.2** Indirect Reasoning

CHAPTER 14   **Improvement of Reasoning**   282
14.1 Definitions   **14.2** Deductive Reasoning in Geometry
14.3 Converse, Inverse, and Contrapositive of a Statement
14.4 Partial Converse and Partial Inverse of a Theorem
14.5 Necessary and Sufficient Conditions

CHAPTER 15   **Constructions**   291
15.1 Introduction   **15.2** Duplicating Segments and Angles
15.3 Constructing Bisectors and Perpendiculars
15.4 Constructing a Triangle   **15.5** Constructing Parallel
Lines   **15.6** Circle Constructions   **15.7** Inscribing and
Circumscribing Regular Polygons   **15.8** Constructing Similar
Triangles

CHAPTER 16   **Proofs of Important Theorems**   306
16.1 Introduction   **16.2** The Proofs

CHAPTER 17   **Transformational Geometry**   318
17.1 Introduction to Transformations   **17.2** Reflections
17.3 Reflections and Analytic Geometry   **17.4** Translations
17.5 Rotations   **17.6** Dilations   **17.7** Properties of
Transformations

Formulas for Reference   335

Appendix   **Introduction to the Graphing Calculator**   338

Answers to Supplementary Problems   342

Index   363

CHAPTER 1

Fundamentals of Algebra: Laws and Operations

1.1 RELATING FUNDAMENTALS OF ARITHMETIC AND ALGEBRA

In this first chapter, we are going to lead you from arithmetic to algebra. Underlying algebra as well as arithmetic are the fundamental ideas that you will need in geometry. As we recall each of the fundamental ideas, we will develop each of them in greater depth and scope.

1.1A Variables and Constants

A *variable* is a letter or other symbol which holds a place for, or represents, any number in a specified set of numbers. The set whose numbers may replace a variable is called the *replacement set* or the *domain* of the variable.

Thus, if x represents, 1, 10, or any other whole number, then x is a variable and the set of whole numbers is the replacement set or domain of x. Numbers such as 1 and 10 that may be replaced are the *values* of the variable.

A *constant* is a letter or other symbol that represents only one number. For example, 5 and π are constants. If a variable represents a set having a single member, such as {5}, then the variable may be regarded as a constant.

A *variable expression* is an expression that contains a variable. Thus, if n is a variable, then

$$7n, \quad n+7, \quad \frac{1}{2}n-5, \quad \frac{3n+1}{5n+1}$$

are variable expressions.

To *evaluate* a variable expression is to find its value for given values of the variable. Thus, using the replacement set {1, 2, 5, 10} as the set of given values of x, the set of values of the expression $2x + 1$ is {3, 5, 11, 21}.

A *formula* is an equation in which a variable is expressed in terms of other variables. Thus, $p = 4s$ is a formula in which the variable p is expressed in terms of the variable s.

1

1.1B Replacing Verbal Statements with Algebraic Equations

In the following examples, note how variable expressions may be used to replace lengthy verbal statements:

VERBAL STATEMENTS	ALGEBRAIC EQUATIONS
1. Seven times a number reduced by the same number equals six times the number.	1. $7n - n = 6n$
2. The sum of twice a number and three times the same number equals five times that number.	2. $2n + 3n = 5n$
3. The perimeter of a square equals four times the length of one of its sides.	3. $p = 4s$

Omitting the multiplication sign, as in $7n$, is the preferred method of indicating multiplication. However, the multiplication sign may not be omitted when showing the multiplication of two numbers.

1.1C Properties of 0 and 1: Additive Identity and Multiplicative Identity

If the domain of the variable n is the set of integers, then each of the following important properties of 0 and 1 is true for each integer. In later work, these properties will be extended to apply to any "real number."

Properties of 0 and 1

VERBAL STATEMENTS	ALGEBRAIC EQUATIONS
Multiplicative Property of Zero	
1. If any number is multiplied by 0, the product is 0.	1. $n \cdot 0 = 0$
Additive Property of Zero	
2. If 0 is added to any number, the sum is the number. Because identically the same number remains when 0 is added to it, 0 is called the *additive identity*.	2. $n + 0 = n$
Multiplicative Property of One	
3. If any number is multiplied by 1, the product is the number. Because identically the same number remains when it is multiplied by 1, 1 is called the *multiplicative identity*.	3. $n \cdot 1 = n$

SOLVED PROBLEMS

1.1 REPLACING A VERBAL STATEMENT BY AN ALGEBRAIC EQUATION

Using variable expressions, replace each verbal statement by an algebraic equation: (a) If six times a number is reduced by the same number, the result must be five times the number. (b) The sum of twice a number, three times the same number and four times the same number is equivalent to nine times the number. (c) Increasing a number by itself and 20 is the same as doubling the number and adding 20.

Illustrative Solution

(c) Increasing a number by itself and 20   is the same as   doubling the number   and adding 20.

$$n + n + 20 \qquad = \qquad 2n \qquad +20$$

Ans. $n + n + 20 = 2n + 20$

Ans. (a) $6n - n = 5n$ (b) $2n + 3n + 4n = 9n$

1.2   REPLACING A VERBAL RULE BY A FORMULA

Using the initial letters of words as variables, replace each verbal rule by a formula: (a) The perimeter of a square is four times the length of a side. (b) The perimeter of a rectangle is twice the length added to twice the width. (c) The area of a rectangle is the product of its length and width. (d) The selling price of an article is the sum of its cost and its profit.

Illustrative Solution

(a)   Using p for the perimeter of a square and s for the length of a side, the formula is $p = 4s$.

Ans.   (b) $p = 2l + 2w$ (c) $a = lw$ (d) $s = c + p$

1.3   EVALUATING VARIABLE EXPRESSIONS USING A REPLACEMENT SET

Using the replacement set $\{1, 2, 5, -10\}$ as the set of values of n, evaluate the following expressions and list these values in a set:

(a) $7n$ (b) $n + 7$ (c) $5 - \dfrac{1}{2}n$ (d) $\dfrac{3n + 1}{5n - 1}$

Illustrative Solution

(d)   If $n = 1$, $3n + 1 = 3 \cdot 1 + 1 = 4$ and $5n - 1 = 5 \cdot 1 - 1 = 4$. Hence,

$$\frac{3n + 1}{5n - 1} = \frac{4}{4} = 1$$

If $n = 2, 5$, and -10, then

$$\frac{3n + 1}{5n - 1} = \frac{7}{9} \quad \frac{3n + 1}{5n - 1} = \frac{16}{24} = \frac{2}{3} \quad \frac{3n + 1}{5n - 1} = \frac{29}{51}$$

respectively. List these values in a set: $\{1, 7/9, 2/3, 31/49\}$ *Ans.*

Ans.   (a) $\{7, 14, 35, 70\}$ (b) $\{8, 9, 12, 17\}$ (c) $\{4\frac{1}{2}, 4, 2\frac{1}{2}, 0\}$

1.2   COMMUTATIVE LAW OF ADDITION

Addends are numbers being added. Their *sum* is the answer obtained. Thus, in $5a + 3a = 8a$, the addends are $5a$ and $3a$. Their sum is $8a$.

Numerical addends are numbers used as addends. Thus, in $3 + 4 + 6 = 13$, 3, 4, and 6 are numerical addends.

Literal addends are variables which represent numbers being added. Thus, in $a + b = 8$, a and b are literal addends.

Commutative Law of Addition

Interchanging addends does not change their sum.

$$a + b = b + a$$
$$a + b + c = b + c + a$$

Thus, $2 + 3 = 3 + 2$ and $3 + 4 + 6 = 4 + 6 + 3$.

An important application of the Commutative Law of Addition in algebra is rearranging addends in a preferred order. Thus, $b + c + a$ becomes $a + b + c$ if the literal addends are to be arranged alphabetically. Also, $3 + x$ becomes $x + 3$ if the literal addend is to precede the numerical addend.

SOLVED PROBLEMS

1.4   REARRANGING ADDENDS TO OBTAIN A PREFERRED ORDER
    Rearrange the addends so that literal addends are arranged alphabetically and precede numerical addends:

(a) $3 + b$ (c) $d + 10 + e$ (e) $15 + x + 10$ (g) $w + y + x$

(b) $c + a$ (d) $c + 12 + b$ (f) $20 + s + r$ (h) $b + 8 + c + a$

Illustrative Solution

(d) Arrange the literal addends alphabetically; thus, $b + c$. Since the literal addends precede the numerical addend 12, the required result is $b + c + 12$.

Ans.   (a) $b + 3$ (c) $d + e + 10$ (f) $r + s + 20$ (h) $a + b + c + 8$

        (b) $a + c$ (e) $x + 25$ (g) $w + x + y$

1.3   COMMUTATIVE LAW OF MULTIPLICATION

    Factors are numbers being multiplied. Their *product* is the answer obtained. Thus, in $5 \times 3 = 15$, the factors are 5 and 3. Their product is 15.
    Numerical factors are numbers used as factors. Thus, in $2 \times 3 \times 5 = 30$, 2, 3, and 5 are numerical factors.
    Literal factors are variables which represent numbers being multiplied. Thus, in $ab = 20$, a and b are literal factors.

Commutative Law of Multiplication

Interchanging factors does not change their product.

$$ab = ba$$
$$cba = abc$$

Thus, $3a \times 5 = 5 \times 2a = 10a$.
    An important application of the Commutative Law of Multiplication in algebra is rearranging factors in a preferred order. Thus, bca becomes abc if the literal factors are arranged alphabetically. Also, $x3$ becomes $3x$ if the numerical factor is to precede the literal factor.

SOLVED PROBLEMS

1.5   REARRANGING FACTORS TO OBTAIN A PREFERRED ORDER
    Rearrange the factors so that literal factors are arranged alphabetically and follow numerical factors:

(a) $b3$ (c) $d10e$ (e) $-15x10$ (g) wyx

(b) ca (d) $c12b$ (f) $20sr$ (h) $b35ca$

Illustrative Solution

(e) The numerical factors -15 and 10 should precede the literal factor x. Hence, the result is $-15 \cdot 10 \cdot x$ or $-150x$.

Ans.   (a) $3b$ (b) ac (c) $10de$ (d) $12bc$ (f) $20rs$ (g) wxy (h) $35abc$

1.4   SYMBOLIZING THE FUNDAMENTAL OPERATIONS IN ALGEBRA: DIVISION BY ZERO

The symbols for the fundamental operations are as follows:

1. ADDITION: +   3. MULTIPLICATION: ×, ( ), ·, no sign
2. SUBTRACTION: −   4. DIVISION: ÷; :, fraction bar

Thus,

$n + 4$ means "add n and 4" $4 \times n, 4(n), 4 \cdot n$ mean "multiply n and 4"

$n - 4$ means "subtract 4 from n" $n \div 4, n{:}4, \dfrac{n}{4} n/4$ mean "divide n by 4"

Rule: Division by zero is an impossible operation.

Thus, $4 \div 0$ or $x \div 0$ is meaningless. Also, $4/n$ is meaningless if $n = 0$.

SOLVED PROBLEMS

1.6   **SYMBOLS FOR MULTIPLICATION AND DIVISION**
Symbolize each, avoiding multiplication signs where possible:

(a) 8 times -11 (c) b times c (e) 5 multiplied by a and the result divided by b
(b) 8 times x (d) 8 divided by x (f) d divided by the product of 7 and e

Ans.   (a) $8 \times (-11), 8 \cdot (-11), 8(-11)$ or $(8)(-11)$ (c) $b \cdot c$ or bc (avoid $b \times c$)   (e) $\dfrac{5a}{b}$ or $5a \div b$

(b) $8 \cdot x$ or $8x$ (avoid $8 \times x$)   (d) $\dfrac{8}{x}$ or $8 \div x \left(\dfrac{8}{x} \text{ is preferred}\right)$ (f) $\dfrac{d}{7e}$ or $d \div (7e)$

1.7   **DIVISION BY ZERO: FRACTIONS WHOSE DENOMINATOR HAS A SINGLE VARIABLE**
When is each division impossible? Give a reason for your answer.

(a) $\dfrac{5}{a}$ (b) $\dfrac{7}{2b}$ (c) $\dfrac{2}{c-6}$ (d) $\dfrac{10}{3d-9}$ (e) $\dfrac{3x}{35-7e}$

Illustrative Solution

(c) Division is impossible if $c - 6$, the denominator, equals 0. If $c - 6 = 0$, $c = 6$.

Ans. (a) $a = 0$ (b) $b = 0$ (d) $d = 3$ (e) $e = 5$

1.8   **DIVISION BY ZERO: FRACTIONS WHOSE DENOMINATOR HAS MORE THAN ONE VARIABLE**
When is each division impossible? Give a reason for your answer.

(a) $\dfrac{a}{xy}$ (b) $\dfrac{3x}{2ab}$ (c) $\dfrac{10}{x-y}$ (d) $\dfrac{x}{y-2z}$

Illustrative Solution

(b) Division is impossible if $2ab$, the denominator, equals 0. If $2ab = 0$, $a = 0$ or $b = 0$.

Ans. (a) $x = 0$ or $y = 0$ (c) $x = y$ (d) $y = 2z$

1.5 EXPRESSING ADDITION AND SUBTRACTION ALGEBRAICALLY

In algebra, changing verbal expressions into algebraic expressions is of major importance. The operations of addition and subtraction are denoted by words such as the following:

WORDS DENOTING ADDITION		WORDS DENOTING SUBTRACTION	
sum	more than	difference	less than
plus	greater than	minus	smaller than
gain	larger than	lose	fewer than
increase	enlarge	decrease	shorten
rise	grow	drop	depreciate
expand	augment	lower	diminish

The Commutative Law applies to addition but does *not* apply to subtraction. Thus, "the sum of n and 20" may be represented by either $n + 20$ or $20 + n$. But "20 minus a number" may be represented only by $20 - n$ and *not* by $n - 20$.

SOLVED PROBLEMS

1.9 EXPRESSING ADDITION ALGEBRAICALLY

If n represents a number, express algebraically: (a) the sum of the number and 7, (b) the number plus 8, (c) the number increased by 9, (d) 15 plus the number, (e) 20 enlarged by the number, (f) 25 augmented by the number.

Illustrative Solution

(e) Use + for "enlarged" to obtain $20 + n$ or, by interchanging addends, $n + 20$.

Ans. (a) $n + 7$ or $7 + n$ (c) $n + 9$ or $9 + n$ (f) $25 + n$ or $n + 25$

       (b) $n + 8$ or $8 + n$ (d) $15 + n$ or $n + 15$

1.10 EXPRESSING SUBTRACTION ALGEBRAICALLY

If n represents a number, express algebraically: (a) the difference if the number is subtracted from 15, (b) the number diminished by 20, (c) 25 less than the number, (d) 25 less the number, (e) the difference if 15 is subtracted from the number, (f) 50 subtracted from the number, (g) the number subtracted from 50, (h) the number reduced by 75.

Illustrative Solution

(a) Express algebraically as $15 - n$. (Do not use $n - 15$.)

Ans. (b) $n - 20$ (d) $25 - n$ (f) $n - 50$ (h) $n - 75$

       (c) $n - 25$ (e) $n - 15$ (g) $50 - n$

1.11 CHANGING VERBAL EXPRESSIONS INTO ALGEBRAIC EXPRESSIONS

Express algebraically: (a) the no. of km of a weight that is 10 km heavier than w km, (b) the no. of mi in a distance that is 40 mi farther than d mi, (c) the no. of degrees in a temperature 50° hotter than t degrees, (d) the no. of dollars in a price $60 cheaper than p dollars, (e) the no. of mph in a speed 30 mph faster than r mph, (f) the no. of ft in a length of l ft expanded 6 ft, (g) the no. of oz in a weight that is 10 oz lighter than w oz.

Illustrative Solution

(*d*) Since "cheaper" means fewer dollars, use $-$ to express it.   *Ans. p $-$ 60*

(*f*) Since "expanded" means greater length, use $+$ to express it.   *Ans. l $+$ 6*

Ans.   (*a*) $w + 10$ (*b*) $d + 40$ (*c*) $t + 50$ (*e*) $r + 30$ (*g*) $w - 10$

1.6   EXPRESSING MULTIPLICATION AND DIVISION ALGEBRAICALLY

WORDS DENOTING MULTIPLICATION		WORDS DENOTING DIVISION	
multiplied by	double	divided by	ratio
times	triple or treble	quotient	half
product	quadruple		
twice	quintuple		

The Commutative Law applies to multiplication but does *not* apply to division. Thus, "the product of *n* and 10" may be represented by either $n10$ or $10n$ (the latter is preferred). But "a number divided by 20" may be represented only by $n/20$ and *not* by $20/n$.

SOLVED PROBLEMS

1.12   EXPRESSING MULTIPLICATION OR DIVISION
State verbal expressions that may be represented by each of the following:

(*a*) $-5x$ (*b*) $\dfrac{y}{5}$ (*c*) $\dfrac{5w}{7}$

Ans.   (*a*) (1) -5 multiplied by x (*b*) (1) y divided by 5   (*c*) (1) five-sevenths of w

(2) x multiplied by -5 (2) quotient of y and 5   (2) $5w$ divided by 7

(3) -5 times x (3) ratio of y to 5   (3) quotient of $5w$ divided by 7

(4) product of -5 and x (4) one-fifth of y (4) ratio of $5w$ to 7

1.13   EXPRESSING DIVISION ALGEBRAICALLY
If *n* represents a number, express algebraically in the form of a fraction: (*a*) the quotient of the number and 10, (*b*) the ratio of 10 to the number, (*c*) twice the number divided by 7, (*d*) 20 divided by the product of the number and 3.

Illustrative Solution

(*b*) "Ratio" denotes division. Hence, express the ratio of 10 to the number in fraction form as $10/n$. (Do not use $n/10$.)

Ans.   (*a*) $\dfrac{n}{10}$ (*c*) $\dfrac{2n}{7}$ (*d*) $\dfrac{20}{3n}$

1.7 EXPRESSING ALGEBRAICALLY STATEMENTS INVOLVING TWO OR MORE OPERATIONS

In addition to their use in signed numbers, parentheses ( ) are used to indicate that an expression is to be treated as a single number. Thus, "double the sum of 4 and x" is written as $2(4 + x)$.

SOLVED PROBLEMS

1.14 EXPRESSING ALGEBRAICALLY EXPRESSIONS INVOLVING TWO OPERATIONS

Express algebraically: (a) a increased by twice b, (b) twice the sum of a and b, (c) 30 decreased by three times c, (d) three times the difference of 30 and c, (e) 50 minus the product of 10 and p, (f) the product of 50 and the sum of p and 10, (g) 100 increased by the quotient of x and y, (h) the quotient of x and the sum of y and 100, (i) the average of s and 20.

Illustrative Solutions

(b) Express the sum of a and b as $(a + b)$, using parentheses. The correct answer is $2(a + b)$, not $2a + b$.

(i) The average of s and 20 is one-half their sum.   Ans. $\frac{1}{2}(s + 20)$ or $\frac{s + 20}{2}$

Ans. (a) $a + 2b$ (d) $3(30 - c)$ (f) $50(p + 10)$ (h) $\dfrac{x}{7 + 100}$

      (c) $30 - 3c$ (e) $50 - 10p$ (g) $100 + \dfrac{x}{y}$

1.15 EXPRESSING ALGEBRAICALLY MORE DIFFICULT EXPRESSIONS

Express algebraically: (a) half of a, increased by the product of 25 and b; (b) four times c, decreased by one-fifth of d; (c) half the sum of m and twice n; (d) the average of m, r, and 80; (e) 60 diminished by one-third the product of 7 and x; (f) twice the sum of e and 30, diminished by 40; (g) two-thirds the sum of n and three-sevenths of p; (h) the product of a and b, decreased by twice the difference of c and d; (i) the quotient of x and 10, minus four times their sum.

Illustrative Solutions

(d) The average of three numbers is one-third of their sum.   Ans. $(m + r + 80)/3$

(h) Express the difference of c and d as $(c - d)$, using parentheses. However, the product of a and b may be expressed as ab without parentheses.   Ans. $ab - 2(c - d)$

Ans. (a) $\dfrac{2}{2} + 25b$ (c) $\dfrac{m + 2n}{2}$ or $\dfrac{1}{2}(m + 2n)$ (f) $2(e + 30) - 40$ (i) $\dfrac{x}{10} - 4(x + 10)$

      (b) $4c - \dfrac{d}{5}$ (e) $60 - \dfrac{7x}{3}$ (g) $\dfrac{2}{3}\left(n + \dfrac{3p}{7}\right)$

1.16 CHANGING VERBAL EXPRESSIONS INTO ALGEBRAIC EXPRESSIONS

Express algebraically: (a) a speed in mph that is 30 mph faster than twice another of r mph, (b) a weight in lb that is 20 lb lighter than 3 times another of w lb, (c) a temperature in degrees that is 15° colder than two-thirds another of $t°$, (d) a price in cents that is 25¢ cheaper than another of D dollars, (e) a length in inches that in 8 in. longer than another of f ft.

Illustrative Solutions

(*d*)   *D* dollars equals 100*D* cents. "25¢ cheaper" means 25¢ less.   *Ans.* $100D - 25$

(*e*)   *f* feet equals 12*f* inches. "8 in. longer" means 8 inches more.   *Ans.* $12f + 8$

Ans.   (*a*) $2r + 30$ (*b*) $3w - 20$ (*c*) $\dfrac{2t}{3} - 15$

1.8   ASSOCIATIVE LAWS OF ADDITION AND MULTIPLICATION

Associative Law of Addition
The way in which quantities are added in groups of two does not change their sum.
$$a + b + c = (a + b) + c = a + (b + c)$$

Thus, the sum $2x + 3x + 5x$ may be found by obtaining *partial sums* in two ways:

(1)   Add $2x$ and $3x$ to obtain a partial sum of $5x$, then add $5x$ and $5x$:

$$2x + 3x + 5x = (2x + 3x) + 5x = 5x + 5x = 10x.$$

(2)   Add $3x$ and $5x$ to obtain a partial sum of $8x$, then add $2x$ and $8x$.

Associative Law of Multiplication
The way in which quantities are multiplied in groups of two does not change their product.
$$abc = (ab)c = a(bc)$$

Thus, the product $2 \cdot 3 \cdot 5q$ may be found by obtaining *partial products* in two ways:

(1)   Multiply 2 and 3 to obtain a partial product of 6, then multiply 6 and $5q$:

$$2 \cdot 3 \cdot 5q = (2 \cdot 3) \cdot 5q = 6 \cdot 5q = 30q$$

(2)   Multiply 3 and 5 to obtain a partial product of 15, then multiply 2 and 15:

$$2 \cdot 3 \cdot 5 = 2 \cdot (3 \cdot 5) = 2 \cdot 15 = 30$$

Hence, $2 \cdot 3 \cdot 5 = (2 \cdot 3) \cdot 5 = 2 \cdot (3 \cdot 5)$.

A sum may be simplified by using both the Commutative and Associative Laws of Addition:

$$25 + (467 + 175) = 25 + (175 + 467) \text{ (Commutative Law of Addition)}$$
$$= (25 + 175) + 467 \text{ (Associative Law of Addition)}$$
$$= 200 + 467 = 667 \text{ } Ans.$$

A product may be simplified by using both the Commutative and Associative Laws of Multiplication:

$$\left(\frac{1}{4} \times 35\right) \times 400 = \left(35 \times \frac{1}{4}\right) \times 400 \text{ (Commutative Law of Multiplication)}$$

$$= 35 \times \left(\frac{1}{4} \times 400\right) \text{ (Associative Law of Multiplication)}$$

$$= 35 \times 100 = 3500 \text{ } Ans.$$

SOLVED PROBLEMS

1.17   **VERIFYING THE ASSOCIATIVE LAWS FOR THREE NUMBERS**
If $a = 2$, $b = 3$, $c = 5$, and $d = 10$, show that

(a) $(a + b) + c = a + (b + c)$ (d) $(ab)c = a(bc)$

(b) $(a + c) + d = a + (c + d)$ (e) $(ac)d = a(cd)$

(c) $(d + b) + c = d + (b + c)$ (f) $(db)c = d(bc)$

Solutions

(a) $(5) + 5 = 2 + (8)$ and $10 = 10$ (d) $(6)5 = 2(15)$ and $30 = 30$

(b) $(7) + 10 = 2 + (15)$ and $17 = 17$ (e) $(10)10 = 2(50)$ and $100 = 100$

(c) $(13) + 5 = 10 + (8)$ and $18 = 18$ (f) $(30)5 = 10(15)$ and $150 = 150$

1.18   **VERIFYING THE ASSOCIATIVE LAWS FOR FOUR NUMBERS**
If $a = 1$, $b = 5$, $c = 10$, $d = 20$, and $e = 100$, show that

(a) $(a + b + c) + d = a + (b + c + d)$ (d) $(abc)d = a(bcd)$

(b) $(b + c) + (d + e) = b + (c + d + e)$ (e) $(bc)(de) = b(cde)$

(c) $(d + a + c) + e = (d + a) + (c + e)$ (f) $(dac)e = (da)(ce)$

Solutions

(a) $(16) + 20 = 1 + (35)$ and $36 = 36$ (d) $(50)20 = 1(1000)$ and $1000 = 1000$

(b) $(15) + (120) = 5 + (130)$ and $135 = 135$ (e) $(50)(2000) = 5(20,000)$ and $100,000 = 100,000$

(c) $(31) + 100 = (21) + (110)$ and $131 = 131$ (f) $(200)100 = (20)(1000)$ and $20,000 = 20,000$

1.9   ORDER IN WHICH FUNDAMENTAL OPERATIONS ARE PERFORMED

In evaluating an expression, the operations involved must be performed in a certain order. Note in the following, how *multiplication and division must precede addition and subtraction*!

1.9A   To Evaluate a Numerical Expression Not Containing Parentheses

Evaluate: (a) $3 + 4 \times 2$ (b) $5 \times 4 - 18 \div 6$

PROCEDURE   SOLUTIONS

1. Do **multiplications and division (M & D)** in order   1. $3 + 4 \times 2$ $5 \times 4 - 18 \div 6$
    from left to right: $3 + 8$ $20 - 3$

2. Do remaining **additions and subtractions (A & S)** in   2.   11 *Ans.*   17 *Ans.*
    order from left to right:

1.9B   To Evaluate an Algebraic Expression Not Containing Parentheses

Evaluate $x + 2y - \dfrac{z}{5}$ when $x = 5$, $y = 3$, $z = 20$.

PROCEDURE	SOLUTION
1. **Substitute** the value given for each variable:	1. $x + 2y - \dfrac{z}{5}$ $5 + 2(3) - \dfrac{20}{5}$
2. Do **multiplications and divisions (M & D)** in order from left to right:	2. $5 + 6 - 4$
3. Do remaining **additions and subtractions (A & S)** in order from left to right:	3. 7 *Ans.*

1.9C   To Evaluate an Algebraic Expression Containing Parentheses

Evaluate: $2(a + b) + 3a - \dfrac{b}{2}$ if $a = 7$, $b = 2$.

PROCEDURE	SOLUTION
1. **Substitute** the value given for each variable:	1. $2(7 + 2) + 3 \cdot 7 - \dfrac{2}{2}$
2. **Evaluate inside parentheses:**	2. $2 \cdot 9 + 3 \cdot 7 - \dfrac{2}{2}$
3. Do **multiplications and divisions (M & D)** in order from left to right:	3. $18 + 21 - 1$
4. Do remaining **additions and subtractions (A & S)** in order from left to right:	4. 38 *Ans.*

SOLVED PROBLEMS

1.19   EVALUATING NUMERICAL EXPRESSIONS
Evaluate:

(a) $24 \div 4 + 8$,   (b) $24 + 8 \div 4$,   (c) $8 \times 6 - 10 \div 5 + 12$.

Solutions

1. Do M & D:	(a) $24 \div 4 + 8$	(b) $24 + 8 \div 4$	(c) $8 \times 6 - 10 \div 5 + 12$
2. Do A & S:	$6 + 8$	$24 + 2$	$48 - 2 + 12$
	14 *Ans.*	26 *Ans.*	58 *Ans.*

1.20   EVALUATING ALGEBRAIC EXPRESSIONS
Evaluate if $a = 8$, $b = 10$, $x = 3$:

(a) $4b - \dfrac{a}{4}$ (b) $12x + ab$ (c) $\dfrac{3a}{4} + \dfrac{4b}{5} - \dfrac{2x}{3}$

Solutions

(a) $4b - \dfrac{a}{4}$ (b) $12x + ab$ (c) $\dfrac{3a}{4} + \dfrac{4b}{5} - \dfrac{2x}{3}$

1.	Substitute:	$4 \times 10 - \dfrac{8}{4}$	$12 \times 3 + 8 \times 10$	$\dfrac{3}{4} \times 8 + \dfrac{4}{5} \times 10 - \dfrac{2}{3} \times 3$
2.	Do M & D:	$40 - 2$	$36 + 80$	$6 + 8 - 2$
3.	Do A & S:	38 *Ans.*	116 *Ans.*	12 *Ans.*

1.21 EVALUATING NUMERICAL EXPRESSIONS CONTAINING PARENTHESES
Evaluate:

(a) $3(4-2)+12$ (b) $7-\dfrac{1}{2}(14-6)$ (c) $8+\dfrac{1}{3}(4+2)$ (d) $20-5(4-1)$

Solutions

		(a) $3(4-2)+12$	(b) $7-\dfrac{1}{2}(14-6)$	(c) $8+\dfrac{1}{3}(4+2)$	(d) $20-5(4-1)$
1.	Substitute:				
2.	Do ( ):	$3 \cdot 2 + 12$	$7 - \dfrac{1}{2} \cdot 8$	$8 + \dfrac{1}{3} \cdot 6$	$20 - 5 \cdot 3$
3.	Do M & D:	$6 + 12$	$7 - 4$	$8 + 2$	$20 - 15$
4.	Do A & S:	18 *Ans.*	3 *Ans.*	10 *Ans.*	5 *Ans.*

1.22 EVALUATING ALGEBRAIC EXPRESSIONS CONTAINING PARENTHESES
Evaluate if $a = 10$, $b = 2$, $x = 12$:

(a) $3(x+2b)-30$ (b) $8+2\left(\dfrac{a}{b}+x\right)$ (c) $3x-\dfrac{1}{2}(a+b)$

Solutions

		(a) $3(x+2b)-30$	(b) $8+2\left(\dfrac{a}{b}+x\right)$	(c) $3x-\dfrac{1}{2}(a+b)$
1.	Substitute:	$3(12+2\cdot2)-30$	$8+2\left(\dfrac{10}{2}+12\right)$	$3\cdot12-\dfrac{1}{2}(10+2)$
2.	Do ( ):	$3\cdot16-30$	$8+2\cdot17$	$36-\dfrac{1}{2}\cdot12$
3.	Do M & D:	$48-30$	$8+34$	$36-6$
4.	Do A & S:	18 *Ans.*	42 *Ans.*	30 *Ans.*

1.23 EVALUATING WHEN THE VALUE OF ONE OF THE VARIABLES IS ZERO
Evaluate if $w = 4$, $x = 2$, and $y = 0$:

(a) $wx+y$ (b) $w+xy$ (c) $\dfrac{w+y}{x}$ (d) $\dfrac{xy}{w}$ (e) $\dfrac{x}{w+y}$ (f) $\dfrac{wx}{y}$

Solutions

(a) $wx+y$ (b) $w+xy$ (c) $\dfrac{w+y}{x}$ (d) $\dfrac{xy}{w}$ (e) $\dfrac{x}{w+y}$ (f) $\dfrac{wx}{y}$

1. Substitute:	$4 \times 2 + 0$	$4 + 2 \times 0$	$\dfrac{4+0}{2}$	$\dfrac{2 \times 0}{4}$	$\dfrac{2}{4+0}$	has no meaning if $y = 0$. *Ans.*
2. Do M & D:	$8 + 0$	$4 + 0$	$\dfrac{4}{2}$	$\dfrac{0}{4}$	$\dfrac{2}{4}$	
3. Do A & S:	8 *Ans.*	4 *Ans.*	2 *Ans.*	0 *Ans.*	$\dfrac{1}{2}$ *Ans.*	

1.10 TERMS, FACTORS, AND COEFFICIENTS

A *term* is a number, a variable, or the product or quotient of numbers and variables. Thus, 5, $8y$, cd, $3wx$, and $2/3r$ are terms. An *expression* consists of one or more terms connected by plus or minus signs. Thus, the expression $8y + 5$ consists of two terms, $8y$ and 5.

A *factor of a term* is any of the numbers or variables multiplied to form the term. Thus, 8 and y are factors of the term $8y$; 2 is also a factor, though not a *displayed* factor.

Any factor or group of factors of a term is the *coefficient* of the product of the remaining factors. Thus, in $3abc$, 3 is the *numerical coefficient* of abc, while abc is the *literal coefficient* of 3. Note that the literal coefficient is the product of the variable factors.

SOLVED PROBLEMS

1.24 TERMS IN EXPRESSIONS

State the terms in each expression:

(a) $8abc$ (c) $8a + bc$ (e) $3 + bcd$

(b) $-8 + a + bc$ (d) $3b + c + d$ (f) $3(b + c) + d$

Illustrative Solutions

(b) The two plus signs separate $-8 + a + bc$ into 3 terms: -8, a, and bc.

(f) 3 and $(b + c)$ are factors of the term $3(b + c)$. There are 2 terms: $3(b + c)$ and d.

Ans. (a) $8abc$ (c) $8a, bc$ (d) $3b, c, d$ (e) $3, bcd$

1.25 FACTORS OF TERMS

State the factors of the following terms, disregarding 1 and the term itself.

(a) 21   (c) rs (e) $\dfrac{1}{3}m$ (g) $\dfrac{n+3}{5}$

(b) 121   (d) $5cd$ (f) $\dfrac{n}{-5}$ (h) $3(x + 2)$

Illustrative Solutions

(a) Since $21 = 3 \cdot 7$, 3 and 7 are factors.

(d) $5cd$ is the product of the factors 5, c, and d.

(g) $\dfrac{n+3}{5}$ is the product of the factors $\dfrac{1}{5}$, and $(n + 3)$

Ans. (b) 11, 11   (c) r, s (e) $1/3, m$ (f) $-1/5, n$ (h) $3, (x + 2)$

1.26   NUMERICAL AND LITERAL COEFFICIENT

State each numerical and literal coefficient:

(a) y (b) $\dfrac{4x}{5}$ (c) $\dfrac{w}{7}$ (d) $.7abc$ (e) $8(a+b)$

Solutions

	(a) y	(b) $\dfrac{4x}{5}$	(c) $\dfrac{w}{7}$	(d) $.7abc$	(e) $8(a+b)$
Numerical Coefficient:	1	$\dfrac{4}{5}$	$\dfrac{1}{7}$	.7	8
Literal Coefficient:	y	x	w	abc	$(a+b)$

1.11   REPEATED MULTIPLYING OF A FACTOR: BASE, EXPONENT, AND POWER

$$\text{BASE}^{EXPONENT} = \text{POWER}$$

In $2 \cdot 2 \cdot 2 \cdot 2 \cdot 2$, the factor 2 is being multiplied repeatedly. This may be written in a shorter form as 2^5, where the repeated factor 2 is the *base* while the small 5 written above and to the right of 2 is the *exponent*. The answer 32 is called the fifth *power* of 2.

An exponent is a number which indicates how many times another number, the base, is being used as a repeated factor. The power is the answer obtained. Thus, in the expression $x \cdot x \cdot x \cdot x = x^4$, 4 is the exponent and x is the base.

1.11A   Variable Bases: Squares and Cubes

The area of a square with side s is found by multiplying s by s. This may be written as $A = s^2$ and read "Area equals s-square." Here, A is the second power of s. See Fig. 1-1(a).

The volume of a cube with side s is found by multiplying s three times; that is, $s \cdot s \cdot s$. This may be written as $V = s^3$ and read "Volume equals s-cube." Here, V is the third power of s. See Fig. 1-1(b).

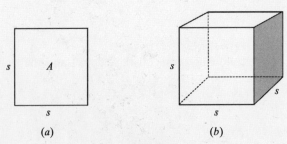

Fig. 1-1

1.11B   Reading Powers

b^2 is read as "b-square," "b to the second power," "b-second" or "b to the second." x^3 is read as "x-cube," "x to the third power," "x-third" or "x to the third."

SOLVED PROBLEMS

1.27   WRITING AS BASES AND EXPONENTS
Write each, using bases and exponents:

(a) $bbccc$ (c) $(3y)(3y)(3y)$ (e) $7rr(s-8)$ (g) $\dfrac{7w}{xx}$ (i) $2(a+b)(a+b)$

(b) $bbbbb$ (d) $12bccd$ (f) $\dfrac{yy}{x}$ (h) $\dfrac{2ttt}{5vvvv}$

Illustrative Solutions

(c)   Since $(3y)$ is repeated as a factor 3 times, $(3y)(3y)(3y) = (3y)^3$.

(i)   Since $(a+b)$ is repeated as a factor twice, $2(a+b)(a+b) = 2(a+b)^2$.

Ans.   (a) b^2c^3 (d) $12bc^2d$ (f) $\dfrac{y^2}{x}$ (h) $\dfrac{2t^3}{5v^4}$

(b) b^5 (e) $7r^2(s-8)$ (g) $\dfrac{7w}{x^2}$

1.28   WRITING WITHOUT EXPONENTS
Write each without exponents:

(a) 2^6 (d) x^5 (g) $(2x)^3$ (j) $\dfrac{a^5}{b^2}$

(b) $3 \cdot 4^2$ (e) $10y^4z^2$ (h) $6(5y)^2$ (k) $\dfrac{2(a+b)^2}{c^5}$

(c) $5 \cdot 7^3 \cdot 8$ (f) $8rs^2t^3$ (i) $4(a-b)^2$ (l) $\dfrac{a^2+b^2}{c^3-d^3}$

Illustrative Solutions:

(e)   In $10y^4z^2$, y is repeated as a factor 4 times and z twice.   *Ans.* $10yyyyzz$

(k)   In $\dfrac{2(a+b)^2}{c^5}$, $(a+b)$ is repeated as a factor twice and c 5 times.   *Ans.* $\dfrac{2(a+b)(a+b)}{ccccc}$

Ans.   (a) $2 \cdot 2 \cdot 2 \cdot 2 \cdot 2 \cdot 2$ (d) $xxxxx$ (h) $6(5y)(5y)$ (l) $\dfrac{aa+bb}{ccc-ddd}$

(b) $3 \cdot 4 \cdot 4$ (f) $8rssttt$ (i) $4(a-b)(a-b)$

(c) $5 \cdot 7 \cdot 7 \cdot 7 \cdot 8$ (g) $(2x)(2x)(2x)$ (j) $\dfrac{aaaaa}{bb}$

1.29   EVALUATING POWERS OF VARIABLE BASES
Evaluate if $a = 5$, $b = 1$, and $c = 10$:

(a) a^3 (d) $2a^2$ (g) $\left(\dfrac{1}{2}c\right)^2$ (j) a^2+c^2 (m) $c^2(a+b)$

(b) b^4 (e) $(2a)^2$ (h) $\left(\dfrac{b}{3}\right)^2$ (k) $(c+3b)^2$ (n) $c(a-b^2)$

(c) c^2 (f) $(a+2)^2$ (i) $\dfrac{4a^2}{c}$ (l) $5(a^2-b^2)$ (o) $3b(a^3-c^2)$

Solutions

(a) $5 \cdot 5 \cdot 5 = 125$ (d) $2 \cdot 25 = 50$ (g) $5^2 = 25$ (j) $25 + 100 = 125$ (m) $100 \cdot 6 = 600$

(b) $1 \cdot 1 \cdot 1 \cdot 1 = 1$ (e) $10^2 = 100$ (h) $\left(\dfrac{1}{3}\right)^2 = \dfrac{1}{9}$ (k) $13^2 = 169$ (n) $10 \cdot 4 = 40$

(c) $10 \cdot 10 = 100$ (f) $7^2 = 49$ (i) $\dfrac{100}{10} = 10$ (l) $5 \cdot 24 = 120$ (o) $3 \cdot 25 = 75$

Supplementary Problems

The number in parentheses associated with each supplementary problem refers to a set of solved problems of the same kind. Refer to those for help.

1. Using variable expressions, replace each verbal statement by an algebraic equation.   (1.1)

(a) Three times a number added to eight times the same number is equivalent to eleven times the number. (b) The difference between ten times a number and one-half of the same number is the same as nine and one-half times the number. (c) The perimeter of an equilateral triangle is equal to three times the length of one of the sides. (d) The area of a square is found by multiplying the length of a side by itself.

2. Using the initial letters of words as variables, replace each verbal rule by a formula.   (1.2)

(a) The perimeter of a regular hexagon is six times a side. (b) The area of a triangle is one-half the product of the base and height. (c) The profit made on an article is the difference between the selling price and the cost. (d) The semiperimeter of a rectangle is the sum of the length and width. (e) The volume of a rectangular box is the product of the length, the width, and the height.

3. Using the replacement set $\{2, 3, 6, 10\}$ as the set of values of n, evaluate the following expressions and list these values in a set: (1.3)

(a) $5n$ (b) $2n + 5$ (c) $10 - \dfrac{1}{2}n$ (d) $\dfrac{2n - 1}{2n + 1}$ (e) $150\% n$

4. Rearrange the addends so that literal addends are arranged alphabetically and precede numerical addends: (1.4)

(a) $10 + d$ (c) $15 + g + f$ (e) $d + b + e + a$ (g) $5 + m + 4 + j + 11$

(b) $y + x$ (d) $17 + r + q + 13$ (f) $s + 12 + p + 48$ (h) $v + w + 16 + t + 50$

5. Rearrange the factors so that literal factors are arranged alphabetically and follow numerical factors: (1.5)

(a) $r8$ (c) $qp7$ (e) $x5y7w$ (g) $2h5k10$ (i) $cd13ab$

(b) cab (d) $4v3t$ (f) $def13c$ (h) $r11sm4$

6. Symbolize each, avoiding multiplication signs where possible: (1.6)

(a) the product of 8, m and n (e) f divided by 7

(b) 5 times p times q (f) 25 divided by x

(c) two-thirds of c (g) the product of u and v divided by 9

(d) one-half of b multiplied by h

7. When is each division impossible?   (1.7, 1.8)

(a) $\dfrac{7}{d}$ (c) $\dfrac{3}{4x}$ (e) $\dfrac{b}{7-c}$ (g) $\dfrac{50}{w-y}$ (i) $\dfrac{45}{x-2y}$

(b) $\dfrac{r}{t}$ (d) $\dfrac{5}{a-8}$ (f) $\dfrac{10}{2x-4}$ (h) $\dfrac{100}{pq}$

8. If n represents a number, express algebraically: (1.9, 1.10)

(a) 25 more than the number   (g) 30 less than the number

(b) 30 greater than the number   (h) 35 fewer than the number

(c) the sum of the number and 35   (i) 40 less the number

(d) the number increased by 40   (j) 45 decreased by the number

(e) 45 plus the number   (k) 50 minus the number

(f) 50 added to the number   (l) 55 subtracted from the number

9. Express algebraically: (a) the no. of kg of a weight that is 15 kg lighter than w kg, (b) the no. of ft in a length that is 50 ft shorter than l ft, (c) the no. of sec in a time interval that is 1 minute less than t sec, (d) the no. of cents in a price that is \$1 more than p cents, (e) the no. of ft per sec (fps) in a speed that is 20 fps slower than r fps, (f) the no. of ft in a distance that is 10 yd farther than d ft, (g) the no. of sq ft in an area that is 30 sq ft greater than A sq ft, (h) the no. of degrees in a temperature that is 40° colder than $t°$, (i) the no. of floors in a building that is 8 floors higher than f floors, (j) the no. of yr in an age 5 yr younger than a yr.   (1.11)

10. Express algebraically: (1.12, 1.13)

(a) x times 3   (c) product of 12 and y (e) 10 divided by y

(b) one-eighth of b (d) three-eighths of r (f) quotient of y and 10

11. Express algebraically: (1.14, 1.15)

(a) b decreased by one-half c (f) twice d, less 25

(b) one-third of g, decreased by 5   (g) 8 more than the product of 5 and x

(c) four times r, divided by 9   (h) four times the sum of r and 9

(d) the average of m and 60   (i) the average of 60, m, p, and q

(e) three-quarters of x, less y (j) the ratio of b to three times c

12. Express algebraically: (a) a distance in m that is 25 m shorter than three times another of d m, (b) a weight in oz that is 5 oz more than twice another of w oz, (c) a temperature in degrees that is 8° warmer than five times another of $T°$, (d) a price in dollars that is \$50 higher than one-half another of p dollars, (e) a price in cents that is 50¢ cheaper than one-third another of p cents, (f) a length in cm that is 2 cm longer than y m.   (1.16)

13. If $a = 2$, $b = 5$, $c = 10$, $d = 100$, and $e = 1,000$, show that   (1.17, 1.18)

(a) $(a+b)+c = a+(b+c)$ (d) $(a+c+e)+(d+b) = (a+c)+(e+d)+b$

(b) $(a+b)+(c+d) = (a+b+c)+d$ (e) $(ab)c = a(bc)$

(c) $(b+d)+(a+e) = b+(d+a+e)$ (f) $(ab)(cd) = (abc)d$

14. Evaluate: (1.19)

(a) $40 - 2 \times 5$ (c) $40 \div 2 + 5$ (e) $16 \div 2 - \frac{1}{2} \cdot 10$ (g) $40 \times 2 - 40 \div 2$

(b) $3 \times 8 - 2 \times 5$ (d) $3 + 8 - 2 \times 5$ (f) $3 + 8 \times 2 \times 5$ (h) $3 + 8 \times 2 - 5 \div 10$

15. Evaluate if $a = 5$, $b = 6$ and $c = 10$: (1.20)

(a) $a + b - c$ (e) $\frac{3c}{a}$ (i) $\frac{a+b}{c-9}$ (m) $a + \frac{c-b}{2}$

(b) $a + 2b$ (f) $3 + \frac{c}{a}$ (j) $\frac{ab}{c}$ (n) $a + c - \frac{b}{2}$

(c) $a + \frac{b}{2}$ (g) $\frac{4}{5}c$ or $\frac{4c}{5}$ (k) $5a + 4b - 2c$ (o) $\frac{a+c-b}{3}$

(d) $\frac{a+b}{2}$ (h) $\frac{2}{3}b + \frac{3}{2}c$ (l) $6c - 2ab$

16. Evaluate: (1.21)

(a) $5(8 + 2)$ (d) $8(2 \cdot 5 - 3)$ (g) $(3 \cdot 6 + 2)5$ (j) $(4 + 4)4 - 4$

(b) $5(8 - 2)$ (e) $8 \cdot 2(5 - 3)$ (h) $3(6 + 2)5$ (k) $(4 + 4)(4 - 4)$

(c) $8 + 2(5 - 3)$ (f) $3(6 + 2 \cdot 5)$ (i) $4(4 \cdot 4 - 4)$ (l) $4 + 4(4 - 4)$

17. Evaluate if $a = 4$, $b = 3$, and $c = 5$: (1.22)

(a) $a(b + c)$ (d) $\frac{1}{2}(a + b + c)$ (g) $3(b + 2)c$ (j) $3(a + 2c) - b$

(b) $b(c - a)$ (e) $\frac{1}{2}(a + b) + c$ (h) $(3b + 2)c$ (k) $3(a + 2c - b)$

(c) $c(a - b)$ (f) $3(b + 2c)$ (i) $3a + 2(c - b)$ (l) $3(a + 2)(c - b)$

18. Evaluate if $x = 3$, $y = 2$, and $z = 0$: (1.23)

(a) $x + y + z$ (e) $y(x + z)$ (i) $xy + z$ (m) $\frac{x}{y + z}$

(b) $x - y - z$ (f) $\frac{z}{x}$ (j) $x + yz$ (n) $x + \frac{z}{y}$

(c) $x(y + z)$ (g) $\frac{x}{z}$ (k) $xz + yz$ (o) $\frac{y + z}{x}$

(d) $z(x + y)$ (h) xyz (l) $\frac{z}{x + y}$

19. State the terms in each expression: (1.24)

(a) $5xyz$ (c) $5 + x + y + z$ (e) $3ab + c$

(b) $5 + xyz$ (d) $3a + bc$ (f) $3a(b + c)$

20. State the factors of the following, disregarding 1 and the term itself: (1.25)

(a) 77 (b) 25 (c) pq (d) $\frac{3}{4}x$ (e) $\frac{w}{10}$ (f) $8(x - 5)$ (g) $\frac{y - 2}{4}$

21. State each numerical and literal coefficient: (1.26)

(a) w (b) $\frac{1}{8}x$ (c) $\frac{n}{10}$ (d) $.03ab$ (e) $\frac{3y}{10}$ (f) $\frac{2a}{3b}$ (g) $\frac{3}{5}(a - b)$

22. Write each, using bases and exponents: (1.27)

(a) $7 \cdot 3 \cdot 3$ (b) $7xyyy$ (c) $\dfrac{7x}{yyy}$ (d) $(7x)(7x)$ (e) $(a+5)(a+5)$ (f) $\dfrac{2rrw}{5stvv}$

23. Write each without exponents: (1.28)

(a) $4 \cdot 7^2$ (b) $\dfrac{1}{2}y^4$ (c) $\dfrac{5a}{b^4}$ (d) $(ab)^3$ (e) $(x+2)^2$ (f) $\dfrac{a^2 - b^3}{c + d^2}$

24. Evaluate if $a = 3$ and $b = 2$:   (1.29)

(a) $a^2 b$ (d) $a + b^2$ (g) $a^3 b$ (j) $(a - b)^3$

(b) ab^2 (e) $(a + b)^2$ (h) $(ab)^3$ (k) $a^2 b^3$

(c) $(ab)^2$ (f) $a^2 + b^2$ (i) $a^3 - b^3$ (l) $a^3 b^2$

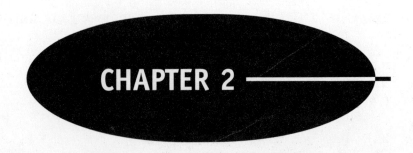

CHAPTER 2

Fundamentals of Algebra: Equations and Formulas

2.1 VARIABLES AND EQUATIONS

The question "What number increased by 7 equals 10?" may be answered by expressing the problem in the form of an equation, $n + 7 = 10$. In this equation, n is a variable whose replacement set is assumed to be the set of all numbers.

The equation $n + 7 = 10$ is solved by comparing it with the addition fact $3 + 7 = 10$; thus,

$$\left.\begin{array}{r} n + 7 = 10 \\ 3 + 7 = 10 \end{array}\right\} \rightarrow \text{Conclusion: } n = 3 \ \textit{Ans.}$$

A *solution* or *root* of an equation is a value of the variable for which the equation becomes a true statement. A solution or root of an equation is said to *satisfy* the equation.

Thus, 3 is a solution or root of the equation $n + 7 = 10$ and satisfies the equation.

2.1A Solving an Equation: Solution Set or Truth Set

Solving an equation is the process of finding the roots or solutions of the equation. The *solution set* or *truth set* of an equation is the set of roots of the equation.

Thus, $n + 7 = 10$ is solved when the root 3 is found. The solution set of $n + 7 = 10$ is {3}.

2.1B Identities and Equivalent Expressions

We have found that the equation $n + 7 = 10$ has one root, 3, and no other. However, there are equations that are satisfied by any number! For example, $3n - 2n = n$ is such an equation. Such an equation is an *identity* or *unconditional equation*, and both the left side and the right side are *equivalent expressions*. To show that $3n - 2n = n$ is an identity, simply combine like terms to obtain $n = n$.

Substitution Rule for Equivalent Expressions

In any process, an expression may be replaced by an equivalent expression.

Thus, to solve $3n - 2n + 7 = 10$, substitute n for $3n - 2n$ to obtain $n + 7 = 10$. We now solve $n + 7 = 10$ to obtain 3, the root or solution of $3n - 2n + 7 = 10$.

2.1C Properties of an Equality

Three important properties of an equality are the following:

1. **Reflexive Property:** $a = a$.
   According to the reflexive property, an expression equals itself. Thus, $n + 7 = n + 7$

2. **Symmetric Property:** If $a = b$, then $b = a$.
   According to the symmetric property, we can interchange the sides of an equation. Thus, if $10 = n + 7$, then $n + 7 = 10$.

3. **Transitive Property:** If $a = b$ and $b = c$, then $a = c$.
   According to the transitive property, two expressions equal to a third expression are equal to each other. Thus, if $n + 7 = 10$ and $3 + 7 = 10$, then $n + 7 = 3 + 7$.

2.1D Checking or Verifying an Equation

Checking or *verifying* an equation is the process of testing to see if a given number is a root of an equation.

Thus, the equation $3n - 2n + 7 = 10$ may be checked to see if 3 or 5 is a root. (Read $\stackrel{?}{=}$ as "should equal." Read $\neq$ as "does not equal.")

Check for $n = 3$	**Check for $n = 5$**
$3n - 2n + 7 = 10$	$3n - 2n + 7 = 10$
$3(3) - 2(3) + 7 \stackrel{?}{=} 10$	$3(5) - 2(5) + 7 \stackrel{?}{=} 10$
$9 - 6 + 7 \stackrel{?}{=} 10$	$15 - 10 + 7 \stackrel{?}{=} 10$
$10 = 10 ✓$	$12 \neq 10$

The check shows that 3 is a root of $3n - 2n + 7 = 10$ but 5 is not.

SOLVED PROBLEMS

2.1 **CHECKING AN EQUATION TO DETERMINE A ROOT**
By checking, determine which is a root of the equation:

(a) Check $2n + 3n = 25$ for $n = 5$ and $n = 6$ (b) Check $8x - 14 = 6x$ for $x = 6$ and $x = 7$

Solutions

Check: $n = 5$	$n = 6$	**Check:** $x = 6$	$x = 7$
$2n + 3n = 25$	$2n + 3n = 25$	$8x - 14 = 6x$	$8x - 14 = 6x$
$2(5) + 3(5) \stackrel{?}{=} 25$	$2(6) + 3(6) \stackrel{?}{=} 25$	$8(6) - 14 \stackrel{?}{=} 6(6)$	$8(7) - 14 \stackrel{?}{=} 6(7)$
$10 + 15 \stackrel{?}{=} 25$	$12 + 18 \stackrel{?}{=} 25$	$48 - 14 \stackrel{?}{=} 36$	$56 - 14 \stackrel{?}{=} 42$
$25 = 25$	$30 \neq 25$	$34 \neq 36$	$42 = 42$

Ans. 5 is a root of $2n + 3n = 25$ *Ans.* 7 is a root of $8x - 14 = 6x$

2.2 **CHECKING AN IDENTITY TO SHOW THAT ANY GIVEN VALUE IS A ROOT**
By checking the identity $4(x + 2) = 4x + 8$, show that it is satisfied by (a) $x = 10$, (b) $x = 6$, (c) $x = 4\frac{1}{2}$, (d) $x = 3.2$.

Solutions

(a) $4(x + 2) = 4x + 8$ (b) $4(x + 2) = 4x + 8$ (c) $4(x + 2) = 4x + 8$ (d) $4(x + 2) = 4x + 8$

$\quad\quad 4(10 + 2) \overset{?}{=} 4(10) + 8$ $\quad 4(6 + 2) \overset{?}{=} 4(6) + 8$ $\quad 4(4\frac{1}{2} + 2) \overset{?}{=} 4(4\frac{1}{2}) + 8$ $\quad 4(3.2 + 2) \overset{?}{=} 4(3.2) + 8$

$\quad\quad\quad 4(12) \overset{?}{=} 40 + 8$ $\quad\quad 4(8) \overset{?}{=} 24 + 8$ $\quad\quad 4(6\frac{1}{2}) \overset{?}{=} 18 + 8$ $\quad\quad 4(5.2) \overset{?}{=} 12.8 + 8$

$\quad\quad\quad\quad 48 = 48$ $\quad\quad\quad 32 = 32$ $\quad\quad\quad\quad 26 = 26$ $\quad\quad\quad 20.8 = 20.8$

2.2 TRANSLATING VERBAL PROBLEMS INTO EQUATIONS

In algebra, a simple verbal problem having one unknown is solved when the unknown is found. In the process, it is necessary to "translate" a verbal sentence into an equation. The first step in problems of this type is to let the unknown be represented by a variable.

Thus, if n represents the unknown in "Twice what number equals 12?", we obtain "$2n = 12$."

SOLVED PROBLEMS

2.3   TRANSLATING INTO EQUATIONS

Translate into an equation, letting n represent the unknown number. (a) 4 less than what number equals 8? (b) One-half of what number equals 10? (c) Ten times what number equals 20? (d) What number increased by 12 equals 17? (e) Twice what number added to 8 is 16? (f) 15 less than three times what number is 27? (g) The sum of what number and twice the same number is 18? (h) What number and 4 more equals five times the number? (i) Twice the sum of a certain number and five is 24. What is the number?

Illustrative Solution

(e) Use $2n$ for "twice what number."   *Ans.* $8 + 2n = 16$ or $2n + 8 = 16$.

Ans. (a) $n - 4 = 8$ (c) $10n = 20$ (f) $3n - 15 = 27$ (h) $n + 4 = 5n$

$\quad\quad\quad (b)$ $\dfrac{n}{2} = 10$ (d) $n + 12 = 17$ (g) $n + 2n = 18$ (i) $2(n + 5) = 24$

2.4   MATCHING SENTENCES AND EQUATIONS

Match a sentence in Column 1 with an equation in Column 2.

Column 1	**Column 2**
1. The product of 8 and a number is 40.	(a) $n - 8 = 40$
2. A number increased by 8 is 40.	(b) $8(n + 8) = 40$
3. 8 less than a number equals 40.	(c) $8n = 40$
4. Eight times a number, less 8, is 40.	(d) $n/8 = 40$
5. Eight times the sum of a number and 8 is 40.	(e) $8n - 8 = 40$
6. One-eighth of a number is 40.	(f) $n + 8 = 40$

Ans.   1. and (c) 2. and (f) 3. and (a) 4. and (e) 5. and (b) 6. and (d)

2.5   REPRESENTING UNKNOWNS

Represent the unknown by a variable and obtain an equation for each problem. (a) A man worked for 5 hours and earned \$87.50. What was his hourly wage? (b) How old is Henry now, if ten years ago, he was 23 years old? (c) After gaining 12 lb, Mary weighed 120 lb. What was her previous weight? (d) A baseball team won four times as many games as it lost. How many games did it lose, if it played a total of 100 games?

Solutions

(a) Let w = his hourly wage in dollars.
Then, $5w = 87.50$

(b) Let H = Henry's age now.
Then, $H - 10 = 23$

(c) Let M = Mary's previous weight in lb.
Then, $M + 12 = 120$

(d) Let n = no. of games lost, and
$4n$ = no. of games won.
Then, $n + 4n = 100$

2.3 SOLVING SIMPLE EQUATIONS USING INVERSE OPERATIONS

In mathematics, you may think of inverse operations as operations that undo each other. To understand the functions of inverse operations, think how often you do something and then undo it. You earn money, then you spend it; you get up, then you lie down; you open a door, then you close it; you go out, then you go in. Of course, not everything that you do can be undone. Consider putting an egg together after you have fried it.

Rule 1: Addition and subtraction are inverse operations.

Rule 2: Multiplication and division are inverse operations.

Figure 2-1(a) illustrates how addition undoes subtraction, and how subtraction undoes addition. Figure 2-1(b) illustrates how multiplication undoes division, and how division undoes multiplication.

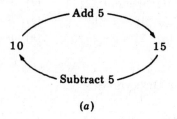   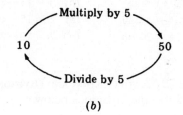

(a)   (b)

Fig. 2-1

(1) Addition and subtraction are shown to be inverse operations since adding 5 to 10, then subtracting 5 from the result leads back to 10; or, subtracting 5 from 15, then adding 5 to the result leads back to 15.

Suppose that in addition to having $10 you earn $5, making a total of $15. If now you spend $5, you will have the original amount of $10.

(2) Multiplication and division are shown to be inverse operations since multiplying 10 by 5, then dividing the result by 5 leads back to 10; or, dividing 50 by 5, then multiplying the result by 5 leads back to 50.

Suppose that the number of persons in a group of 10 is multiplied by 5, making a new group of 50 persons. If now the number is divided by 5, we will find that the number of persons is the original number 10.

An understanding of inverse operations helps us understand the solution of equations. To solve an equation, think of it as asking a question, as in each of the following examples.

EQUATION	QUESTION ASKED BY EQUATION	FINDING ROOT OF EQUATION
1. $n + 4 = 12$	What number plus 4 equals 12?	$n = 12 - 4 = 8$
2. $n - 4 = 12$	What number minus 4 equals 12?	$n = 12 + 4 = 16$
3. $4n = 12$	What number multiplied by 4 equals 12?	$n = 12 \div 4 = 3$
4. $n/4 = 12$	What number divided by 4 equals 12?	$n = 12 \cdot 4 = 48$

Note the inverse operations involved in the above cases.

1. The equation $n + 4 = 12$, involving *addition*, is solved by *subtracting* 4 from 12.

2. The equation $n - 4 = 12$, involving *subtraction*, is solved by *adding* 4 to 12.

3. The equation $4n = 12$, involving *multiplication*, is solved by *dividing* 4 into 12.

4. The equation $n/4 = 12$, involving *division*, is solved by *multiplying* 4 by 12.

SOLVED PROBLEMS

2.6   RULE 1: ADDITION AND SUBTRACTION ARE INVERSE OPERATIONS
Solve the equations below.

Equations Involving Addition to Unknown	Solutions by Subtraction	Equations Involving Subtraction from Unknown	Solutions by Addition
(a) $x + 3 = 8$	(a) $x = 8 - 3$ or 5	(e) $x - 10 = 2$	(e) $x = 2 + 10$ or 12
(b) $5 + y = 13$	(b) $y = 13 - 5$ or 8	(f) $w - 20 = 12$	(f) $w = 12 + 20$ or 32
(c) $15 = a + 10$	(c) $a = 15 - 10$ or 5	(g) $18 = a - 13$	(g) $a = 18 + 13$ or 31
(d) $28 = 20 + b$	(d) $b = 28 - 20$ or 8	(h) $21 = b - 2$	(h) $b = 21 + 2$ or 23

2.7   RULE 2: MULTIPLICATION AND DIVISION ARE INVERSE OPERATIONS
Solve the equations below.

Equations Involving Multiplication of Unknown	Solutions by Division	Equation Involving Division of Unknown	Solutions by Multiplication
(a) $3x = 12$	(a) $x = \dfrac{12}{3}$ or 4	(e) $\dfrac{x}{3} = 12$	(e) $x = 12 \cdot 3$ or 36
(b) $12y = 3$	(b) $y = \dfrac{3}{12}$ or $\dfrac{1}{4}$	(f) $\dfrac{y}{12} = 3$	(f) $y = 3 \cdot 12$ or 36
(c) $35 = 7a$	(c) $a = \dfrac{35}{7}$ or 5	(g) $4 = \dfrac{a}{7}$	(g) $a = 4 \cdot 7$ or 28
(d) $7 = 35b$	(d) $b = \dfrac{7}{35}$ or $\dfrac{1}{5}$	(h) $7 = \dfrac{b}{4}$	(h) $b = 7 \cdot 4$ or 28

2.8   SOLVING BY USING INVERSE OPERATIONS
Solve each equation, showing the inverse operation used to solve.

(a) $x + 5 = 20$ (c) $5x - 20$ (e) $10 + y = 30$ (g) $\dfrac{y}{10} = 30$

(b) $x - 5 = 20$ (d) $\dfrac{x}{5} = 20$ (f) $10y = 30$ (h) $14 = a + 7$

(i) $14 = a - 7$ (k) $14 = \dfrac{a}{7}$ (m) $8b = 2$ (o) $24 = 6 + c$

(j) $14 = 7a$ (l) $b - 8 = 2$ (n) $\dfrac{b}{8} = 2$

Solutions

(a) $x = 20 - 5$ or 15 (f) $y = \dfrac{30}{10}$ or 3 (k) $a = 14(7)$ or 98

(b) $x = 20 + 5$ or 25 (g) $y = 30(10)$ or 300 (l) $b = 2 + 8$ or 10

(c) $x = \dfrac{20}{5}$ or 4 (h) $a = 14 - 7$ or 7 (m) $b = \dfrac{2}{8}$ or $\dfrac{1}{4}$

(d) $x = 20(5)$ or 100 (i) $a = 14 + 7$ or 21 (n) $b = 2(8)$ or 16

(e) $y = 30 - 10$ or 20 (j) $a = \dfrac{14}{7}$ or 2 (o) $c = 24 - 6$ or 18

2.4 RULES FOR SOLVING EQUATIONS

Equivalent equations are equations having the same solution set; that is, the same root or roots. In the previous section, equations were solved by changing, or transforming, them into equivalent equations using inverse operations.

Thus, $n + 4 = 12$ and $n = 12 - 4$ are equivalent equations.

2.4A Rules of Equality for Solving Equations

1. **Addition Rule:** To change an equation into an equivalent equation, the same number may be added to both sides.

2. **Subtraction Rule:** To change an equation into an equivalent equation, the same number may be subtracted from both sides.

3. **Multiplication Rule:** To change an equation into an equivalent equation, both sides may be multiplied by the same number.

4. **Division Rule:** To change an equation into an equivalent equation, both sides may be divided by any number except zero.

These four rules of equality may be summed up in one rule:

The Rule of Equality for Fundamental Operations

To change an equation into an equivalent equation, perform the same fundamental operation on both sides using the same number, excepting division by zero.

To understand these rules of equality, think of an equality as a scale in balance (Fig. 2-2). If only one side of a balanced scale is changed, the scale becomes unbalanced. To balance the scale, exactly the same change must be made on the other side. Similarly, if only one side of an equation is changed, the result need not be an equivalent equation. To obtain an equivalent equation, the same fundamental operation should be performed on both sides using the same number.

BALANCED SCALES

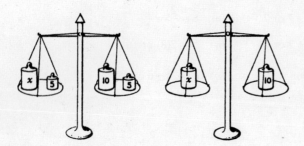

$\left\{\begin{array}{l}\text{If 5 is subtracted from} \\ \text{both sides of a balanced scale, the} \\ \text{scale is still in balance.}\end{array}\right.$

EQUALITIES

If $x + 5 = 15$

then $x + 5 - 5 = 15 - 5$

and $x = 10$

$\left\{\begin{array}{l}\text{If 5 is subtracted from both} \\ \text{sides of an equation, the result} \\ \text{is an equivalent equation.}\end{array}\right.$

Fig. 2-2

SOLVED PROBLEMS

2.9   USING RULES OF EQUALITY

State the equality rule and the change needed to solve the equation:

(a) $\begin{aligned} x + 15 &= 21 \\ -15 &= -15 \\ \hline x &= 6 \end{aligned}$ (b) $\begin{aligned} 40 &= r - 8 \\ +8 &= +8 \\ \hline 48 &= r \end{aligned}$ (c) $\begin{aligned} 25 &= 5m \\ \frac{25}{5} &= \frac{5m}{5} \\ 5 &= m \end{aligned}$ (d) $\begin{aligned} \frac{n}{8} &= 3 \\ 8 \cdot \frac{n}{8} &= 8 \cdot 3 \\ n &= 24 \end{aligned}$ (e) $\begin{aligned} 24x &= 8 \\ \frac{24x}{24} &= \frac{8}{24} \\ x &= \frac{1}{3} \end{aligned}$

Solutions

(a)   Using the subtraction rule of equality, 15 is subtracted from each side.

(b)   Using the addition rule of equality, 8 is added to each side.

(c)   Using the division rule of equality, each side is divided by 5.

(d)   Using the multiplication rule of equality, each side is multiplied by 8.

(e)   Using the division rule of equality, each side is divided by 24.

2.5   USING DIVISION TO SOLVE AN EQUATION

2.5A   To Solve an Equation Using the Division Rule of Equality

Solve: (a) $2n = 16$ (b) $16n = 2$

PROCEDURE   SOLUTIONS

1. Divide both sides of the equation by the   **1.** $\mathbf{D}_2$ $\dfrac{2n}{2} = \dfrac{16}{2}$ $\mathbf{D}_{16}$ $\dfrac{16n}{16} = \dfrac{2}{16}$
   coefficient or multiplier of the variable:
                                                            Ans. $n = 8$ *Ans.* $n = \dfrac{1}{8}$

2. Check the original equation: **2.** $2n = 16$ $16n = 2$

                                                            $2(8) \overset{?}{=} 16$ $16\left(\dfrac{1}{8}\right) \overset{?}{=} 2$

                                                            $16 = 16$ ✔ $2 = 2$ ✔

Note 1:   **D** is a convenient symbol for "dividing both sides."
            $\mathbf{D}_2$ means "divide both sides by 2."

Note 2:   A common factor may be eliminated in $\dfrac{\overset{1}{\cancel{2}}n}{\cancel{2}}$ and $\dfrac{\overset{1}{\cancel{16}}n}{\cancel{16}}$.

SOLVED PROBLEMS

2.10   **S**OLVING **E**QUATIONS WITH **N**ATURAL **N**UMBER **C**OEFFICIENTS
   Solve: (a) $7x = 35$, (b) $35y = 7$, (c) $33 = 11z$, (d) $11 = 33w$.

Solutions

(a) $7x = 35$ (b) $35y = 7$ (c) $33 = 11z$ (d) $11 = 33w$

$\mathbf{D}_7$ $\dfrac{7x}{7} = \dfrac{35}{7}$ $\mathbf{D}_{35}$ $\dfrac{35y}{35} = \dfrac{7}{35}$ $\mathbf{D}_{11}$ $\dfrac{33}{11} = \dfrac{11z}{11}$ $\mathbf{D}_{11}$ $\dfrac{11}{33} = \dfrac{33w}{33}$

Ans. $x = 5$ *Ans.* $y = \dfrac{1}{5}$ *Ans.* $3 = z$ *Ans.* $\dfrac{1}{3} = w$

Check: **Check:**   **Check:**   **Check:**
   $7x = 35$ $35y = 7$ $33 = 11z$ $11 = 33w$

$7(5) \overset{?}{=} 35$ $35\left(\dfrac{1}{5}\right) \overset{?}{=} 7$ $33 \overset{?}{=} 11(3)$ $11 \overset{?}{=} 33\left(\dfrac{1}{3}\right)$

$35 = 35$ $7 = 7$ $33 = 33$ $11 = 11$

2.11   **D**IVISION IN **E**QUATIONS WITH **D**ECIMAL **C**OEFFICIENTS
   Find each solution set: (a) $.3a = 9$, (b) $1.2b = 48$, (c) $15 = .05c$.

Solutions

(a) $.3a = 9$ (b) $1.2b = 48$ (c) $15 = .05c$

$\mathbf{D}_{.3}$ $\dfrac{.3a}{.3} = \dfrac{9}{.3}$ $\mathbf{D}_{1.2}$ $\dfrac{1.2b}{1.2} = \dfrac{48}{1.2}$ $\mathbf{D}_{.05}$ $\dfrac{15}{.05} = \dfrac{.05c}{.05}$

         $a = 30$ $b = 40$ $300 = c$
Ans. $\{30\}$ *Ans.* $\{40\}$ *Ans.* $\{300\}$

Check: **Check:**   **Check:**
   $.3a = 9$ $1.2b = 48$ $15 = .05c$

$.3(30) \overset{?}{=} 9$ $1.2(40) \overset{?}{=} 48$ $15 \overset{?}{=} .05(300)$

   $9 = 9$ $48 = 48$ $15 = 15$

2.12 SOLVING EQUATIONS WITH PER CENTS AS COEFFICIENTS

Solve: (a) $22\%s = 88$, (b) $75\%t = 18$, (c) $72 = 2\%n$.

Solutions

(a) $22\%s = 88$
Since $22\% = .22$,

$$\mathbf{D}_{.22} \quad \frac{.22s}{.22} = \frac{88}{.22}$$

Ans. $s = 400$

Check:

$$22\% = 88$$
$$(.22)(400) \overset{?}{=} 88$$
$$88 = 88$$

(b) $75\%t = 18$
Since $75\% = .75$,

$$\mathbf{D}_{.75} \quad \frac{.75t}{.75} = \frac{18}{.75}$$

Ans. $t = 24$

Check:

$$75\%t = 18$$
$$(.75)(24) \overset{?}{=} 18$$
$$18 = 18$$

(c) $72 = 2\%n$
Since $2\% = .02$,

$$\mathbf{D}_{.02} \quad \frac{72}{.02} = \frac{.02n}{.02}$$

Ans. $3600 = n$

Check:

$$72 = 2\%n$$
$$72 \overset{?}{=} (.02)(3600)$$
$$72 = 72$$

2.13 SOLVING EQUATIONS WITH LIKE TERMS ON ONE SIDE

Solve: (a) $60 = 7x - x$, (b) $3x + 5x = 48$, (c) $7x - 2x = 55$.

Solutions

(a) $60 = 7x = x$
$60 = 6x$

$$\mathbf{D}_6 \quad \frac{60}{6} = \frac{6x}{6}$$

Ans. $10 = x$

Check:

$$60 = 7x - x$$
$$60 \overset{?}{=} 70 - 10$$
$$60 = 60$$

(b) $3x + 5x = 48$
$8x = 48$

$$\mathbf{D}_8 \quad \frac{8x}{8} = \frac{48}{8}$$

Ans. $x = 6$

Check:

$$3x + 5x = 48$$
$$18 + 30 \overset{?}{=} 48$$
$$48 = 48$$

(c) $7x - 2x = 55$
$5x = 55$

$$\mathbf{D}_5 \quad \frac{5x}{5} = \frac{55}{5}$$

Ans. $x = 11$

Check:

$$7x - 2x = 55$$
$$77 - 22 \overset{?}{=} 55$$
$$55 = 55$$

2.14 DIVISION RULE IN PROBLEM SOLVING

John worked 7 hr and earned $85.40. What was his hourly wage?

Solution

Let h = his hourly wage in $.
Then, $7h = 85.40$

$$\mathbf{D}_7 \quad \frac{7h}{7} = \frac{85.40}{7}$$

$$h = 12.20$$

Ans. John's hourly wage was $12.20

Check (the problem)

In 7 hr, John should earn $85.40
Hence,
$$7(\$12.20) \overset{?}{=} \$85.40$$
$$\$85.40 = \$85.40$$

2.15 DIVISION RULE IN PROBLEM SOLVING

Mr. Black's commission rate was 3%. If he earned $66 in commission, how much did he sell?

Solution

Let s = Mr. Black's sales in \$.

Then, $3\% s = .03s = 66$

$\mathbf{D}_{.03}$ $\dfrac{.03s}{.03} = \dfrac{66}{.03}$

$s = 2200$

Check (the problem):

At 3%, Mr. Black's commission should be \$66. Hence,

3% of $\$2200 \overset{?}{=} \$66$

$(.03)(\$2200) = \$66$

$\$66 = \$66$

Ans. Mr. Black's sales were \$2200.

2.6 USING MULTIPLICATION TO SOLVE AN EQUATION

2.6A To Solve an Equation Using the Multiplication Rule of Equality

Solve: (a) $\dfrac{w}{3} = 5$ (b) $10 = \dfrac{x}{7}$

PROCEDURE   SOLUTIONS

1. Multiply both sides of the equation by the divisor of the variable:

1. $\mathbf{M}_3$ $3 \cdot \dfrac{w}{3} = 3 \cdot 5$ $\mathbf{M}_7$ $7 \cdot 10 = 7 \cdot \dfrac{x}{7}$

Ans. $w = 15$ *Ans.* $70 = x$

2. Check the original equation:

2. $\dfrac{w}{3} = 5$ $10 = \dfrac{x}{7}$

$\dfrac{15}{3} \overset{?}{=} 5$ $10 \overset{?}{=} \dfrac{70}{7}$

$5 = 5$ $10 = 10$

Note 1:   **M** is a convenient symbol for "multiplying both sides."
       $\mathbf{M}_3$ means "multiply both sides by 3."

Note 2:   **A** common factor may be eliminated in $\overset{1}{\cancel{3}} \cdot \dfrac{w}{\cancel{3}}$ and $\overset{1}{\cancel{7}} \cdot \dfrac{x}{\cancel{7}}$.

2.6B To Solve an Equation whose Variable has a Fractional Coefficient

To divide by a fraction, invert the fraction and multiply.

Solve: (a) $\dfrac{2}{3}x = 8$ (b) $\dfrac{5}{3}y = 25$

PROCEDURE   SOLUTIONS

1. Multiply both sides of the equation by the fractional coefficient inverted:

1. $\mathbf{M}_{3/2}$ $\dfrac{3}{2} \cdot \dfrac{2}{3}x = \dfrac{3}{2} \cdot 8$ $\mathbf{M}_{3/5}$ $\dfrac{3}{5} \cdot \dfrac{5}{3}y = \dfrac{3}{5} \cdot 25$

Ans. $x = 12$ *Ans.* $y = 15$

2. Check the original equation:

2. $\dfrac{2}{3}x = 8$ $\dfrac{5}{3}y = 25$

$\dfrac{2}{3} \cdot 12 \overset{?}{=} 8$ $\dfrac{5}{3} \cdot 15 \overset{?}{=} 25$

$8 = 8$ $25 = 25$

SOLVED PROBLEMS

2.16   SOLVING EQUATIONS WITH NATURAL NUMBER DIVISORS
Solve:

(a) $\dfrac{x}{8} = 4$ (b) $\dfrac{1}{3}y = 12$ (c) $20 = \dfrac{z}{10}$ (d) $.2 = \dfrac{w}{40}$

Solutions

(a) $\dfrac{x}{8} = 4$ (b) $\dfrac{1}{3}y = 12$ (c) $20 = \dfrac{z}{10}$ (d) $.2 = \dfrac{w}{40}$

$\mathbf{M_8}$ $8 \cdot \dfrac{x}{8} = 8 \cdot 4$ $\mathbf{M_3}$ $3 \cdot \dfrac{1}{3}y = 3 \cdot 12$ $\mathbf{M_{10}}$ $10 \cdot 20 = 10 \cdot \dfrac{z}{10}$ $\mathbf{M_{40}}$ $40(.2) = 40 \cdot \dfrac{w}{40}$

Ans. $x = 32$ *Ans.* $y = 36$ *Ans.* $200 = z$ *Ans.* $8 = w$

Check: **Check:**   **Check:**   **Check:**

$$\frac{x}{8} = 4 \qquad\qquad \frac{1}{3}y = 12 \qquad\qquad 20 = \frac{z}{10} \qquad\qquad .2 = \frac{w}{40}$$

$$\frac{32}{8} \overset{?}{=} 4 \qquad\qquad \frac{1}{3}(36) \overset{?}{=} 12 \qquad\qquad 20 \overset{?}{=} \frac{200}{10} \qquad\qquad .2 \overset{?}{=} \frac{8}{40}$$

$$4 = 4 \qquad\qquad\qquad 12 = 12 \qquad\qquad\qquad 20 = 20 \qquad\qquad\qquad .2 = .2$$

2.17   SOLVING EQUATIONS WITH DECIMAL DIVISORS
Find each solution set:

(a) $\dfrac{a}{.5} = 4$ (b) $\dfrac{b}{.08} = 400$ (c) $1.5 = \dfrac{c}{1.2}$

Solutions

(a) $\dfrac{a}{.5} = 4$ (b) $\dfrac{b}{.08} = 400$ (c) $1.5 = \dfrac{c}{1.2}$

$\mathbf{M_{.5}}$ $.5\left(\dfrac{a}{.5}\right) = .5(4)$ $\mathbf{M_{.08}}$ $.08\left(\dfrac{b}{.08}\right) = .08(400)$ $\mathbf{M_{1.2}}$ $1.2(1.5) = 1.2\left(\dfrac{c}{1.2}\right)$

       $a = 2$ $b = 32$ $1.8 = c$

Ans.   {2}   *Ans.*   {32}   *Ans.*   {1.8}

2.18   SOLVING EQUATIONS WITH FRACTIONAL COEFFICIENTS
Solve:

(a) $\dfrac{2}{5}x = 10$ (b) $1\tfrac{1}{3}w = 30$ (c) $c - \dfrac{1}{4}c = 24$

Solutions

(a) $\dfrac{2}{5}x = 10$ (b) $1\tfrac{1}{3}w = 30$ (c) $c - \dfrac{1}{4}c = 24$

                             $\dfrac{4}{3}w = 30$ $\dfrac{3}{4}c = 24$

$\mathbf{M_{5/2}}$ $\dfrac{5}{2} \cdot \dfrac{2}{5}x = \dfrac{5}{2} \cdot 10$ $\mathbf{M_{3/4}}$ $\dfrac{3}{4} \cdot \dfrac{4}{3}w = \dfrac{3}{4} \cdot 30$ $\mathbf{M_{4/3}}$ $\dfrac{4}{3} \cdot \dfrac{3}{4}c = \dfrac{4}{3} \cdot 24$

Ans. $x = 25$ *Ans.* $w = 22\tfrac{1}{2}$ *Ans.* $c = 32$

2.19   SOLVING EQUATIONS WITH PER CENTS AS COEFFICIENTS

Solve: (a) $66\frac{2}{3}\%s = 22$, (b) $87\frac{1}{2}\%t = 35$, (c) $120\%w = 72$.

Solutions

Replace a per cent by a fraction if the per cent equals an easy fraction.

(a) $66\frac{2}{3}\%s = 22$ (b) $87\frac{1}{2}\%t = 35$ (c) $120\%w = 72$

$$\frac{2}{3}s = 22 \qquad\qquad \frac{7}{8}t = 35 \qquad\qquad \frac{6}{5}w = 72$$

$\mathbf{M}_{3/2}$ $\dfrac{3}{2}\cdot\dfrac{2}{3}s = \dfrac{3}{2}(22)$ $\mathbf{M}_{8/7}$ $\dfrac{8}{7}\cdot\dfrac{7}{8}t = \dfrac{8}{7}(35)$ $\mathbf{M}_{5/6}$ $\dfrac{5}{6}\cdot\dfrac{6}{5}w = \dfrac{5}{6}(72)$

Ans. $s = 33$ *Ans.* $t = 40$ *Ans.* $w = 60$

2.20   MULTIPLICATION RULE IN PROBLEM SOLVING

Mr. White receives 5% on a stock investment. If his interest at the end of one year was $140, how large was his investment?

Solution

Let s = the invested sum $.

Then, $5\%s = \dfrac{s}{20} = 140$

$\mathbf{M}_{20}$ $20\cdot\dfrac{s}{20} = 20(140)$

                          $s = 2800$

Ans.   The investment was $2800.

Check (the problem):

5% of the investment should be $140.
Hence,

$$5\%(\$2800) \overset{?}{=} \$140$$
$$\$140 = \$140$$

2.7   USING SUBTRACTION TO SOLVE AN EQUATION

2.7A   To Solve an Equation Using the Subtraction Rule of Equality (Section 4)

Solve: (a) $w + 12 = 19$ (b) $28 = 11 + x$.

PROCEDURE   SOLUTIONS

1. Subtract from both sides the   **1.** $w + 12 = \ \ 19$ $28 = \ \ 11 + x$
number added to the variable: $\mathbf{S}_{12}$ $\underline{-12 = -12}$ $\mathbf{S}_{11}$ $\underline{-11 = -11}$
                                   Ans. $w \ \ \ = \ \ \ 7$ *Ans.* $17 = \ \ \ \ \ x$

2. Check the original equation: **2.** $w + 12 = 19$ $28 = 11 + x$
                                                     $7 + 12 \overset{?}{=} 19$ $28 \overset{?}{=} 11 + 17$
                                                     $19 = 19$ $28 = 28$

Note: **S** is a convenient symbol for "subtracting from both sides."
          $\mathbf{S}_{11}$ means "subtract 11 from both sides."

SOLVED PROBLEMS

2.21   SUBTRACTION RULE IN EQUATIONS CONTAINING NATURAL NUMBERS

Find each solution set: (a) $r + 8 = 13$, (b) $15 + t = 60$, (c) $110 = s + 20$.

Solutions

(a)　$r + 8 = 13$
S_8　$\underline{-8 = -8}$
　　$r = 5$

Ans. {5}

Check:

$$r + 8 = 13$$
$$5 + 8 \overset{?}{=} 13$$
$$13 = 13$$

(b)　$15 + t = 60$
S_{15}　$\underline{-15 = -15}$
　　$t = 45$

Ans. {45}

Check:

$$15 + t = 60$$
$$15 + 45 \overset{?}{=} 60$$
$$60 = 60$$

(c)　$100 = s + 20$
S_{20}　$\underline{-20 = -20}$
　　$90 = s$

Ans. {90}

Check:

$$110 = s + 20$$
$$110 \overset{?}{=} 90 + 20$$
$$110 = 110$$

2.22 **SUBTRACTION RULE IN EQUATIONS CONTAINING FRACTIONS OR DECIMALS**

Solve: (a) $b + \frac{1}{3} = 3\frac{2}{3}$, (b) $2\frac{3}{4} + c = 8\frac{1}{2}$, (c) $20.8 = d + 6.9$.

Solutions

(a)　$b + \frac{1}{3} = 3\frac{2}{3}$
$S_{1/3}$　$\underline{-\frac{1}{3} = -\frac{1}{3}}$
Ans. $b = 3\frac{1}{3}$

Check:

$$b + \frac{1}{3} = 3\frac{2}{3}$$
$$3\frac{1}{3} + \frac{1}{3} \overset{?}{=} 3\frac{2}{3}$$
$$3\frac{2}{3} = 3\frac{2}{3}$$

(b)　$2\frac{3}{4} + c = 8\frac{1}{2}$
$S_{2\frac{3}{4}}$　$\underline{-2\frac{3}{4} = -2\frac{3}{4}}$
Ans.　$c = 5\frac{3}{4}$

Check:

$$2\frac{3}{4} + c = 8\frac{1}{2}$$
$$2\frac{3}{4} + 5\frac{3}{4} \overset{?}{=} 8\frac{1}{2}$$
$$8\frac{1}{2} = 8\frac{1}{2}$$

(c)　$20.8 = d + 6.9$
$S_{6.9}$　$\underline{-6.9 = -6.9}$
Ans.　$13.9 = d$

Check:

$$20.8 = d + 6.9$$
$$20.8 \overset{?}{=} 13.9 + 6.9$$
$$20.8 = 20.8$$

2.23 **SUBTRACTION RULE IN PROBLEM SOLVING**

After an increase of 22¢, the price of eggs rose to $1.22. What was the original price?

Solution

Let p = original price in ¢.

Then,　$p + 22 = 122$
S_{22}　$\underline{-22 = -22}$
　　$p = 100$

Check (the problem):

After increasing 22¢, the new
price should be $1.22. Hence,
$$\$1.00 + 22¢ \overset{?}{=} \$1.22$$
$$\$1.22 = \$1.22$$

Ans.　The original price was $1.00

2.24 **SUBTRACTION RULE IN PROBLEM SOLVING**

Harold's height is 5 ft 3 in. If he is 9 in. taller than John, how tall is John?

Solution

Let J = John's height in ft.

Then,　$J + \frac{3}{4} = 5\frac{1}{4}$　(9 in. = $\frac{3}{4}$ ft)

$S_{3/4}$　$\underline{-\frac{3}{4} = -\frac{3}{4}}$
　　$J = 4\frac{1}{2}$

Check (the problem):

9 in. more than John's height
should equal 5 ft 3 in.
$$4\frac{1}{2} \text{ ft} + \frac{3}{4} \text{ ft} \overset{?}{=} 5\frac{1}{4} \text{ ft}$$
$$5\frac{1}{4} \text{ ft} = 5\frac{1}{4} \text{ ft}$$

Ans.　John is $4\frac{1}{2}$ ft or 4 ft 6 in. tall.

2.8   USING ADDITION TO SOLVE AN EQUATION

2.8A   To Solve an Equation Using the Addition Rule of Equality (Section 4)

Solve: (a) $n - 19 = 21$ (b) $17 = m - 8$

PROCEDURE   SOLUTIONS

1. Add to both sides the number
    subtracted from the variable:

1.		$n - 19 =$	21		$17 = m - 8$
$\mathbf{A}_{19}$		$+19 =$	$+19$	$\mathbf{A}_s$	$+ 8 = \quad + 8$
Ans.		$n \quad =$	40	*Ans.*	$25 = m$

2. Check the original equation:

2.	$n - 19 = 21$	$17 = m - 8$
	$40 - 19 \overset{?}{=} 21$	$17 \overset{?}{=} 25 - 8$
	$21 = 21$	$17 = 17$

Note: **A** is a convenient symbol for "adding to both sides."
        $\mathbf{A}_{19}$ means "add 19 to both sides."

SOLVED PROBLEMS

2.25   ADDITION RULE IN EQUATIONS CONTAINING NATURAL NUMBERS
Find each solution set: (a) $w - 10 = 19$, (b) $x - 19 = 10$, (c) $7 = y - 82$, (d) $82 = z - 7$.

Solutions

(a) $w - 10 =$	19	(b) $x - 19 =$	10	(c) $7 = y - 82$	(d) $82 = z - 7$
$\mathbf{A}_{10} \quad +10 =$	$+10$	$\mathbf{A}_{19} \quad +19 =$	$+19$	$\mathbf{A}_{82} \quad +82 = \quad +82$	$\mathbf{A}_7 \quad +7 = \quad +7$
$w \quad =$	29	$x \quad =$	29	$89 = y$	$89 = z$

Ans. {29}   *Ans.* {29}   *Ans.* {89}   *Ans.* {89}

Check: **Check:**   **Check:**   **Check:**

$w - 10 = 19$ $x - 19 = 10$ $7 = y - 82$ $82 = z - 7$

$29 - 10 \overset{?}{=} 19$ $29 - 19 \overset{?}{=} 10$ $7 \overset{?}{=} 89 - 82$ $82 \overset{?}{=} 89 - 7$

$19 = 19$ $10 = 10$ $7 = 7$ $82 = 82$

2.26   ADDITION RULE IN EQUATIONS CONTAINING FRACTIONS OR DECIMALS
Solve: (a) $h - \frac{3}{8} = 5\frac{1}{4}$, (b) $j - 20\frac{7}{12} = 1\frac{1}{12}$, (c) $12.5 = m - 2.9$.

Solutions

(a) $h - \frac{3}{8} = 5\frac{1}{4}$	(b) $j - 20\frac{7}{12} = 1\frac{1}{12}$	(c) $12.5 = m - 2.9$
$\mathbf{A}_{3/8} \quad +\frac{3}{8} = \frac{3}{8}$	$\mathbf{A}_{20\frac{7}{12}} \quad +20\frac{7}{12} = 20\frac{7}{12}$	$\mathbf{A}_{2.9} \quad +2.9 = \quad +2.9$
Ans. $h \quad = 5\frac{5}{8}$	*Ans.* $j \quad = 21\frac{2}{3}$	*Ans.* $15.4 = m$

2.27   ADDITION RULE IN PROBLEM SOLVING

A drop of 8° brought the temperature to 64°. What was the original temperature?

Solution

Let t = original temperature in °.

Then, $\quad t - 8 = 64$

$\mathbf{A}_8 \qquad \underline{+8 = +8}$

$\qquad\quad t \quad = \overline{72}$

Check (the problem):

The original temperature, dropped 8°, should become 64°. Hence,

$$72° - 8° \stackrel{?}{=} 64°$$
$$64° = 64°$$

Ans.   The original temperature was 72°.

2.28   ADDITION RULE IN PROBLEM SOLVING

After giving 15 marbles to Sam, Joe has 43 left. How many did Joe have originally?

Solutions

Let m = original no. of marbles.

Then, $\quad m - 15 = \quad 43$

$\mathbf{A}_{15} \qquad \underline{+15 = +15}$

$\qquad\quad m \quad = \quad 58$

Check (the problem):

The original number of marbles, less 15, should be 43. Hence,

$$58 \text{ marbles} - 15 \text{ marbles} \stackrel{?}{=} 43 \text{ marbles}$$
$$43 \text{ marbles} = 43 \text{ marbles}$$

Ans.   Joe had 58 marbles at first.

2.9   USING TWO OR MORE OPERATIONS TO SOLVE AN EQUATION

In equations where two operations are performed upon the variable, two inverse operations may be needed to solve the equation.

Thus, in $2x + 7 = 19$, the two operations upon the variable are *multiplication and addition*. To solve, use *division and subtraction*, performing subtraction first.

Also, in $(x/3) - 5 = 2$, the two operations upon the variable are *division and subtraction*. To solve, use *multiplication and addition*, performing addition first.

2.9A   To Solve Equations Using Two Inverse Operations

Solve: $\qquad\qquad\qquad\qquad$ (a) $2x + 7 = 19$ $\qquad\qquad$ (b) $\dfrac{x}{3} - 5 = 2$

PROCEDURE $\qquad\qquad\qquad\qquad\qquad\qquad\qquad\qquad\qquad$ SOLUTIONS

1. Perform addition to undo subtraction, or subtraction to undo addition:

    1. $\quad 2x + 7 = \quad 19$

    $\mathbf{S}_7 \qquad \underline{-7 = -7}$

    $\qquad 2x \quad = \quad 12$

    $\mathbf{A}_5 \qquad \dfrac{x}{3} - 5 = \quad 2$

    $\qquad \underline{+5 = +5}$

    $\qquad \dfrac{x}{3} \quad = \quad 7$

2. Perform multiplication to undo division, or division to undo multiplication:

    2. $\mathbf{D}_2$ $\quad \dfrac{2x}{2} = \dfrac{12}{2}$

    Ans. $\qquad x = 6$

    $\mathbf{M}_3 \qquad 3 \cdot \dfrac{x}{3} = 3 \cdot 7$

    Ans. $\qquad x = 21$

3. Check the original equation:

    3. $\quad 2x + 7 = 19$

    $\qquad 2(6) + 7 \stackrel{?}{=} 19$

    $\qquad\qquad 19 = 19$

    $\dfrac{x}{3} - 5 = 2$

    $\dfrac{21}{3} - 5 \stackrel{?}{=} 2$

    $\qquad 2 = 2$

SOLVED PROBLEMS

2.29 **USING TWO INVERSE OPERATIONS TO SOLVE AN EQUATION**
Solve:

(a) $2x + 7 = 11$ (b) $3x - 5 = 7$ (c) $\dfrac{x}{3} + 5 = 7$ (d) $\dfrac{x}{5} - 3 = 7$

Solutions

(a) $2x + 7 = 11$ (b) $3x - 5 = 7$ (c) $\dfrac{x}{3} + 5 = 7$ (d) $\dfrac{x}{5} - 3 = 7$

S_7 $\dfrac{-7 = -7}{2x = 4}$ A_5 $\dfrac{+5 = +5}{3x = +12}$ S_5 $\dfrac{-5 = -5}{\dfrac{x}{3} = 2}$ A_3 $\dfrac{+3 = 3}{\dfrac{x}{5} = 10}$

D_2 $\dfrac{2x}{2} = \dfrac{4}{2}$ D_3 $\dfrac{3x}{3} = \dfrac{12}{3}$ M_3 $3 \cdot \dfrac{x}{3} = 3 \cdot 2$ M_5 $5 \cdot \dfrac{x}{5} = 5 \cdot 10$

Ans. $x = 2$ *Ans.* $x = 4$ *Ans.* $x = 6$ *Ans.* $x = 50$

(a) **Check:**   (b) **Check:**   (c) **Check:**   (d) **Check:**

$2x + 7 = 11$ $3x - 5 = 7$ $\dfrac{x}{3} + 5 = 7$ $\dfrac{x}{5} - 3 = 7$

$2(2) + 7 \overset{?}{=} 11$ $3(4) - 5 \overset{?}{=} 7$ $\dfrac{6}{3} + 5 \overset{?}{=} 7$ $\dfrac{50}{5} - 3 \overset{?}{=} 7$

$4 + 7 \overset{?}{=} 11$ $12 - 5 \overset{?}{=} 7$ $2 + 5 \overset{?}{=} 7$ $10 - 3 \overset{?}{=} 7$

$11 = 11$ $7 = 7$ $7 = 7$ $7 = 7$

2.30 **SOLVING EQUATIONS WITH LIKE TERMS ON THE SAME SIDE**
Solve: (a) $8n + 4n - 3 = 9$, (b) $13n + 4 + n = 39$, (c) $10 = 7 + n - (n/2)$.

Solutions

Combine like terms first.

(a) $\begin{aligned} 8n + 4n - 3 &= 9 \\ 12n - 3 &= 9 \end{aligned}$ (b) $\begin{aligned} 13n + 4 + n &= 39 \\ 14n + 4 &= 39 \end{aligned}$ (c) $10 = 7 + n - \dfrac{n}{2}$

A_3 $\dfrac{+3 = 3}{12n = 12}$ S_4 $\dfrac{-4 = -4}{14n = 35}$ $10 = 7 + \dfrac{n}{2}$

D_{12} $\dfrac{12n}{12} = \dfrac{12}{12}$ D_{14} $\dfrac{14n}{14} = \dfrac{35}{14}$ S_7 $\dfrac{-7 = -7}{3 = \dfrac{n}{2}}$

Ans. $n = 1$ *Ans.* $n = 2\frac{1}{2}$ M_2 $2 \cdot 3 = 2 \cdot \dfrac{n}{2}$

Ans. $6 = n$

2.31 **SOLVING EQUATIONS WITH LIKE TERMS ON OPPOSITE SIDES**
Find each solution set: (a) $5n = 40 - 3n$, (b) $4u + 5 = 5u - 30$, (c) $3r + 10 = 2r + 20$.

Solutions

First, add or subtract to collect like terms on the same side.

(a) $5n = 40 - 3n$
$\mathbf{A}_{3n}$ $\underline{+3n = \quad +3n}$
    $8n = 40$

$\mathbf{D}_8$ $\dfrac{8n}{8} = \dfrac{40}{8}$

    $n = 5$

Ans.   {5}

(b) $4u + 5 = 5u - 30$
$\mathbf{A}_{30}$ $\underline{+30 = \quad +30}$
    $4u + 35 = 5u$

$\mathbf{S}_{4u}$ $\dfrac{-4u \quad = -4u}{35 = u}$

Ans.   {35}

(c) $3r + 10 = 2r + 20$
$\mathbf{S}_{10}$ $\dfrac{-10 = \quad -10}{3r = 2r + 10}$

$\mathbf{S}_{2r}$ $\dfrac{-2r \quad = -2r}{r = 10}$

Ans.   {10}

2.32 SOLVING EQUATIONS IN WHICH THE VARIABLE IS A DIVISOR
Solve:

(a) $\dfrac{8}{x} = 2$ (b) $12 = \dfrac{3}{y}$ (c) $\dfrac{7}{x} = \dfrac{1}{5}$ (d) $\dfrac{1}{7} = \dfrac{3}{x}$

Solutions

First multiply both sides by the variable.

(a) $\dfrac{8}{x} = 2$

$\mathbf{M}_x$ $\left(\dfrac{8}{x}\right)x = 2x$

    $8 = 2x$

$\mathbf{D}_2$ $\dfrac{8}{2} = \dfrac{2x}{2}$

Ans. $4 = x$

(b) $12 = \dfrac{3}{y}$

$\mathbf{M}_y$ $12y = \left(\dfrac{3}{y}\right)y$

    $12y = 3$

$\mathbf{D}_{12}$ $\dfrac{12y}{12} = \dfrac{3}{12}$

Ans. $y = 1/4$

(c) $\dfrac{7}{x} = \dfrac{1}{5}$

$\mathbf{M}_x$ $\dfrac{7}{x} \cdot x = \dfrac{1}{5}x$

    $7 = \dfrac{x}{5}$

$\mathbf{M}_5$ $5(7) = 5\left(\dfrac{x}{5}\right)$

Ans. $35 = x$

(d) $\dfrac{1}{7} = \dfrac{3}{x}$

$\mathbf{M}_x$ $\dfrac{1}{7}x = \dfrac{3}{x} \cdot x$

    $\dfrac{x}{7} = 3$

$\mathbf{M}_7$ $7 \cdot \dfrac{x}{7} = 7 \cdot 3$

Ans. $x = 21$

2.33 SOLVING EQUATIONS WHOSE VARIABLE HAS A FRACTIONAL COEFFICIENT

(a) Solve: $\dfrac{3}{8}x = 9$

(b) Solve: $25 = \dfrac{5}{4}x$

Solutions
Using One Operation

(a) $\dfrac{3}{8}x = 9$

$\mathbf{M}_{8/3}$ $\dfrac{8}{3} \cdot \dfrac{3}{8}x = \dfrac{8}{3} \cdot 9$

Ans. $x = 24$

Using Two Operations

(a) $\dfrac{3}{8}x = 9$

$\mathbf{M}_8$ $8 \cdot \dfrac{3}{8}x = 8 \cdot 9$

    $3x = 72$

$\mathbf{D}_3$ $\dfrac{3x}{3} = \dfrac{72}{3}$

Ans. $x = 24$

Solutions
Using One Operation

(b) $25 = \dfrac{5}{4}x$

$\mathbf{M}_{4/5}$ $\dfrac{4}{5} \cdot 25 = \dfrac{4}{5} \cdot \dfrac{5}{4}x$

Ans. $20 = x$

Using Two Operations

(b) $25 = \dfrac{5}{4}x$

$\mathbf{M}_4$ $4(25) = 4 \cdot \dfrac{5}{4}x$

    $100 = 5x$

$\mathbf{D}_5$ $\dfrac{100}{5} = \dfrac{5x}{5}$

Ans. $20 = x$

2.34 SOLVING MORE DIFFICULT EQUATIONS
Solve:

(a) $\dfrac{3}{4}y - 5 = 7$ (b) $8 + \dfrac{2}{7}b = 20$ (c) $48 - \dfrac{5}{3}w = 23$

Solutions

(a) $\dfrac{3}{4}y - 5 = 7$

$\mathbf{A_5}$ $\quad\underline{\quad +5 = +5\quad}$

$\dfrac{3}{4}y = 12$

$\mathbf{M_{4/3}}$ $\quad\dfrac{4}{3}\cdot\dfrac{3}{4}y = \dfrac{4}{3}\cdot 12$

Ans. $\quad y = 16$

(b) $8 + \dfrac{2}{7}b = 20$

$\mathbf{S_8}$ $\quad\underline{-8\qquad\; = -8}$

$\dfrac{2}{7}b = 12$

$\mathbf{M_{7/2}}$ $\quad\dfrac{7}{2}\cdot\dfrac{2}{7}b = \dfrac{7}{2}\cdot 12$

Ans. $\quad b = 42$

(c) $48 - \dfrac{5}{3}w = 23$

$\mathbf{A_{\frac{5}{3}w}}$ $\quad\underline{+\dfrac{5}{3}w = +\dfrac{5}{3}w}$

$48 = \dfrac{5}{3}w + 23$

$\mathbf{S_{23}}$ $\quad\underline{-23 = \qquad -23}$

$25 = \dfrac{5}{3}w$

$\mathbf{M_{3/5}}$ $\quad\dfrac{3}{5}\cdot 25 = \dfrac{3}{5}\cdot\dfrac{5}{3}w$

Ans. $\quad 15 = w$

2.35   Using Two Operations in Problem Solving

How many boys are there in a class of 36 pupils if the number of girls is (a) 6 more? (b) three times as many?

Solutions

(a)　　Let b = number of boys
　　Then $b + 6$ = number of girls
　　　　$b + (b + 6) = 36$
$\mathbf{S_6}$　　　　$2b + 6 = 36$
$\mathbf{D_2}$　　　　$2b = 30$
　　　　　　$b = 15$

Ans. There are 15 boys.

(b)　　Let b = number of boys
　　Then $3b$ = number of girls
　　　　$b + 3b = 36$
$\mathbf{D_4}$　　　　$4b = 36$
　　　　　$b = 9$

Ans. There are 9 boys.

2.36   Using Two Operations in Problem Solving

Paul has $26.00 in his bank. By adding equal deposits each week for 20 weeks, he hopes to have $78.00. How much should each weekly deposit be?

Solution

Let d = no. of dollars in each deposit
Then, $\quad 20d + 26 = 78$
$\mathbf{S_{260}}$ $\quad\underline{\quad\;\; -26 = -26}$
$\mathbf{D_{20}}$ $\quad 20d \quad = 52$
$\qquad\quad d \quad = 2.60$

Ans. He must deposit $2.60 a week.

Check (the problem):
　20 deposits and $26 should equal
　the total of $78. Hence,
　$20(\$2.60) + \$26 \overset{?}{=} \$78$
　$\$52 + \$26 \overset{?}{=} \$78$
　$\$78 = \$78$

2.37   Using Two Operations in Problem Solving

Mr. Richards sold his boat for $9000. His loss amounted to two-fifths of his cost. What did the boat cost him?

Solution

Let c = cost in $

Then
$$9000 = c - \frac{2}{5}c$$

$$9000 = \frac{3}{5}c$$

$\mathbf{M}_{5/3}$ $\frac{5}{3} \cdot 9000 = \frac{5}{3} \cdot \frac{3}{5}c$

$$15,000 = c$$

Ans.   The cost was $15,000.

Check (the problem):

If $15,000 is the cost, the loss is $\frac{2}{5} \cdot$ $15,000 or $6,000. The selling price of $9000 should be the cost minus the loss. Hence,

$9000 \overset{?}{=} $15,000 - $6000

$9000 = $9000

2.10   DERIVING FORMULAS

2.10A   To Derive a Formula for Related Quantities

Derive a formula relating the *distance* (D) traveled in a *time* (T) at a *rate of speed* (R).

PROCEDURE	SOLUTION
1. Obtain sets of values for these quantities, using convenient numbers:	**1.** **Sets of Values** At 50 mph for 2 hr, 100 mi will be traveled. At 25 mph for 10 hr, 250 mi will be traveled. At 40 mph for 3 hr, 120 mi will be traveled.
2. Tabulate those sets of values:	**2.** **Table of Values.** (Place units above quantities.)

(mph) Rate (R)	(hr) Time (T)	(mi) Distance (D)
50	2	$50 \cdot 2 = 100$
25	10	$25 \cdot 10 = 250$
40	3	$40 \cdot 3 = 120$

3. State the rule that follows:	**3.** **Rule:** The product of the rate and time equals the distance.
4. State the formula that expresses the rule:	**4.** **Formula:** $= RT = D$

Note: If D is in mi and T in hr, then R must be in mi per hr (mph). In general, *rate* must be in distance units per time unit.

2.10B   Obtaining Formulas from a More General Formula

A formula such as $RT = D$ relates *three* quantities: time, rate, and distance. Each of these quantities may vary in value; that is, they may have many values. However, in a problem, situation, or discussion, one of these quantities may have a fixed value. When such is the case, this constant value may be used to obtain a formula relating the other *two* quantities.

Thus, $D = RT$ leads to $D = 30T$ if the rate of speed is fixed at 30 mph, 30 km/hr, etc.

SOLVED PROBLEMS

2.38   DERIVING A COIN FORMULA

Derive a formula for the number of nickels (n) equivalent (equal in value) to q quarters.

Solution

1.   Sets of Values

2 quarters equals 10 nickels
4 quarters equals 20 nickels
10 quarters equals 50 nickels
q quarters equals $5n$ nickels

$\longrightarrow$

2.   Table of Values

No. of Quarters (q)	No. of Nickels (n)
2	$5 \cdot 2$ or 10
4	$5 \cdot 4$ or 20
10	$5 \cdot 10$ or 50
q	$5q$

3.   Rule: The number of nickels equivalent to a number of quarters is five times that number.

4.   Formula: $n = 5q$ *Ans.*

2.39   DERIVING A COIN FORMULA

Derive a formula for the value in cents (c) of d dimes and n nickels

Solution

1.   Sets of Values

3 dimes and 4 nickels equals 50¢
4 dimes and 2 nickels equals 50¢
5 dimes and 3 nickels equals 65¢
d dimes and n nickels equals $(10d + 5n)$¢

$\longrightarrow$

2.   Table of Values

No. of Dimes (d)	No. of Nickels (n)	(cents) Value of Dimes & Nickels (c)
3	4	$10 \cdot 3 + 5 \cdot 4$ or 50
4	2	$10 \cdot 4 + 5 \cdot 2$ or 50
5	3	$10 \cdot 5 + 5 \cdot 3$ or 65
d	n	$10d + 5n$

3.   Rule: The value in cents of dimes and nickels is ten times the number of dimes plus five times the number of nickels.

4.   Formula: $c = 10d + 5n$ *Ans.*

2.40   DERIVING COIN FORMULAS

Derive a formula for (a) the number of pennies (p) equivalent to q quarters, (b) the number of nickels (n) equivalent to d dimes, (c) the number of quarters (q) equivalent to D dollars, (d) the number of pennies (p) equivalent to n nickels and q quarters, (e) the number of nickels (n) equivalent to q quarters and d dimes.

Illustrative Solution

(b) Since 1 nickel is 5¢ and 1 dime is 10¢, if nickels are traded for dimes, the number of nickels must be twice the number of dimes.   *Ans.* $n = 2d$.

Ans.   (a) $p = 25q$ (c) $q = 4D$ (d) $p = 5n + 25q$ (e) $n = 5q + 2d$

2.41 DERIVING TIME FORMULAS

Derive a formula for (*a*) the number of seconds (*s*) in *m* minutes, (*b*) the number of hours (*h*) in *d* days, (*c*) the number of weeks (*w*) in *d* days, (*d*) the number of days (*d*) in *w* weeks and 5 days, (*e*) the number of minutes (*m*) in *h* hours and 30 sec.

Illustrative Solution

(*c*) Since 1 week = 7 days, the number of weeks, *w*, equals 1/7 of the number of days, *d*. *Ans.* $w = (1/7)d$ or $w = d/7$.

Ans.   (*a*) $s = 60m$ (*b*) $h = 24d$ (*d*) $d = 7w + 5$ (*e*) $m = 60h + .5$ or $m = 60h + \frac{1}{2}$

2.42 DERIVING LENGTH FORMULAS

Derive a formula for (*a*) the number of in. (*i*) in *f* ft, (*b*) the number of ft (*f*) in *y* yd, (*c*) the number of yd (*y*) in *i* in., (*d*) the number of ft (*f*) in *m* mi and 50 yd, (*e*) the number of cm (*c*) in *m* meters.

Illustrative Solution

(*b*) Since 1 yd = 3 ft, the number of ft in *y* yd must be three times *y*.   *Ans.* $f = 3y$

Ans.   (*a*) $i = 12f$ (*c*) $y = i/36$ (*d*) $f = 5280m + 150$ (*e*) $c = \dfrac{m}{100}$

2.43 OBTAINING FORMULAS FROM $D = RT$

From $D = RT$, derive a formula relating (*a*) the distance in mi and time in hr when the rate is 35 mph, (*b*) the distance in ft and the time in sec when sound travels at 1100 ft sec, (*c*) the distance in mi and the time in sec if light travels at 186,000 mi per sec, (*d*) the distance in mi and the rate in mph when the time of travel is 1 hr and 30 min, (*e*) the rate in km per hr and the time in hr when the distance traveled is 125 km.

Illustrative Solutions

(*d*) If the distance is in miles and the rate in miles per hour (mph), the time of travel *T*, is in hours. Since $T = 1\frac{1}{2}$ hours, substitute $1\frac{1}{2}$ for *T* in $D = RT$.   *Ans.* $D = 1\frac{1}{2}R$. (*e*) Since the distance traveled, *D*, is 125, substitute 125 for *D* in $D = RT$.   *Ans.* $125 = RT$.

Ans.   (*a*) $D = 35T$ (*b*) $D = 1100T$ (*c*) $D = 186,000T$

2.11 TRANSFORMING FORMULAS

The *subject of a formula* is the variable that is expressed in terms of the other variables. Thus, in $p = 4s$, *p* is the subject of the formula.

Transforming a formula is the process of changing the subject of the formula. Thus, $p = 4s$ becomes $p/4 = s$ when both sides are divided by 4. In the transforming of the formula, the subject has changed from *p* to *s*.

In *Solving a formula* for a variable, the formula is transformed in order that the variable be made the subject. Thus, to solve $D = 5T$ for *T*, transform it into $T = D/5$.

2.11A Use of Inverse Operations to Transform Formulas

1. Use *division* to undo *multiplication*.
    Thus, $c = 25q$ becomes $c/25 = q$ by division.

2. Use *multiplication* to undo *division*.
   Thus, $w = d/7$ becomes $7w = d$ by multiplication.

3. Use *subtraction* to undo *addition*.
   Thus, $S = P + C$ becomes $S - P = C$ by subtraction.

4. Use *addition* to undo *subtraction*.
   Thus, $S = C - L$ becomes $S + L = C$ by addition.

Formulas may be transformed by transposing terms. In transposing a term, change its sign. Thus, $a + b = 180$ becomes $a = 180 - b$ when $+b$ is transposed. Actually, $+b$ has been subtracted from both sides to undo addition.

SOLVED PROBLEMS

2.44 TRANSFORMATIONS REQUIRING DIVISION
Solve: (a) $D = RT$ for R, (b) $D = RT$ for T, (c) $V = LWH$ for L, (d) $c = 10d$ for d, (e) $C = 2\pi r$ for r.

Solutions

(a) $D = RT$

$\mathbf{D}_T \quad \dfrac{D}{T} = \dfrac{RT}{T}$

Ans. $\dfrac{D}{T} = R$

(b) $D = RT$

$\mathbf{D}_R \quad \dfrac{D}{R} = \dfrac{RT}{R}$

Ans. $\dfrac{D}{R} = T$

(c) $V = LWH$

$\mathbf{D}_{WH} \quad \dfrac{V}{WH} = \dfrac{LWH}{WH}$

Ans. $\dfrac{V}{WH} = L$

(d) $c = 10d$

$\mathbf{D}_{10} \quad \dfrac{c}{10} = \dfrac{10d}{10}$

Ans. $\dfrac{c}{10} = d$

(e) $C = 2\pi r$

$\mathbf{D}_{2\pi} \quad \dfrac{C}{2\pi} = \dfrac{2\pi r}{2\pi}$

Ans. $\dfrac{C}{2\pi} = r$

2.45 TRANSFORMATIONS REQUIRING MULTIPLICATION
Solve:

(a) $\dfrac{i}{12} = f$ for i (b) $f = \dfrac{n}{d}$ for n (c) $\dfrac{V}{LW} = H$ for V (d) $\dfrac{b}{2} = \dfrac{A}{h}$ for A

Solutions

(a) $\dfrac{i}{12} = f$

$\mathbf{M}_{12} \quad 12\left(\dfrac{i}{12}\right) = 12f$

Ans. $i = 12f$

(b) $f = \dfrac{n}{d}$

$\mathbf{M}_d \quad df = d\left(\dfrac{n}{d}\right)$

Ans. $df = n$

(c) $\dfrac{V}{LW} = H$

$\mathbf{M}_{LW} \quad LW\left(\dfrac{V}{LW}\right) = LWH$

Ans. $V = LWH$

(d) $\dfrac{b}{2} = \dfrac{A}{h}$

$\mathbf{M}_h \quad \left(\dfrac{b}{2}\right)h = \left(\dfrac{A}{h}\right)h$

Ans. $\dfrac{bh}{2} = A$

2.46 TRANSFORMATIONS REQUIRING ADDITION OR SUBTRACTION (TRANSPOSING)
Solve: (a) $a + b = 90$ for a, (b) $a = b - 180$ for b, (c) $a + c = b + 100d$ for b, (d) $a - b - 25 = c$ for a.

Solutions

(a) $a + b = 90$

Transpose b:
Ans. $a = 90 - b$

(b) $a = b - 180$
Transpose -180:
Ans. $a + 180 = b$

(c) $a + c = b + 100d$
Transpose $100d$:
Ans. $a + c - 100d = b$

(d) $a - b - 25 = c$
Transpose $-b - 25$:
Ans. $a = b + c + 25$

2.47 TRANSFORMATIONS REQUIRING TWO OPERATIONS
Solve:

(a) $P = 2a + b$ for a (b) $c = 10d + 25q$ for q (c) $F = \dfrac{9}{5}C + 32$ for C (d) $V = \dfrac{1}{3}Bh$ for B

Solutions

(a) $P = 2a + b$ (b) $c = 10d + 25q$ (c) $F = \dfrac{9}{5}C + 32$ (d) $V = \dfrac{1}{3}Bh$

Transpose b: **Transpose** $10d$:   **Transpose** 32: $\mathbf{M_3}$ $3V = 3\left(\dfrac{1}{3}Bh\right)$

$\mathbf{D_2}$ $P - b = 2a$ $\mathbf{D_{25}}$ $c - 10d = 25q$ $\mathbf{M_{5 \cdot 9}}$ $F - 32 = \dfrac{9}{5}C$ $\mathbf{D_h}$ $3V = Bh$

Ans. $\dfrac{P - b}{2} = a$ Ans. $\dfrac{c - 10d}{25} = q$ Ans. $\dfrac{5}{9}(F - 32) = C$ Ans. $\dfrac{3V}{h} = B$

2.48 MORE DIFFICULT TRANSFORMATIONS
  Solve:

(a) $A = \dfrac{h}{2}(b + b')$ for h (b) $S = \dfrac{n}{2}(a + l)$ for a (c) $l = a + (n-1)d$ for n

Solutions

(a) $A = \dfrac{h}{2}(b + b')$ (b) $S = \dfrac{n}{2}(a + l)$ (c) $l = a + (n-1)d$

$\mathbf{M_2}$ $2A = 2\left(\dfrac{h}{2}\right)(b + b')$ $\mathbf{M_2}$ $2S = n(a + l)$ **Transpose** a:
$\mathbf{D_d}$ $l - a = (n-1)d$

$\mathbf{D_{(b+b')}}$ $\dfrac{2A}{b + b'} = \dfrac{h(b + b')}{b + b'}$ $\mathbf{D_n}$ $\dfrac{2S}{n} = a + l$ $\dfrac{l - a}{d} = n - 1$

      Transpose $+l$:   **Transpose** -1:

Ans. $\dfrac{2A}{b + b'} = h$ Ans. $\dfrac{2S}{n} - l = a$ Ans. $\dfrac{l - a}{d} + 1 = n$

2.12 FINDING THE VALUE OF A VARIABLE IN A FORMULA

2.12A When the Variable is the Subject of the Formula

To evaluate the subject of a formula, replace the other variables by their given values and evaluate the numerical expression that results.
Thus, in $A = bh$ if $b = 10$ and $h = 5$, then $A = 10(5)$ or 50.

2.12B When the Variable is not the Subject of the Formula

When a variable is to be found and it is not the subject of the formula, two methods may be used:

Method 1: *Substitute first*, then solve.

Method 2: *Transform the formula first* to make the variable to be found the subject of the formula, then substitute and evaluate.

Thus, in $p = 3s$ if $p = 27$, the value of s may be found

(1) by substituting first: $27 = 3s$, $s = 9$.

(2) by transforming first: transform $p = 3s$ into $s = \dfrac{p}{3}$. Then $s = \dfrac{27}{3}$ or 9.

SOLVED PROBLEMS

2.49   To Find the Value of a Variable that is the Subject of a Formula

(a) Find V if $V = lwh$ and $l = 10$, $w = 2$, $h = 3.2$.

(b) Find S if $S = \dfrac{n}{2}(a + l)$ and $n = 8$, $a = 5$, $l - 12$.

(c) Find A if $A = p + prt$ and $p = 800$, $r = .04$, $t = 3$.

(d) Find S if $S = \dfrac{1}{2}gt^2$ and $g = 32$, $t = 5$.

Solutions

(a) $V = lwh$ (b) $S = \dfrac{n}{2}(a + l)$ (c) $A = p + prt$ (d) $S = \dfrac{1}{2}gt^2$

    $V = 10(2)(3.2)$ $A = 800 + 800(.04)3$

Ans. $V = 64$ $S = \dfrac{8}{2}(5 + 12)$ *Ans.* $A = 896$ $S = \dfrac{1}{2} \cdot 32 \cdot 5^2$

                            Ans. $S = 68$ *Ans.* $S = 400$

2.50   To Find the Value of a Variable that is not the Subject of a Formula

(a) Find h if $A = bh$,   (b) Find a if $p = 2a + b$,   (c) Find h if $V = \dfrac{1}{3}Bh$,

    $b = 13$ and $A = 156$. $b = 20$ and $p = 74$. $B = 240$ and $V = 960$.

Solutions

(a) **(1) By substitution first:**   (b) **(1) By substitution first:**   (c) **(1) By substitution first:**

        $A = bh$ $p = 2a + b$ $V = \dfrac{1}{3}Bh$

      $156 = 13h$ $74 = 2a + 20$

Ans. $12 = h$ *Ans.* $27 = a$ $960 = \dfrac{1}{3}(240)h$

                                                       Ans. $12 = h$

(2) By transformation first: **(2) By transformation first:**   **(2) By transformation first:**

        $A = nh$ $p = 2a + b$ $V = \dfrac{1}{3}Bh$

Transform: $\dfrac{A}{b} = h$ **Transform:** $\dfrac{p - b}{2} = a$ **Transform:** $\dfrac{3V}{B} = h$

Substitute: $\dfrac{156}{13} = h$ **Substitute:** $\dfrac{74 - 20}{2} = a$ **Substitute:** $\dfrac{3(960)}{240} = h$

Ans. $12 = h$ *Ans.* $27 = a$ *Ans.* $12 = h$

2.51   More Difficult Evaluations Using Transformations
    (a) Find h if $V = \pi r^2 h$, $\pi = 3.14$, $V = 9420$ and $r = 10$. (b) Find t if $A = p + prt$, $A = 864$, $p = 800$ and $r = 2$.

Solutions

(a)
$$V = \pi r^2 h$$

Transform:
$$\frac{V}{\pi r^2} = h$$

Substitute:
$$\frac{9420}{(3.14)(100)} = h$$

$$\frac{9420}{314} = h$$

Ans.
$$30 = h$$

(b)
$$A = p + prt$$

Transform:
$$\frac{A - p}{pr} = t$$

Substitute:
$$\frac{864 - 800}{800(2)} = t$$

$$\frac{64}{1600} = t$$

Ans.
$$.04 = t$$

2.52 FINDING AN UNKNOWN IN A PROBLEM

A train takes 3 hours and 15 minutes to go a distance of 247 miles. Find its average speed.

Solution

Here, $D = RT$, $T = 3\frac{1}{4}$ hr and $D = 247$. To find R:

(1) By substitution first:

$$247 = \frac{13}{4}R$$

$$\mathbf{M}_{4/13} \quad \frac{4}{13} \cdot \overset{19}{\cancel{247}} = R$$

$$76 = R$$

(2) By transformation first:

$$D = RT$$

Transform:
$$\frac{D}{T} = R$$

$$247 \div \frac{13}{4} = R$$

$$247 \cdot \frac{4}{13} = R$$

Ans. Average speed is 76 mph.

2.13 USING A CALCULATOR TO PERFORM THE FUNDAMENTAL OPERATIONS OF ARITHMETIC

The calculator can be used to perform various calculations with extreme ease. We assume, for this text, that the student will use a basic calculator which accommodates the four basic operations, as well as signed numbers, squares, and square roots. Figure 2-3 illustrates such a calculator.

SOLVED PROBLEMS

2.53 SIMPLE CALCULATIONS

Evaluate each using a calculator:

(a) $430y + 920y$ (b) $671r + 284r$ (c) $\dfrac{5063x^3}{29x} \ (x \neq 0)$

Solutions

(a) Press: $\boxed{4}\boxed{3}\boxed{0}\boxed{+}\boxed{9}\boxed{2}\boxed{0}\boxed{=}$

Answer on screen 1,350; Answer = 1350y

(b) Press: $\boxed{6}\boxed{7}\boxed{1}\boxed{*}\boxed{2}\boxed{8}\boxed{4}\boxed{=}$

Answer on screen 190,564; Answer = $190{,}564r^2$

Fig. 2-3

(c)   Press: $\boxed{5}\,\boxed{0}\,\boxed{6}\,\boxed{3}\,\boxed{\div}\,\boxed{2}\,\boxed{9}\,\boxed{=}$

Answer on screen 174.5862; Answer $= 174.5862x^2$

2.54   USING THE CALCULATOR TO SOLVE ALGEBRAIC PROBLEMS

(a) Find w if $V = lwh$ and $V = 1.386$, $l = .87$, and $h = .06$.

(b) Find l if $S = \dfrac{n}{2}(a + l)$ and $S = 184$, $n = 16.8$, and $a = 9$.

Solutions

(a) $V = lwh$

   $1.386 = .87(w)(.06)$

Use the calculator to conjugate $(.87)(.06)$. The answer is .0522. Use the calculator to complete the problem:

   $1.386 = .0522w$

   $w = \dfrac{1.386}{.0522} = 26.551724$ (this is the answer to 8 significant digits)

(b) $S = \dfrac{n}{2}(a + l)$

   $184 = \dfrac{16.8}{2}(a + l)$

Use the calculator to conjugate $\dfrac{16.8}{2}\,(=8.4)$ and $\dfrac{184}{8.4}$.

   $\dfrac{184}{8.4} = 21.904761$ (this is the answer to 8 significant digits)

   $21.904761 = 9 + l$

   $l = 12.904761$

2.14 GRAPHS

A *number scale* is a line on which distances from a point are marked off in equal units, positively in one direction and negatively in the other. The *origin* is the zero point from which distances are measured. Figure 2-4 shows a horizontal number scale.

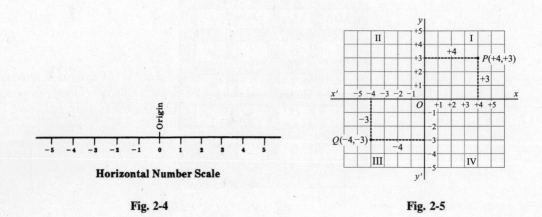

Horizontal Number Scale

Fig. 2-4

Fig. 2-5

The *graph* shown in Fig. 2-5 is formed by combining two number scales at right angles to each other so that their zero points coincide. The horizontal number scale is called the *x-axis*, and the vertical number scale is the *y-axis*. The point where the two scales cross each other is, again, called the *origin*.

A *point is located on a graph by its coordinates*, which are its distances from the axes. The *abscissa* or *x-coordinate* of a point is its distance from the *y*-axis. The *ordinate* or *y-coordinate* of a point is its distance from the *x*-axis.

When the coordinates of a point are stated, the *x*-coordinate precedes the *y*-coordinate. Thus the coordinates of point P in Fig. 2-5 are written $(4, 3)$; those for Q are $(-4, -3)$. Note the parentheses.

The *quadrants* of a graph are the four parts cut off by the axes. These are numbered I, II, III, and IV in a counterclockwise direction, as shown.

SOLVED PROBLEMS

2.55 LOCATING POINTS ON A GRAPH

Give the coordinates of the following points in Fig. 2-6.

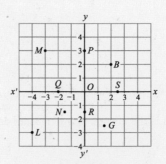

Fig. 2-6

(a) B	(c) O	(e) N	(g) P	(i) R
(b) M	(d) L	(f) G	(h) Q	(j) S

Answers

(a) (2,2)	(c) (0,0)	(e) $(-1\frac{1}{2}, -1\frac{1}{2})$	(g) (0,3)	(i) $(0, -1\frac{1}{2})$
(b) (−3,3)	(d) (−4,−3)	(f) $(1\frac{1}{2}, -2\frac{1}{2})$	(h) (−2,0)	(j) $(2\frac{1}{2}, 0)$

2.56   COORDINATES OF POINTS IN THE FOUR QUADRANTS

What are the signs of the coordinates of (a) a point in quadrant I; (b) a point in quadrant II; (c) a point in quadrant III; (d) a point in quadrant IV? Show which coordinate has a sign and which zero value for a point between quadrants (e) IV and I; (f) I and II; (g) II and III; (h) III and IV.

Solutions

(a) (+, +)	(c) (−, −)	(e) (+, 0)	(g) (−, 0)
(b) (−, +)	(d) (+, −)	(f) (0, +)	(h) (0, −)

2.57   GRAPHING A QUADRILATERAL

If the vertices of a rectangle have the coordinates $A(3,1)$, $B(-5,1)$, $C(-5,-3)$, and $D(3,-3)$, find its perimeter and area.

Solution

The base and height of the rectangle are 8 and 4 (see Fig. 2-7). Hence the perimeter is $2b + 2h = 2(8) + 2(4) = 24$, and the area is $bh = (8)(4) = 32$.

2.58   GRAPHING A TRIANGLE

If the vertices of a triangle have the coordinates $A(4\frac{1}{2}, -2)$, $B(-2\frac{1}{2}, -2)$ and $C(1,5)$, find its area.

Solution

The length of the base is $BA = 7$ (see Fig. 2-8). The height is $CD = 7$. Then $A = \frac{1}{2}bh = \frac{1}{2}(7)(7) = 24\frac{1}{2}$.

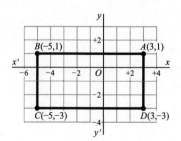

Fig. 2-7

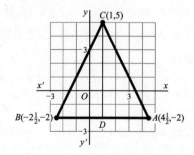

Fig. 2-8

2.15 MIDPOINT OF A SEGMENT

The coordinates (x_m, y_m) of the midpoint M of the line segment joining $P(x_1, y_1)$ to $Q(x_2, y_2)$ are

$$x_m = \tfrac{1}{2}(x_1 + x_2) \qquad \text{and} \qquad y_m = \tfrac{1}{2}(y_1 + y_2)$$

In Fig. 2-9, segment y_m is the median of trapezoid $CPQD$, whose bases are y_1 and y_2. Since the length of a median is one-half the sum of the bases, $y_m = \tfrac{1}{2}(y_1 + y_2)$. Similarly, segment x_m is the median of trapezoid $ABQP$, whose bases are x_1 and x_2; hence $x_m = \tfrac{1}{2}(x_1 + x_2)$.

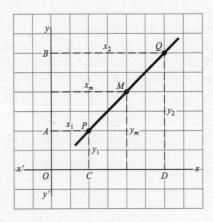

Fig. 2-9

SOLVED PROBLEMS

2.59   **APPLYING THE MIDPOINT FORMULA**

If M is the midpoint of $\overline{PQ}$, find the coordinates of (a) M if the coordinates of P and Q are $P(3,4)$ and $Q(5,8)$; (b) Q if the coordinates of P and M are $P(1,5)$ and $M(3,4)$.

Solutions

(a) $x_m = \tfrac{1}{2}(x_1 + x_2) = \tfrac{1}{2}(3 + 5) = 4$; $y_m = \tfrac{1}{2}(y_1 + y_2) = \tfrac{1}{2}(4 + 8) = 6$.

(b) $x_m = \tfrac{1}{2}(x_1 + x_2)$, so $3 = \tfrac{1}{2}(1 + x_2)$ and $x_2 = 5$; $y_m = \tfrac{1}{2}(y_1 + y_2)$, so $4 = \tfrac{1}{2}(5 + y_2)$ and $y_2 = 3$.

2.60   **DETERMINING IF SEGMENTS BISECT EACH OTHER**

The vertices of a quadrilateral are $A(0,0)$, $B(0,3)$, $C(4,3)$, and $D(4,0)$.

(a)   Show that $ABCD$ is a rectangle.

(b)   Show that the midpoint of $\overline{AC}$ is also the midpoint of $\overline{BD}$.

(c)   Do the diagonals bisect each other? Why?

Solutions

(a)   From Fig. 2-10, $AB = CD = 3$ and $BC = AD = 4$; hence $ABCD$ is a parallelogram. Since $\angle BAD$ is a right angle, $ABCD$ is a rectangle.

(b)   The coordinates of the midpoint of $\overline{AC}$ are $x = \tfrac{1}{2}(0 + 4) = 2$, $y = \tfrac{1}{2}(0 + 3) = 1\tfrac{1}{2}$. The coordinates of the midpoint of $\overline{BD}$ are $x = \tfrac{1}{2}(0 + 4) = 2$, $y = \tfrac{1}{2}(3 + 0) = 1\tfrac{1}{2}$. Hence $(2, 1\tfrac{1}{2})$ is the midpoint of both $\overline{AC}$ and $\overline{BD}$.

(c)   Yes, since the midpoints of both diagonals are the same point.

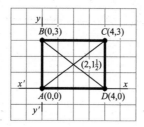

Fig. 2-10

2.16   DISTANCE BETWEEN TWO POINTS

PRINCIPLE 1:   *The distance between two points having the same ordinate (or y-value) is the absolute value of the difference of their abscissas.* (Hence, the distance between two points must be *positive*.)

Thus the distance between the points $P(6,1)$ and $Q(9,1)$ is $9 - 6 = 3$.

PRINCIPLE 2:   *The distance between two points having the same abscissa (or x-value) is the absolute value of the difference of their ordinates.*

Thus the distance between the points $P(2,1)$ and $Q(2,4)$ is $4 - 1 = 3$.

PRINCIPLE 3:   *The distance d between the points $P_1(x_1, y_1)$ and $P_2(x_2, y_2)$ is*

$$d = \sqrt{(x_2 - x_1)^2 + (y_2 - y_1)^2} \quad \text{or} \quad d = \sqrt{(\Delta x)^2 + (\Delta y)^2}$$

The difference $x_2 - x_1$ is denoted by the symbol Δx; the difference $y_2 - y_1$ is denoted by Δy. Delta (Δ) is the fourth letter of the Greek alphabet, corresponding to our d. The differences Δx and Δy may be positive or negative.

SOLVED PROBLEMS

2.61   USING THE DISTANCE FORMULA

Use the distance formula to find the distance between $A(2,5)$ and $B(6,8)$.

Solutions

The distance from $A(2,5)$ to $B(6,8)$ is found as follows:

$$\frac{(x, y)}{}$$

$$B(6,8) \rightarrow x_2 = 6, \ y_2 = 8$$
$$A(2,5) \rightarrow x_1 = 2, \ y_1 = 5$$

$$d^2 = (x_2 - x_1)^2 + (y_2 - y_1)^2$$
$$d^2 = (6 - 2)^2 + (8 - 5)^2 = 4^2 + 3^2 = 25 \quad \text{and} \quad d = 5$$

2.62   APPLYING THE DISTANCE FORMULA TO A TRIANGLE

(*a*)   Find the lengths of the sides of a triangle whose vertices are $A(1,1)$, $B(1,4)$, and $C(5,1)$.

(*b*)   Show that the triangle whose vertices are $G(2,10)$, $H(3,2)$, and $J(6,4)$ is a right triangle.

Solutions

See Fig. 2-11.

(*a*) $AC = 5 - 1 = 4$ and $AB = 4 - 1 = 3$; $BC = \sqrt{(5-1)^2 + (1-4)^2} = \sqrt{4^2 + (-3)^2} = 5$.

(*b*) $(GJ)^2 = (6-2)^2 + (4-10)^2 = 52$; $(HJ)^2 = (6-3)^2 + (4-2)^2 = 13$; $(GH)^2 = (2-3)^2 + (10-2)^2 = 65$. Since $(GJ)^2 + (HJ)^2 = (GH)^2$, $\triangle GHJ$ is a right triangle.

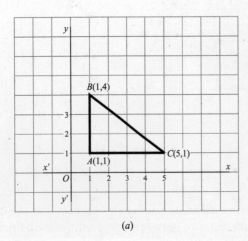
(a)

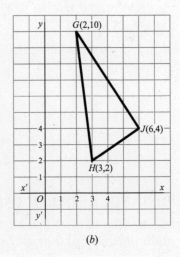
(b)

Fig. 2-11

2.63   **APPLYING THE DISTANCE FORMULA TO A PARALLELOGRAM**

The coordinates of the vertices of a quadrilateral are $A(2,2)$, $B(3,5)$, $C(6,7)$, and $D(5,4)$. Show that $ABCD$ is a parallelogram.

Solution

See Fig. 2-12, where we have

$$AB = \sqrt{(3-2)^2 + (5-2)^2} = \sqrt{1^2 + 3^2} = \sqrt{10}$$
$$CD = \sqrt{(6-5)^2 + (7-4)^2} = \sqrt{1^2 + 3^2} = \sqrt{10}$$
$$BC = \sqrt{(6-3)^2 + (7-5)^2} = \sqrt{3^2 + 2^2} = \sqrt{13}$$
$$AD = \sqrt{(5-2)^2 + (4-2)^2} = \sqrt{3^2 + 2^2} = \sqrt{13}$$

Thus $\overline{AB} \cong \overline{CD}$ and $\overline{BC} \cong \overline{AD}$. Since opposite sides are congruent $ABCD$ is a parallelogram.

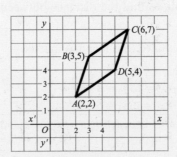

Fig. 2-12

2.64   **APPLYING THE DISTANCE FORMULA TO A CIRCLE**

A circle is tangent to the x-axis and has its center at $(6,4)$. Where is the point $(9,7)$ with respect to the circle?

Solution

Since the circle is tangent to the x-axis, $\overline{AQ}$ in Fig. 2-13 is a radius. By Principle 2, $AQ = 4$.

By Principle 3, $BQ = \sqrt{(9-6)^2 + (7-4)^2} = \sqrt{3^2 + 3^2} = \sqrt{18}$. Since $\sqrt{18}$ is greater than 4, $\overline{BQ}$ is greater than a radius so B is outside the circle.

2.17 SLOPE OF A LINE

PRINCIPLE 1: *If a line passes through the points $P_1(x_1, y_1)$ and $P_2(x_2, y_2)$, then*

$$Slope \ of \ \overleftrightarrow{P_1 P_2} = \frac{y_2 - y_1}{x_2 - x_1} = \frac{\Delta y}{\Delta x}$$

PRINCIPLE 2: *The line whose equation is $y = mx + b$ has slope m.*

PRINICPLE 3: *The slope of a line equals the tangent of its inclination.*

The inclination i of a line is the angle above the x-axis that is included between the line and the positive direction of the x-axis (see Fig. 2-14). In the figure,

$$\text{Slope of } \overleftrightarrow{P_1 P_2} = \frac{y_2 - y_1}{x_2 - x_1} = \frac{\Delta y}{\Delta x} = m = \tan i$$

The slope is independent of the order in which the end points are selected. Thus,

$$\text{Slope of } \overleftrightarrow{P_1 P_2} = \frac{y_2 - y_1}{x_2 - x_1} = \frac{y_1 - y_2}{x_1 - x_2} = \text{slope of } P_2 P_1$$

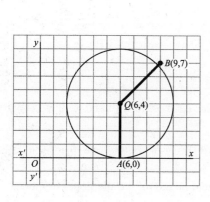

Fig. 2-13

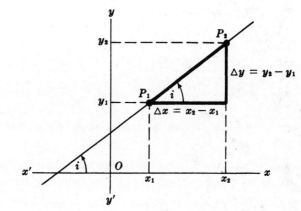

Fig. 2-14

2.17A Positive and Negative Slopes

PRINCIPLE 4: *If a line slants upward from left to right, its inclination i is an acute angle and its slope is positive* (Fig. 2-15).

PRINCIPLE 5: *If a line slants downward from left to right, its inclination is an obtuse angle and its slope is negative* (Fig. 2-16).

PRINCIPLE 6: *If a line is parallel to the x-axis, its inclination is 0° and its slope is 0* (Fig. 2-17).

PRINCIPLE 7: *If a line is perpendicular to the x-axis, its inclination is 90° and it has no slope* (Fig. 2-18).

2.17B Slopes of Parallel and Perpendicular Lines

PRINCIPLE 8: *Parallel lines have the same slope.*

In Fig. 2-19, $l \| l'$; hence corresponding angles i and i' are equal, and $\tan i = \tan i'$ or $m = m'$, where m and m' are the slopes of l and l'.

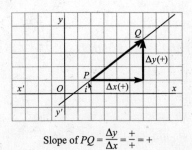

Slope of $PQ = \dfrac{\Delta y}{\Delta x} = \dfrac{+}{+} = +$

Fig. 2-15

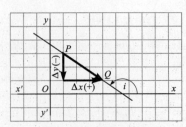

Slope of $PQ = \dfrac{\Delta y}{\Delta x} = \dfrac{-}{+} = -$

Fig. 2-16

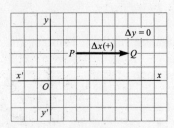

Slope of $PQ = \dfrac{\Delta y}{\Delta x} = \dfrac{0}{+} = 0$

Fig. 2-17

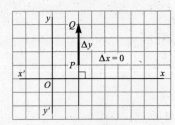

Slope of $PQ = \dfrac{\Delta y}{\Delta x} = \dfrac{+}{0}$ (meaningless)

Fig. 2-18

PRINCIPLE 9: *Lines having the same slope are parallel to each other.* (This is the converse of Principle 8.)

PRINCIPLE 10: *Perpendicular lines have slopes that are negative reciprocals of each other.* (Negative reciprocals are numbers, such as $\frac{2}{5}$ and $-\frac{5}{2}$, whose product is -1.)

Thus in Fig. 2-20, if $l \perp l'$, then $m = -1/m'$ or $mm' = -1$, where m and m' are the slopes of l and l'.

PRINCIPLE 11: *Lines whose slopes are negative reciprocals of each other are perpendicular.* (This is the converse of Principle 10.)

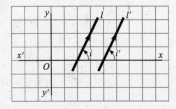

Fig. 2-19

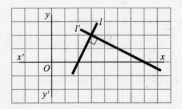

Fig. 2.20

2.17C   Collinear Points

Collinear points are points which lie on the same straight line. Thus A, B, and C are collinear points here:

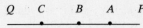

PRINCIPLE 12: *The slope of a straight line is constant all along the line.*

Thus if $\overrightarrow{PQ}$ above is a straight line, the slope of the segment from A to B equals the slope of the segment from C to Q.

PRINCIPLE 13: *If the slope of a segment between a first point and a second equals the slope of the segment between either point and a third, then the points are collinear.*

SOLVED PROBLEMS

2.65 SLOPE AND INCLINATION OF A LINE

    (*a*) Find the slope of the line through $(-2, -1)$ and $(4, 3)$.

    (*b*) Find the slope of the line whose equation is $3y - 4x = 15$.

    (*c*) Find the inclination of the line whose equation is $y = x + 4$.

Solutions

    (*a*) By Principle 1, $m = \dfrac{y_2 - y_1}{x_2 - x_1} = \dfrac{3 - (-1)}{4 - (-2)} = \dfrac{4}{6} = \dfrac{2}{3}$

    (*b*) We may rewrite $3y - 4x = 15$ as $y = \frac{4}{3}x + 5$, from which $m = \frac{4}{3}$.

    (*c*) Since $y = x + 4$, we have $m = 1$; thus $\tan i = 1$ and $i = 45°$.

2.66 SLOPES OF PARALLEL OR PERPENDICULAR LINES

    Find the slope of $\overrightarrow{CD}$ if (*a*) $\overrightarrow{AB} \parallel \overrightarrow{CD}$ and the slope of $\overrightarrow{AB}$ is $\frac{2}{3}$; (*b*) $\overrightarrow{AB} \perp \overrightarrow{CD}$ and the slope of $\overrightarrow{AB}$ is $\frac{3}{4}$.

Solutions

    (*a*) By Principle 8, slope of $\overrightarrow{CD}$ = slope of $\overrightarrow{AB} = \frac{2}{3}$.

    (*b*) By Principle 10, slope of $\overrightarrow{CD} = -\dfrac{1}{\text{slope of } AB} = -\dfrac{1}{\frac{3}{4}} = -\dfrac{4}{3}$.

2.67 APPLYING PRINCIPLES 9 AND 11 TO TRIANGLES AND QUADRILATERALS

    Complete each of the following statements:

    (*a*) In quadrilateral $ABCD$, if the slopes of $\overrightarrow{AB}$, $\overrightarrow{BC}$, $\overrightarrow{CD}$, and $\overrightarrow{DA}$ are $\frac{1}{2}$, -2, $\frac{1}{2}$, and -2, respectively, the quadrilateral is a __?__ .

    (*b*) In triangle LMP, if the slopes of $\overrightarrow{LM}$ and $\overrightarrow{MP}$ are 5 and $-\frac{1}{5}$, then LMP is a __?__ triangle.

Solutions

    (*a*) Since the slopes of the opposite sides are equal, $ABCD$ is a parallelogram. In addition, the slopes of adjacent sides are negative reciprocals; hence those sides are $\perp$ and $ABCD$ is a rectangle.

    (*b*) Since the slopes of $\overrightarrow{LM}$ and $\overrightarrow{MP}$ are negative reciprocals, $\overline{LM} \perp \overline{MP}$ and the triangle is a right triangle.

2.68 APPLYING PRINCIPLE 12

    (*a*) $\overrightarrow{AB}$ has a slope of 2 and points A, B, and C are collinear. What are the slopes of $\overline{AC}$ and $\overline{BC}$?

    (*b*) Find y if $G(1, 4)$, $H(3, 2)$, and $J(9, y)$ are collinear.

Solutions

(a) By Principle 12, $\overline{AC}$ and $\overline{BC}$ have a slope of 2.

(b) By Principle 12, slope of $\overset{\leftrightarrow}{GJ}$ = slope of $\overset{\leftrightarrow}{GH}$. Hence $\dfrac{y-4}{9-1} = \dfrac{2-4}{3-1}$, so that $\dfrac{y-4}{8} = \dfrac{-2}{2} = -1$ and $y = -4$.

Supplementary Problems

1. By checking, determine which value is a root of the equation: (a) $3x + 4x = 42$ for $x = 4$, 6, and 8; (b) $3n + 14 = 47$ for $n = 9$, 10, and 11; (c) $6y - 48 = 2y$ for $y = 8$, 10, and 12. (2.1)

2. By checking, show that x may have any of the following values in the identity $2(x - 3) = 2x - 6$: (a) $x = 10$, (b) $x = 6$, (c) $x = 4\frac{1}{2}$, (d) $x = 3.1$. (2.2)

3. Translate into an equation, letting n represent the number: (a) What number diminished by 8 equals 13? (b) Two-thirds of what number equals 10? (c) Three times the sum of a number and six is 33. What is the number? (d) What number increased by 20 equals three times the same number? (e) What number increased by 5 equals twice the same number, decreased by 4? (2.3)

4. Match the statements in Column 1 with the equations in Column 2. (2.4)

Column 1	**Column 2**
1. The sum of 8 and twice a number is 18.	(a) $\frac{1}{8}n + 2 = 18$
2. Twice a number, less 8, is 18.	(b) $8(n - 2) = 18$
3. Twice the sum of a number and 8 is 18.	(c) $\frac{1}{2}(8 - n) = 18$
4. Eight times the difference of a number and 2 is 18.	(d) $2(n + 8) = 18$
5. One-half the difference of 8 and a number is 18.	(e) $2n + 8 = 18$
6. 2 more than one-eighth of a number is 18.	(f) $\frac{1}{2}n - 8 = 18$
7. 8 less than half a number is 18.	(g) $2n - 8 = 18$

5. Letting n represent the number of games lost, obtain an equation for each problem: (a) A team won three times as many games as it lost. It played a total of 52 games. (b) A team won 20 games more than it lost. It played a total of 84 games. (c) A team won 15 games less than twice the number lost. It played a total of 78 games. (2.5)

6. Solve each equation: (2.6)

(a) $a + 5 = 9$ (e) $x + 11 = 21 + 8$ (i) $45 = m - 13$

(b) $7 + b = 15$ (f) $27 + 13 = 18 + y$ (j) $22 = n - 50$

(c) $20 = c + 12$ (g) $h - 6 = 14$ (k) $x - 42 = 80 - 75$

(d) $75 = 55 + d$ (h) $k - 14 = 6$ (l) $100 - 31 = y - 84$

7. Solve each equation: (2.7)

(a) $4p = 48$ (d) $4n = 2$ (g) $\dfrac{t}{5} = 6$ (i) $\dfrac{x}{15} = 4$ (k) $\dfrac{a}{10} = \dfrac{2}{5}$

(b) $10r = 160$ (e) $12w = 4$ (j) $\dfrac{y}{12} = \dfrac{3}{2}$ (l) $\dfrac{1}{3}b = \dfrac{5}{6}$

                         (h) $\dfrac{u}{65} = 1$

(c) $25s = 35$ (f) $24x = 21$

8. Solve each equation: (2.8)

(a) $n + 8 = 24$ (e) $3 + y = 15$ (i) $16 = y - 20$ (m) $x + \dfrac{1}{3} = 9$

(b) $n - 8 = 24$ (f) $15 = y - 3$ (j) $16 = \dfrac{y}{20}$ (n) $x - \dfrac{1}{3} = 9$

(c) $8n = 24$ (g) $15 = 3y$ (k) $\dfrac{y}{20} = 16$ (o) $\dfrac{1}{3}x = 9$

(d) $\dfrac{n}{8} = 24$ (h) $15 = \dfrac{y}{3}$ (l) $16 + y = 20$ (p) $\dfrac{x}{9} = \dfrac{1}{3}$

9. Solve each equation: (2.8)

(a) $x + 11 = 14$ (f) $h - 3 = 7\frac{1}{2}$ (k) $11r = 55$ (p) $6\frac{1}{2} = \dfrac{l}{2}$

(b) $11 + y = 24$ (g) $35 = m - 20\frac{1}{3}$ (l) $44s = 44$ (q) $1.7 = \dfrac{n}{3}$

(c) $22 = 13 + a$ (h) $17\frac{3}{4} = n - 2\frac{1}{4}$ (m) $10t = 5$ (r) $100 = \dfrac{h}{.7}$

(d) $45 = b + 33$ (i) $x + 1.2 = 5.7$ (n) $8x = 3$ (s) $24 = \dfrac{t}{.5}$

(e) $z - 9 = 3$ (j) $10.8 = y - 3.2$ (o) $3y = 0$ (t) $.009 = \dfrac{x}{1000}$

10. State the equality rule used in each: (2.9)

(a) $6r = 30$ (c) $30 = \dfrac{r}{6}$ (e) $100x = 5$

$\quad \dfrac{6r}{6} = \dfrac{30}{6}$ $\quad 6(30) = 6\dfrac{r}{6}$ $\quad \dfrac{100x}{100} = \dfrac{5}{100}$

$\quad r = 5$ $\quad 180 = r$ $\quad x = \dfrac{1}{20}$

(b) $\quad 30 = r - 6$ (d) $\quad 30 = 6 + r$ (f) $100 = \dfrac{y}{5}$

$\quad \dfrac{+6 = \quad +6}{36 = r}$ $\quad \dfrac{-6 = -6}{24 = r}$ $\quad 5(100) = 5\left(\dfrac{y}{5}\right)$

$\quad 500 = y$

11. Solve each equation: (2.10)

(a) $12x = 60$ (c) $24 = 2z$ (e) $6r = 9$ (g) $10 = 4t$

(b) $60y = 12$ (d) $2 = 24w$ (f) $9s = 6$ (h) $4 = 10u$

12. Find each solution set: (2.11)

(a) $.7a = 21$ (c) $24 = .06c$ (e) $.1h = 100$ (g) $25.2 = .12k$

(b) $1.1b = 55$ (d) $18 = .009d$ (f) $.6j = .96$ (h) $7.5 = .015m$

13. Solve each equation: (2.12)

(a) $10\%s = 7$ (c) $18 = 3\%n$ (e) $5\%m = 13$ (g) $.23 = 1\%y$

(b) $25\%t = 3$ (d) $14 = 70\%w$ (f) $17\%x = 6.8$ (h) $3.69 = 90\%z$

14. Solve each equation: (2.13)

   (a) $14 = 3x - x$ (c) $8z - 3z = 45$ (e) $24 = 4\frac{1}{2}x - \frac{1}{2}x$ (g) $7\frac{1}{2}z - 7z = 28$

   (b) $7y + 3y = 50$ (d) $132 = 10w + 3w - w$ (f) $4y + 15y = 57$ (h) $15w - 3w - 2w = 85$

15. Harry earned \$28.89. What was his hourly wage if he worked: (a) 3 hr, (b) 9 hr, (c) $\frac{1}{2}$ hr?   (2.14)

16. Mr. Brown's commission rate is 5%. How much did he sell if his commissions were: (a) \$85, (b) \$750, (c) \$6.20?   (2.15)

17. Solve each equation: (2.16)

   (a) $\dfrac{x}{3} = 2$ (c) $16 = \dfrac{z}{5}$ (e) $\dfrac{a}{2} = 3$ (g) $.6 = \dfrac{c}{10}$

   (b) $\dfrac{1}{7}y = 12$ (d) $3 = \dfrac{1}{50}w$ (f) $\dfrac{1}{30}b = 20$

18. Find each solution set: (2.17)

   (a) $\dfrac{a}{.7} = 10$ (b) $\dfrac{b}{.02} = 600$ (c) $30 = \dfrac{c}{2.4}$ (d) $11 = \dfrac{d}{.05}$ (e) $\dfrac{m}{.4} = 220$ (f) $\dfrac{n}{.01} = 3$

19. Solve each equation: (2.18)

   (a) $\dfrac{3}{4}x = 21$ (b) $\dfrac{4}{3}y = 32$ (c) $\dfrac{3x}{2} = 9$ (d) $45 = \dfrac{5}{9}y$ (e) $2\frac{1}{5}z = 55$ (f) $2c + \dfrac{1}{2}c = 10$

20. Solve each equation: (2.19)

   (a) $37\frac{1}{2}\%s = 15$ (b) $60\%t = 60$ (c) $16\frac{2}{3}\%n = 14$ (d) $150\%r = 15$ (e) $83\frac{1}{3}\%w = 35$

   Hint: $37\frac{1}{2}\% = 3/8$ $60\% = 3/5$ $16\frac{2}{3}\% = 1/6$ $150\% = 1\frac{1}{2}$ or $3/2$ $83\frac{1}{3}\% = 5/6$

21. On a trip, John covered a distance of 35 km. What was the total distance of the trip if the distance traveled was: (a) 5/6 of the total distance, (b) 70% of the total distance?   (2.20)

22. Mr. Reynolds receives 7% per year on a stock investment. How large is his investment if, at the end of one year, his interest is: (a) \$28, (b) \$350, (c) \$42.70?   (2.20)

23. Find each solution set: (2.21)

   (a) $r + 25 = 70$ (c) $18 = s + 3$ (e) $x + 130 = 754$ (g) $259 = s + 237$

   (b) $31 + t = 140$ (d) $842 = 720 + u$ (f) $116 + y = 807$ (h) $901 = 857 + w$

24. Solve each equation: (2.22)

   (a) $b + 2/3 = 7\frac{2}{3}$ (c) $35.4 = d + 23.2$ (e) $f + 5/8 = 3\frac{1}{2}$ (g) $7.28 = m + .79$

   (b) $1\frac{1}{2} + c = 8\frac{3}{4}$ (d) $87.4 = 80.6 + e$ (f) $8\frac{1}{6} + g = 10\frac{5}{6}$ (h) $15.87 = 6.41 + n$

25. The price of eggs rose 29¢. What was the original price if the new price is: (a) \$1.29, (b) \$1.41?   (2.23)

26. Will is 8 in. taller than George. How tall is George if Will's height is: (a) 5 ft 2 in., (b) 4 ft 3 in.?   (2.24)

27. Find each solution set: (2.25)

   (a) $w - 8 = 22$ (c) $40 = y - 3$ (e) $m - 140 = 25$ (g) $158 = p - 317$

   (b) $x - 22 = 8$ (d) $3 = z - 40$ (f) $n - 200 = 41$ (h) $256 = r - 781$

28. Solve each equation: (2.26)

(a) $h - \frac{7}{8} = 8\frac{3}{4}$ (c) $28.4 = m - 13.9$ (e) $p - 1\frac{5}{12} = 1\frac{7}{12}$ (g) $.03 = s - 2.07$

(b) $j - 34\frac{1}{2} = 65$ (d) $.37 = n - 8.96$ (f) $r - 14\frac{2}{3} = 5\frac{1}{3}$ (h) $5.84 = t - 3.06$

29. What was the original temperature if a drop of $12°$ brought the temperature to: (a) $75°$, (b) $14\frac{1}{2}°$, (c) $6\frac{1}{4}°$?   (2.27)

30. How many marbles did Sam have originally if after giving 35 marbles to Jim, he found that the number of marbles he had left was: (a) 5, (b) 12, (c) 15, (d) 35, (e) 75?   (2.28)

31. Solve each equation: (2.29)

(a) $2x + 5 = 9$ (e) $2x - 5 = 9$ (i) $\frac{x}{4} + 3 = 7$ (m) $\frac{x}{4} - 3 = 7$

(b) $4x + 11 = 21$ (f) $4x - 11 = 21$ (j) $\frac{x}{5} + 2 = 10$ (n) $\frac{x}{5} - 2 = 10$

(c) $20 = 3x + 8$ (g) $60 = 10x - 20$ (k) $17 = \frac{x}{2} + 15$ (o) $3 = \frac{x}{12} - 7\frac{1}{4}$

(d) $13 = 6 + 7x$ (h) $11 = 6x - 16$ (l) $25 = \frac{x}{10} + 2$ (p) $5\frac{1}{2} = \frac{x}{8} - 4$

32. Solve each equation: (2.30)

(a) $10n + 5n - 6 = 9$ (d) $35 = 6p + 8 + 3p$ (g) $40 = 25t + 22 - 13t$

(b) $7m + 10 - 2m = 45$ (e) $19n - 10 + n = 80$ (h) $145 = 10 + 7.6s - 3.1s$

(c) $25 = 19 + 20n - 18n$ (f) $3\frac{1}{2}r + r + 2 = 20$

33. Find each solution set: (2.31)

(a) $5r = 2r + 27$ (c) $10r - 11 = 8r$ (e) $13b = 15 + 3b$ (g) $9u = 16u - 105$

(b) $2r = 90 - 7r$ (d) $18 - 5a = a$ (f) $100 + 3\frac{1}{2}t = 23\frac{1}{2}t$ (h) $5x + 3 - 2x = x + 8$

34. Solve each equation: (2.32)

(a) $\frac{40}{x} = 5$ (c) $14 = \frac{28}{y}$ (e) $\frac{32}{n} = 8$ (g) $4 = \frac{15}{w}$

(b) $\frac{5}{x} = 40$ (d) $28 = \frac{14}{y}$ (f) $\frac{3}{n} = 2$ (h) $15 = \frac{90}{w}$

35. Find each solution set: (2.33, 2.34)

(a) $\frac{7}{8}w = 21$ (d) $1\frac{1}{2}w = 15$ (g) $\frac{4}{5}n + 6 = 22$ (j) $10 = \frac{2}{9}r + 8$

(b) $\frac{3}{4}x = 39$ (e) $2\frac{1}{3}b = 35$ (h) $10 + \frac{6}{5}m = 52$ (k) $6 = 16 - \frac{5t}{3}$

(c) $\frac{5}{4}y = 15$ (f) $2c + 2\frac{1}{2}c = 54$ (i) $30 - \frac{3}{2}p = 24$ (l) $3s + \frac{s}{3} - 7 = 5$

36. Find each solution set: (2.29 to 2.34)

(a) $20 = 3x - 10$ (f) $\dfrac{15}{x} = \dfrac{5}{4}$ (k) $\dfrac{x}{2} + 27 = 30$ (p) $12b - 5 = .28 + b$

(b) $20 = \dfrac{x}{3} - 10$ (g) $5c = 2c + 4.5$ (l) $8x + 3 = 43$ (q) $6d - .8 = 2d$

(c) $15 = \dfrac{3}{4}y$ (h) $.30 - g = .13$ (m) $21 = \dfrac{7}{5}w$ (r) $40 - .5h = 5$

(d) $17 = 24 - z$ (i) $\frac{3}{4}n + 11\frac{1}{2} = 20\frac{1}{2}$ (n) $60 = 66 - 12w$ (s) $40 - \dfrac{3}{5}m = 37$

(e) $3 = \dfrac{39}{x}$ (j) $6w + 5w - 8 = 8w$ (o) $10b - 3b = 49$ (t) $12t - 2t + 10 = 9t + 12$

37. How many girls are there in a class of 30 pupils if: (a) the number of boys is 10, (b) the number of boys is four times as many?   (2.35)

38. Charles has $37 in his bank and hopes to increase this to $100 by making equal deposits each week. How much should he deposit if he deposits money for: (a) 14 wk, (b) 5 wk?   (2.36)

39. Mr. Barr sold his house for $120,000. How much did the house cost him if his loss was: (a) 1/3 of the cost, (b) 20% of the cost?   (2.37)

40. Derive a formula for: (a) the no. of pennies (p) equivalent to d dimes, (b) the no. of dimes (d) equivalent to p pennies, (c) the no. of nickels (n) equivalent to D dollars, (d) the no. of half dollars (h) equivalent to q quarters, (e) the no. of quarters (q) equivalent to d dimes.   (2.38)

41. Derive a formula for: (a) the value in cents (c) of d dimes and q quarters, (b) the value in cents (c) of n nickels and D dollars, (c) the no. of nickels (n) equivalent to d dimes and h half dollars, (d) the no. of dimes (d) equivalent to n nickels and p pennies, (e) the no. of quarters (q) equivalent to D dollars and n nickels.   (2.39, 2.40)

42. Derive a formula for: (a) the no. of sec (s) in h hr; (b) the no. of hr (h) in m min; (c) the no. of hr (h) in w wk; (d) the no. of da (d) in M mo of 30 days; (e) the no. of da (d) in M mo of 30 days, w wk and 5 da; (f) the no. of min (m) in h hr and 30 sec; (g) the no. of da (d) in y yr of 365 da and 3 wk.   (2.41)

43. Derive a formula for: (a) the no. of in (i) in y yd, (b) the no. of yd (y) in f ft, (c) the no. of meters (m) in c centimeters.   (2.42)

44. From $D = RT$, obtain a formula relating: (a) distance in mi and rate in mph for a time of 5 hr, (b) distance in mi and rate in mph for a time of 30 min, (c) time in hr and rate in mph for a distance of 25 mi, (d) time in sec and rate in ft per sec for a distance of 100 yd, (e) distance in ft and time in min for a rate of 20 ft per min, (f) distance in ft and time in min for a rate of 20 ft per sec.   (2.43)

45. Solve: (2.44)

(a) $d = 2r$ for r (e) $c = \pi d$ for d (i) $V = LWH$ for H

(b) $p = 5s$ for s (f) $c = \pi d$ for π (j) $V = 2\pi r^2 h$ for h

(c) $D = 30T$ for T (g) $NP = C$ for N (k) $9C = 5(F - 32)$ for C

(d) $25W = A$ for W (h) $I = PR$ for R (l) $2A = h(b + b')$ for h

46. Solve: (2.45)

(a) $\dfrac{p}{10} = s$ for p (e) $\pi = \dfrac{c}{2r}$ for c (i) $\dfrac{V}{3LW} = H$ for V

(b) $R = \dfrac{D}{15}$ for D (f) $\dfrac{M}{D} = F$ for M (j) $\dfrac{T}{14RS} = \dfrac{1}{2}$ for T

(c) $W = \dfrac{A}{8}$ for A (g) $P = \dfrac{A}{2F}$ for A (k) $\dfrac{L}{KA} = V^2$ for L

(d) $w = \dfrac{d}{7}$ for d (h) $\dfrac{T}{Q} = 5R$ for T (l) $\dfrac{V}{\pi r^2} = \dfrac{h}{3}$ for V

47. Solve: (2.46)

(a) $a + b = 60$ for a (d) $3m = 4n + p$ for p (g) $5a + b = c - d$ for b

(b) $3c + g = 85$ for g (e) $10r = s - 5t$ for s (h) $5a - 4c = 3e + f$ for f

(c) $h - 10r = l$ for h (f) $4g + h - 12 = j$ for h (i) $\dfrac{b}{2} - 10 + c = 100p$ for c

48. Solve: (2.47)

(a) $4P - 3R = 40$ for P (d) $A = \dfrac{1}{2}bh$ for b (g) $\dfrac{R}{2} - 4S = T$ for R

(b) $36 - 5i = 12f$ for i (e) $V = \dfrac{1}{3}\pi r^2 h$ for h (h) $8h - \dfrac{k}{5} = 12$ for k

(c) $\dfrac{P}{2} + R = S$ for P (f) $A = \dfrac{1}{2}h(b + b')$ for h (i) $20p - \dfrac{2}{3}q = 8t$ for q

49. Solve: (2.48)

(a) $l = a + (n - 1)d$ for d (b) $s = \dfrac{n}{2}(a + l)$ for l (c) $F = \dfrac{9}{5}C + 32$ for C

50. Find: (2.49)

(a) l if $l = prt$ and $p = 3000$, $r = .05$, $t = 2$ (e) C if $C = \frac{5}{9}(F - 32)$ and $F = 212$

(b) t if $t = \dfrac{l}{pr}$ and $l = 40$, $p = 2000$, $r = .01$ (f) S if $S = \frac{1}{2}gt^2$ and $g = 32$, $t = 8$

(c) F if $F = \frac{9}{5}C + 32$ and $C = 55$ (g) g if $g = \dfrac{2S}{t^2}$ and $S = 800$, $t = 10$

(d) F if $F = \frac{9}{5}C + 32$ and $C = -40$ (h) S if $S = \dfrac{a - lr}{1 - r}$ and $a = 5$, $l = 40$, $r = -1$

51. Find: (2.50, 2.51)

(a) R if $D = RT$ and $D = 30$, $T = 4$ (e) c if $p = a + 2b + c$ and $p = 33$, $a = 11$, $b = 3\frac{1}{2}$

(b) b if $A = bh$ and $A = 22$, $h = 2.2$ (f) h if $2A = h(b + b')$ and $A = 70$, $b = 3.3$, $b' = 6.7$

(c) a if $P = 2a + b$ and $P = 12$, $b = 3$ (g) B if $V = \frac{1}{3}Bh$ and $V = 480$, $h = 12$

(d) w if $P = 2l + 2w$ and $P = 68$, $l = 21$ (h) a if $l = a + (n - 1)d$ and $l = 140$, $n = 8$, $d = 3$

52. (a) A train takes 5 hours and 20 minutes to go a distance of 304 kilometers. Find its average speed. (b) A rectangle has a perimeter of 2 m and a length of .4 m. Find its width. (2.52)

53. Evaluate using a calculator: (2.53)

    (*a*) $(4.61x^5)(3.006x^6y)$

    (*b*) $\dfrac{.0068q^5}{.09q^2}$

    (*c*) $(27y)(407y^{10})$

54. Using a calculator, find s where $A = stuv$ and $A = 10$, $t = .15$, $v = .013$, and $u = 11$.   (2.54)

55. State the coordinates of each lettered point in Fig. 2-21.   (2.55)

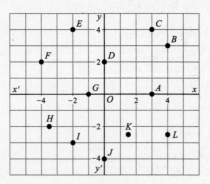

Fig. 2-21

56. Plot each of the following points: (2.56)

    $A(-2,-3)$ $C(0,-1)$ $E(3,-4)$ $G(0,3)$

    $B(-3,2)$ $D(-3,0)$ $F(1\frac{1}{2},2\frac{1}{2})$ $H(3\frac{1}{2},0)$

57. Plot the following points: $A(2,3)$, $B(-3,3)$, $C(-3,-2)$, $D(2,-2)$. Then find the perimeter and area of square *ABCD*.   (2.57)

58. Plot the following points: $A(4,3)$, $B(-1,3)$, $C(-3,-3)$, $D(2,-3)$. Then find the area of parallelogram *ABCD* and triangle *BCD*.   (2.57, 2.58)

59. Find the midpoint of the segment joining   (2.59)

    (*a*) $(0,0)$ and $(8,6)$ (*e*) $(-20,-5)$ and $(0,0)$ (*i*) $(3,4)$ and $(7,6)$

    (*b*) $(0,0)$ and $(5,7)$ (*f*) $(0,4)$ and $(0,16)$ (*j*) $(-2,-8)$ and $(-4,-12)$

    (*c*) $(0,0)$ and $(-8,12)$ (*g*) $(8,0)$ and $(0,-2)$ (*k*) $(7,9)$ and $(3,3)$

    (*d*) $(14,10)$ and $(0,0)$ (*h*) $(-10,0)$ and $(0,-5)$ (*l*) $(2,-1)$ and $(-2,-5)$

60. Find the midpoints of the sides of a triangle whose vertices are   (2.59)

    (*a*) $(0,0)$, $(8,0)$, $(0,6)$ (*c*) $(12,0)$, $(0,-4)$, $(0,0)$ (*e*) $(4,0)$, $(0,-6)$, $(-4,10)$

    (*b*) $(-6,0)$, $(0,0)$, $(0,10)$ (*d*) $(3,5)$, $(5,7)$, $(3,11)$ (*f*) $(-1,-2)$, $(0,2)$, $(1,-1)$

61. Find the midpoints of the sides of the quadrilateral whose successive vertices are   (2.59)

    (*a*) $(0,0)$, $(0,4)$, $(2,10)$, $(6,0)$ (*c*) $(-2,0)$, $(0,4)$, $(6,2)$, $(0,-10)$

    (*b*) $(-3,5)$, $(-1,9)$, $(7,3)$, $(5,-1)$ (*d*) $(-3,-7)$, $(-1,5)$, $(9,0)$, $(5,-8)$

62. Find the midpoints of the diagonals of the quadrilateral whose successive vertices are   (2.59)

    (a) $(0,0), (0,5), (4,12), (8,1)$ (c) $(0,-5), (0,1), (4,9), (4,3)$

    (b) $(-4,-1), (-2,3), (6,1), (2,-8)$

63. Find the center of a circle if the end points of a diameter are   (2.59)

    (a) $(0,0)$ and $(-4,6)$ (c) $(-3,1)$ and $(0,-5)$ (e) (a,b) and $(3a,5b)$

    (b) $(-1,0)$ and $(-5,-12)$ (d) $(0,0)$ and $(2a,2b)$ (f) $(a,2b)$ and $(a,2c)$

64. If M is the midpoint of $\overline{AB}$, find the coordinates of   (2.59)

    (a) M if the coordinates of A and B are $A(2,5)$ and $B(6,11)$

    (b) A if the coordinates of M and B are $M(1,3)$ and $B(3,6)$

    (c) B if the coordinates of A and M are $A(-2,1)$ and $M(2,-1)$

65. The trisection points of $\overline{AD}$ are B and C. Find the coordinates of   (2.59)

    (a) B if the coordinates of A and C are $A(1,2)$ and $C(3,5)$

    (b) D if the coordinates of B and C are $B(0,5)$ and $C(1\frac{1}{2},4)$

    (c) A if the coordinates of B and C are $B(0,6)$ and $C(2,3)$

66. $A(0,0)$, $B(0,5)$, $C(6,5)$, and $D(6,0)$ are the vertices of quadrilateral $ABCD$.   (2.60)

    (a) Prove that $ABCD$ is a rectangle.

    (b) Show that the midpoints of $\overline{AC}$ and $\overline{BD}$ have the same coordinates.

    (c) Do the diagonals bisect each other? Why?

67. The vertices of $\triangle ABC$ are $A(0,0)$, $B(0,4)$, and $C(6,0)$.   (2.60)

    (a) If $\overline{AD}$ is the median to $\overline{BC}$, find the coordinates of D and the midpoint of $\overline{AD}$.

    (b) If $\overline{CE}$ is the median to $\overline{AB}$, find the coordinates of E and the midpoint of $\overline{CE}$.

    (c) Do the medians, $\overline{AD}$ and $\overline{CE}$, bisect each other? Why?

68. Find the distance between each of the following pairs of points: (2.61)

    (a) $(0,0)$ and $(0,5)$ (d) $(-6,-1)$ and $(-6,11)$ (g) $(-3,-4\frac{1}{2})$ and $(-3,4\frac{1}{2})$

    (b) $(4,0)$ and $(-2,0)$ (e) $(5,3)$ and $(5,8.4)$ (h) (a,b) and $(2a,b)$

    (c) $(0,-3)$ and $(0,7)$ (f) $(-1.5,7)$ and $(6,7)$

69. Find the distance separating pairs of the following collinear points: (2.61)

    (a) $(5,-2), (5,1), (5,4)$ (c) $(-4,2), (-3,2), (0,2)$

    (b) $(0,-6), (0,-2), (0,12)$ (d) $(0,b), (a,b), (3a,b)$

70. Find the distance between each of the following pairs of points: (2.61)

    (a) $(0,0)$ and $(5,12)$ (e) $(-3,-6)$ and $(3,2)$ (i) $(3,4)$ and $(4,7)$

    (b) $(-3,-4)$ and $(0,0)$ (f) $(2,3)$ and $(-10,12)$ (j) $(-1,-1)$ and $(1,3)$

    (c) $(0,-6)$ and $(9,6)$ (g) $(2,2)$ and $(5,5)$ (k) $(-3,0)$ and $(0,\sqrt{7})$

    (d) $(4,1)$ and $(7,5)$ (h) $(0,5)$ and $(-5,0)$ (l) $(a,0)$ and $(0,a)$

71. Show that the triangles having the following vertices are isosceles triangles: (2.62)

    (a) $A(3,5)$, $B(6,9)$, and $C(2,6)$ (c) $G(5,-5)$, $H(-2,-2)$, and $J(8,2)$

    (b) $D(2,0)$, $E(6,0)$, and $F(4,4)$ (d) $K(7,0)$, $L(3,4)$, and $M(2,-1)$

72. Which of the triangles having the following vertices are right triangles?   (2.62)

    (a) $A(7,0)$, $B(6,3)$, and $C(12,5)$ (c) $G(1,-1)$, $H(5,0)$, and $J(3,8)$

    (b) $D(2,0)$, $E(5,2)$, and $F(1,8)$ (d) $K(-4,0)$, $L(-2,4)$, and $M(4,-1)$

73. The vertices of $\triangle ABC$ are $A(-2,2)$, $B(4,4)$, and $C(8,2)$. Find the length of the median to (a) $\overline{AB}$; (b) $\overline{AC}$; (c) $\overline{BC}$.   (2.62)

74. (a) The vertices of quadrilateral $ABCD$ are $A(0,0)$, $B(3,2)$, $C(7,7)$, and $D(4,5)$. Show that $ABCD$ is a parallelogram.   (2.63)

    (b) The vertices of quadrilateral $DEFG$ are $D(3,5)$, $E(1,1)$, $F(5,3)$, and $G(7,7)$. Show that $DEFG$ is a rhombus.

    (c) The vertices of quadrilateral $HJKL$ are $H(0,0)$, $J(4,4)$, $K(0,8)$, and $L(-4,4)$. Show that $HJKL$ is a square.

75. Find the radius of a circle that has its center at   (2.64)

    (a) $(0,0)$ and passes through $(-6,8)$ (d) $(2,0)$ and passes through $(7,-12)$

    (b) $(0,0)$ and passes through $(3,-4)$ (e) $(4,3)$ and is tangent to the y-axis

    (c) $(0,0)$ and passes through $(-5,5)$ (f) $(-1,7)$ and is tangent to the line $x=-4$

76. A circle has its center at the origin and a radius of 10. State whether each of the following points is on, inside, or outside of this circle: (a) $(6,8)$; (b) $(-6,8)$; (c) $(0,11)$; (d) $(-10,0)$; (e) $(7,7)$; (f) $(-9,4)$; (g) $(9,\sqrt{19})$.   (2.64)

77. Find the slope of the line through each of the following pairs of points: (2.65)

    (a) $(0,0)$ and $(5,9)$ (e) $(-2,-3)$ and $(7,15)$ (i) $(3,-9)$ and $(0,0)$

    (b) $(0,0)$ and $(9,5)$ (f) $(-2,-3)$ and $(2,1)$ (j) $(0,-2)$ and $(8,10)$

    (c) $(0,0)$ and $(6,15)$ (g) $(3,-4)$ and $(5,6)$ (k) $(-1,-5)$ and $(1,-7)$

    (d) $(2,3)$ and $(6,15)$ (h) $(0,0)$ and $(-4,8)$ (l) $(-3,-4)$ and $(-1,-2)$

78. Find the slope of the line whose equation is   (2.65)

    (a) $y=3x-4$ (e) $y=5x$ (i) $3y=-12x+6$ (m) $\frac{1}{3}y=x-3$

    (b) $y=4x-3$ (f) $y=5$ (j) $3y=12-2x$ (n) $\frac{1}{3}y=2x-6$

    (c) $y=-\frac{1}{2}x+5$ (g) $2y=6x-10$ (k) $y+x=21$ (o) $\frac{1}{4}y=7-x$

    (d) $y=8-7x$ (h) $2y=10x-6$ (l) $2x=12-y$ (p) $\frac{1}{4}y+2x=1$

79. Find the inclination, to the nearest degree, of each of the following lines: (2.65)

    (a) $y=3x-1$ (c) $2y=5x+10$ (e) $5y=5x-3$

    (b) $y=\frac{1}{3}x-1$ (d) $y=\frac{2}{3}x+5$ (f) $y=-3$

80. Find the slope of a line whose inclination is (a) 5°; (b) 17°; (c) 20°; (d) 35°; (e) 45°; (f) 73°; (g) 85°.   (2.65)

81. Find the inclination, to the nearest degree, of a line whose slope is (a) 0; (b) 0.4663; (c) 1, (d) 1.4281; (e) $\frac{1}{8}$; (f) $\frac{1}{2}$; (g) $\frac{3}{4}$; (h) $1\frac{1}{3}$; (i) $2\frac{1}{5}$.   (2.65)

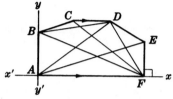

Fig. 2-22

82. In hexagon $ABCDEF$ of Fig. 2-22, $\overline{CD}\|\overline{AF}$. Which sides or diagonals have (a) positive slope; (b) negative slope; (c) zero slope; (d) no slope?   (2.65)

83. Find the slope of a line that is parallel to a line whose slope is (a) 0; (b) has no slope; (c) 5; (d) −5; (e) 0.5; (f) −0.0005.   (2.66)

84. Find the slope of a line that is parallel to the line whose equation is   (2.66)

(a) $y = 0$ (c) $x = 7$ (e) $y = 5x - 2$ (g) $3y - 6x = 12$

(b) $x = 0$ (d) $y = 7$ (f) $x + y = 5$

85. Find the slope of a line that is parallel to a line which passes through (a) $(0,0)$ and $(2,3)$; (b) $(2,-1)$ and $(5,6)$; (c) $(3,4)$ and $(5,2)$; (d) $(1,2)$ and $(0,-4)$.   (2.66)

86. Find the slope of a line that is perpendicular to a line whose slope is   (2.66)

(a) $\frac{1}{2}$ (c) 3   (e) 0.1   (g) $-\frac{4}{5}$ (i) 0

(b) 1   (d) $2\frac{1}{2}$ (f) −1   (h) $-3\frac{1}{4}$ (j) has no slope

87. Find the slope of a line that is perpendicular to a line which passes through (a) $(0,0)$ and $(0,5)$; (b) $(0,0)$ and $(2,1)$; (c) $(0,0)$ and $(3,-1)$; (d) $(1,1)$ and $(3,3)$.   (2.66)

88. In rectangle $DEFG$, the slope of $\overline{DE}$ is $\frac{2}{3}$. What is the slope of (a) $\overline{EF}$; (b) $\overline{FG}$; (c) $\overline{DG}$?   (2.67)

89. In $\square\,ABCD$, the slope of $\overline{AB}$ is 1 and the slope of $\overline{BC}$ is $-\frac{1}{2}$. What is the slope of (a) $\overline{AD}$; (b) $\overline{CD}$; (c) the altitude to $\overline{AD}$; (d) the altitude to $\overline{CD}$?   (2.67)

90. The vertices of $\triangle ABC$ are $A(0,5)$, $B(3,7)$, and $C(5,-1)$. What is the slope of the altitude to (a) $\overline{AB}$; (b) $\overline{BC}$; (e) $\overline{AC}$?   (2.67)

91. Which of the following sets of points are collinear: (a) $(2,1)$, $(4,4)$, $(8,10)$; (b) $(-1,1)$, $(2,4)$, $(6,8)$; (c) $(1,-1)$, $(3,4)$, $(5,8)$?   (2.68)

92. What values of k will make the following trios of points collinear: (a) $A(0,1)$, $B(2,7)$, $C(6,k)$; (b) $D(-1,5)$, $E(3,k)$, $F(5,11)$; (c) $G(0,k)$, $H(1,1)$, $I(3,-1)$?   (2.68)

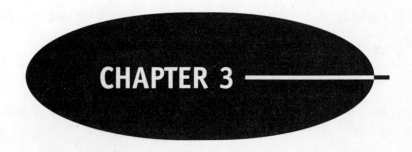

Lines, Angles, and Triangles

3.1 HISTORICAL BACKGROUND OF GEOMETRY

The word *geometry* is derived from the Greek words *geos* (meaning *earth*) and *metron* (meaning *measure*). The ancient Egyptians, Chinese, Babylonians, Romans, and Greeks used geometry for surveying, navigation, astronomy, and other practical occupations.

The Greeks sought to systematize the geometric facts they knew by establishing logical reasons for them and relationships among them. The work of such men as Thales (600 B.C.), Pythagoras (540 B.C.), Plato (390 B.C.), and Aristotle (350 B.C.) in systematizing geometric facts and principles culminated in the geometry text *Elements*, written about 325 B.C. by Euclid. This most remarkable text has been in use for over 2000 years.

3.2 UNDEFINED TERMS OF GEOMETRY: POINT, LINE, AND PLANE

3.2A Point, Line, and Plane are Undefined Terms

These undefined terms underlie the definitions of all geometric terms. They can be given meanings by way of descriptions. However, these descriptions, which follow, are not to be thought of as definitions.

3.2B Point

A point has position only. It has no length, width, or thickness.

A point is represented by a dot. Keep in mind, however, that the dot *represents* a point but *is not* a point, just as a dot on a map may represent a locality but is not the locality. A dot, unlike a point, has size.

A point is designated by a capital letter next to the dot, thus: $_A \cdot P$.

3.2C Line

A line has length but has no width or thickness.

A line may be represented by the path of a piece of chalk on the blackboard or by a stretched rubber band.

A line is designated by the capital letters of any two of its points or by a small letter, thus:

$\overset{\longleftrightarrow}{\underset{A \quad B}{}}$, $\overset{\frown}{\underset{C \quad D}{}}$, $\overset{a}{\nearrow}$, or $\overset{\leftrightarrow}{AB}$.

A line may be straight, curved, or a combination of these. To understand how lines differ, think of a line as being generated by a moving point. A *straight line*, such as $\longleftrightarrow$, is generated by a point moving always in the same direction. A *curved line*, such as $\frown$, is generated by a point moving in a continuously changing direction.

Two lines intersect in a point.

A straight line is unlimited in extent. It may be extended in either direction indefinitely.

A *ray* is the part of a straight line beginning at a given point and extending limitlessly in one

direction: $\overrightarrow{AB}$ and $\overset{B}{\nearrow_{A}}$ designate rays.

In this book, the word *line* will mean "straight line" unless you are told otherwise.

3.2D   Planes

A plane has length and width but no thickness. It may be represented by a blackboard or a side of a box; remember, however, that these are representations of a plane but are not planes.

A plane surface (or plane) is a surface such that a straight line connecting any two of its points lies entirely in it. A plane is a flat surface.

Plane geometry is the geometry of plane figures—those that may be drawn on a plane. Unless you are told otherwise, the word *figure* will mean "plane figure" in this book.

SOLVED PROBLEMS

3.1   ILLUSTRATING UNDEFINED TERMS

Point, line, and plane are undefined terms. State which of these terms is illustrated by (*a*) the top of a desk; (*b*) a projection screen; (*c*) a ruler's edge; (*d*) a stretched thread; (*e*) the tip of a pin.

Solutions

(*a*) Surface; (*b*) surface; (*c*) line; (*d*) line; (*e*) point.

3.3   LINE SEGMENTS

A straight line segment is the part of a straight line between two of its points, including the two points. It is designated by the capital letters of these points or by a small letter. Thus $\overline{AB}$ or r represents the straight line segment $A \overset{r}{\relbar\joinrel\relbar} B$ between A and B.

The expression *straight line segment* may be shortened to *line segment* or to *segment*, if the meaning is clear. Thus, $\overline{AB}$ and *segment AB* both mean "the straight line segment AB."

3.3A   Dividing a Line Segment into Parts

If a line segment is divided into parts:

1. The length of the whole line segment equals the sum of the lengths of its parts. Note that the length of $\overline{AB}$ is designated AB.

2. The length of the whole line segment is greater than the length of any part.
    Suppose $\overline{AB}$ is divided into three parts of lengths a, b, and c, thus: $A\overset{a}{\underset{\bullet}{\rule{1cm}{0.4pt}}}\overset{b}{\underset{\bullet}{\rule{1cm}{0.4pt}}}\overset{c}{\rule{1cm}{0.4pt}}B$. Then $AB = a + b + c$. Also, AB is greater than a; this may be written $AB > a$.

If a line segment is divided into two equal parts:

1. The point of division is the midpoint of the line segment.

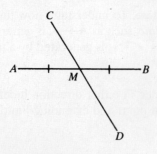

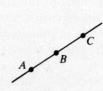

Fig. 3-1   Fig. 3-2

2. A line that crosses at the midpoint is said to bisect the segment.

    Because $AM = MB$ in Fig. 3-1, M is the midpoint of $\overline{AB}$, and $\overline{CD}$ bisects $\overline{AB}$. Equal line segments may be shown by crossing them with the same number of strokes. Note that $\overline{AM}$ and $\overline{MB}$ are crossed with a single stroke.

3. If three points A, B, and C lie on a line, then we say they are *collinear*. If A, B, and C are collinear and $AB + BC = AC$, then B is between A and C (see Fig. 3-2).

3.3B Congruent Segments

Two line segments having the same length are said to be *congruent*. Thus, if $AB = CD$, then $\overline{AB}$ is congruent to $\overline{CD}$, written $\overline{AB} \cong \overline{CD}$.

SOLVED PROBLEMS

3.2   NAMING LINE SEGMENTS AND POINTS
     See Fig. 3-3.

(a) Name each line segment shown.

(b) Name the line segments that intersect at A.

(c) What other line segment can be drawn?

(d) Name the point of intersection of $\overline{CD}$ and $\overline{AD}$.

(e) Name the point of intersection of $\overline{BC}$, $\overline{AC}$, and $\overline{CD}$.

Solutions

(a) $\overline{AB}$, $\overline{BC}$, $\overline{CD}$, $\overline{AC}$, and $\overline{AD}$. These segments may also be named by interchanging the letters; thus BA, CB, DC, CA, and DA are also correct.

(b) $\overline{AB}$, $\overline{AC}$, and $\overline{AD}$ (d) D

(c) $\overline{BD}$ (e) C

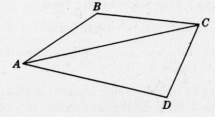

Fig. 3-3

3.3 FINDING LENGTHS AND POINTS OF LINE SEGMENTS

See Fig. 3-4.

(a) State the lengths of $\overline{AB}$, $\overline{AC}$, and $\overline{AF}$.

(b) Name two midpoints.

(c) Name two bisectors.

(d) Name all congruent segments.

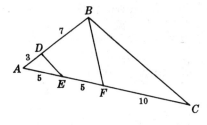

Fig. 3-4

Solutions

(a) $AB = 3 + 7 = 10$; $AC = 5 + 5 + 10 = 20$; $AF = 5 + 5 = 10$.

(b) E is midpoint of $\overline{AF}$; F is midpoint of $\overline{AC}$.

(c) $\overline{DE}$ is bisector of $\overline{AF}$; $\overline{BF}$ is bisector of $\overline{AC}$.

(d) $\overline{AB}$, $\overline{AF}$, and $\overline{FC}$ (all have length 10); $\overline{AE}$ and $\overline{EF}$ (both have length 5).

3.4 CIRCLES

A *circle* is the set of all points in a plane that are the same distance from the *center*. The symbol for circle is ⊙; for circles, ⑤. Thus ⊙O stands for the circle whose center is O.

The *circumference* of a circle is the distance around the circle. It contains 360 degrees (360°).

A *radius* is a segment joining the center of a circle to a point on the circle (see Fig. 3-5). From the definition of a circle, it follows the radii of a circle are congruent. Thus $\overline{OA}$, $\overline{OB}$, and $\overline{OC}$ of Fig. 3-5 are radii of ⊙O and $\overline{OA} \cong \overline{OB} \cong \overline{OC}$.

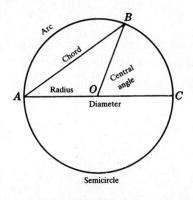

Fig. 3-5

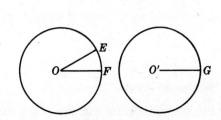

Fig. 3-6

A *chord* is a segment joining any two points on a circle. Thus $\overline{AB}$ and $\overline{AC}$ are chords of ⊙O.

A *diameter* is a chord through the center of the circle; it is the longest chord and is twice the length of a radius. $\overline{AC}$ is a diameter of ⊙O.

An *arc* is a continuous part of a circle. The symbol for arc is ⌒, so that $\overparen{AB}$ stands for arc AB. An arc of measure 1° is 1/360th of a circumference.

A *semicircle* is an arc measuring one-half the circumference of a circle and thus contains 180°. A diameter divides a circle into two semicircles. For example, diameter $\overline{AC}$ cuts ⊙O of Fig. 3-5 into two semicircles.

A *central angle* is an angle formed by two radii. Thus the angle between radii $\overline{OB}$ and $\overline{OC}$ is a central angle. A central angle measuring 1° cuts off an arc of 1°; thus if the central angle between $\overline{OE}$ and $\overline{OF}$ in Fig. 3-6 is 1°, then $\overparen{EF}$ measures 1°.

Congruent circles are circles having congruent radii. Thus if $\overline{OE} \cong \overline{O'G}$, then ⊙$O \cong$ ⊙O'.

SOLVED PROBLEMS

3.4   FINDING LINES AND ARCS IN A CIRCLE

In Fig. 3-7, find (a) OC and AB; (b) the number of degrees in $\widehat{AD}$; (c) the number of degrees in $\widehat{BC}$.

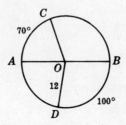

Fig. 3-7

Solutions

(a)   Radius OC = radius OD = 12. Diameter AB = 24.

(b)   Since semicircle ADB contains 180°, $\widehat{AD}$ contains 180° − 100° = 80°.

(c)   Since semicircle ACB contains 180°, $\widehat{BC}$ contains 180° − 70° = 110°.

3.5   ANGLES

An angle is the figure formed by two rays with a common end point. The rays are the *sides* of the angle, while the end point is its *vertex*. The symbol for angle is ∠ or ⊰; the plural is ⊰.

Thus $\vec{AB}$ and $\vec{AC}$ are the sides of the angle shown in Fig. 3-8(a), and A is its vertex.

3.5A   Naming an Angle

An angle may be named in any of the following ways:

1.   With the vertex letter, if there is only one angle having this vertex, as ∠B in Fig. 3-8(b).

2.   With a small letter or a number placed between the sides of the angle and near the vertex, as ∠a or ∠1 in Fig. 3-8(c).

3.   With three capital letters, such that the vertex letter is between two others, one from each side of the angle. In Fig. 3-8(d), ∠E may be named ∠DEG or ∠GED.

Fig. 3-8

3.5B Measuring the Size of an Angle

The size of an angle depends on the extent to which one side of the angle must be rotated, or turned about the vertex, until it meets the other side. We choose degrees to be the unit of measure for angles. The measure of an angle is the number of degrees it contains. We will write $m\angle A = 60°$ to denote that "angle A measures 60°."

The protractor in Fig. 3-9 shows that $\angle A$ measures 60°. If $\overrightarrow{AC}$ were rotated about the vertex A until it met $\overrightarrow{AB}$, the amount of turn would be 60°.

In using a protractor, be sure that the vertex of the angle is at the center and that one side is along the 0°–180° diameter.

The size of an angle *does not* depend on the lengths of the sides of the angle.

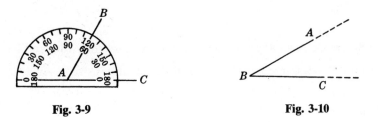

Fig. 3-9   Fig. 3-10

The size of $\angle B$ in Fig. 3-10 would not be changed if its sides $\overrightarrow{AB}$ and $\overrightarrow{BC}$ were made larger or smaller.

No matter how large or small a clock is, the angle formed by its hands at 3 o'clock measures 90°, as shown in Figs. 3-11 and 3-12.

Fig. 3-11   Fig. 3-12

To measure angles with more precision, we divide 1° into 60 equal parts, called *minutes*. Thus, 1° = 60 minutes = 60′. Each minute is divided into 60 equal parts called *seconds* for even greater precision. Thus, we have 1° = 60′ and 1′ = 60″.

3.5C Kinds of Angles

1. *Acute angle*: An acute angle is an angle whose measure is less than 90°.

    Thus, in Fig. 3-13, $a°$ is less than 90°; this is symbolized as $a° < 90°$.

2. *Right angle*: A right angle is an angle that measures 90°.

    Thus, in Fig. 3-14, $m(\text{rt. } \angle A) = 90°$. The square corner denotes a right angle.

3. *Obtuse angle*: An obtuse angle is an angle whose measure is more than 90° and less than 180°.

    Thus, in Fig. 3-15, 90° is less than $b°$ and $b°$ is less than 180°; this is denoted by $90° < b° < 180°$.

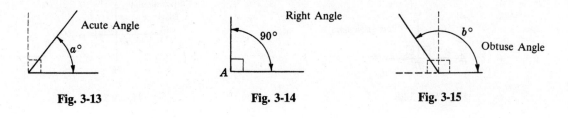

Fig. 3-13   Fig. 3-14   Fig. 3-15

4. *Straight angle*: A straight angle is an angle that measures 180°.

    Thus, in Fig. 3-16, m(st. $\angle B$) = 180°. Note that the sides of a straight angle lie in the same straight line. But do not confuse a straight angle with a straight line!

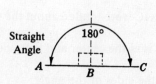

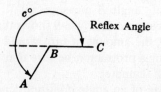

| **Fig. 3-16** | **Fig. 3-17** |

5. *Reflex angle*: A reflex angle is an angle whose measure is more than 180° and less than 360°.

    Thus, in Fig. 3-17, 180° is less than $c°$ and $c°$ is less than 360°; this is symbolized as $180° < c° < 360°$.

3.5D   Additional Angle Facts

1. *Congruent angles* are angles that have the same number of degrees. In other words, if $m\angle A = m\angle B$, then $\angle A \cong \angle B$.

    Thus, in Fig. 3-18, rt.$\angle A \cong$ rt.$\angle B$ since each measures 90°.

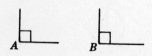

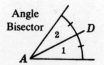

| **Fig. 3-18** | **Fig. 3-19** |

2. A line that *bisects* an angle divides it into two congruent parts.

    Thus, in Fig. 3-19, if $\overrightarrow{AD}$ bisects $\angle A$, then $\angle 1 \cong \angle 2$. (Congruent angles may be shown by crossing their arcs with the same number of strokes. Here the arcs of $\angle$s 1 and 2 are crossed by a single stroke.)

3. *Perpendiculars* are lines or rays or segments that meet at right angles.

    The symbol for perpendicular is $\perp$; for perpendiculars, $\perp$s. In Fig. 3-20, $\overline{CD} \perp \overline{AB}$, so right angles 1 and 2 are formed.

4. A *perpendicular bisector* of a given segment is perpendicular to the segment and bisects it.

    In Fig. 3-21, $\overleftrightarrow{GH}$ is the $\perp$ bisector of $\overline{EF}$; thus $\angle 1$ and $\angle 2$ are right angles and M is the midpoint of $\overline{EF}$.

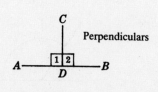

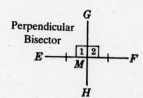

| **Fig. 3-20** | **Fig. 3-21** |

SOLVED PROBLEMS

3.5   NAMING AN ANGLE

Name the following angles in Fig. 3-22: (*a*) two obtuse angles; (*b*) a right angle; (*c*) a straight angle; (*d*) an acute angle at *D*; (*e*) an acute angle at *B*.

Solutions

(*a*) $\angle ABC$ and $\angle ADB$ (or $\angle 1$). The angles may also be named by reversing the order of the letters: $\angle CBA$ and $\angle BDA$.

(*b*) $\angle DBC$ (*d*) $\angle 2$ or $\angle BDC$

(*c*) $\angle ADC$ (*e*) $\angle 3$ or $\angle ABD$

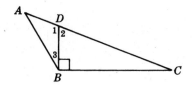

Fig. 3-22

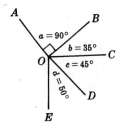

Fig. 3-23

3.6   ADDING AND SUBTRACTING ANGLES

In Fig. 3-23, find (*a*) $m\angle AOC$; (*b*) $m\angle BOE$; (*c*) the measure of obtuse $\angle AOE$.

Solutions

(*a*) $m\angle AOC = m\angle a + m\angle b = 90° + 35° = 125°$

(*b*) $m\angle BOE = m\angle b + m\angle c + m\angle d = 35° + 45° + 50° = 130°$

(*c*) $m\angle AOE = 360° - (m\angle a + m\angle b + m\angle c + m\angle d) = 360° - 220° = 140°$

3.7   FINDING PARTS OF ANGLES

Find (*a*) $\frac{2}{5}$ of the measure of a rt. $\angle$; (*b*) $\frac{2}{3}$ of the measure of a st. $\angle$; (*c*) $\frac{1}{2}$ of 31°; (*d*) $\frac{1}{10}$ of 70°20′.

Solutions

(*a*) $\frac{2}{5}(90°) = 36°$ (*c*) $\frac{1}{2}(31°) = 15\frac{1}{2}° = 15°30′$

(*b*) $\frac{2}{3}(180°) = 120°$ (*d*) $\frac{1}{10}(70°20′) = \frac{1}{10}(70°) + \frac{1}{10}(20′) = 7°2′$

3.8   FINDING ROTATIONS

In a half hour, what turn or rotation is made (*a*) by the minute hand, and (*b*) by the hour hand of a clock? What rotation is needed to turn (*c*) from north to southeast in a clockwise direction; (*d*) from northwest to southwest in a counterclockwise direction (see Fig. 3-24)?

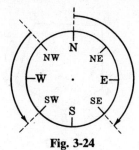

Fig. 3-24

Solutions

(*a*) In 1 hour, a minute hand completes a full circle of 360°. Hence in a half hour it turns 180°.

(*b*) In 1 hour, an hour hand turns $\frac{1}{12}$ of 360° or 30°. Hence in a half hour it turns 15°.

(*c*) Add a turn of 90° from north to east and a turn of 45° from east to southeast to get 90° + 45° = 135°.

(*d*) The turn from northwest to southwest is $\frac{1}{4}(360°) = 90°$.

3.9   FINDING ANGLES

Find the measure of the angle formed by the hands of the clock in Fig. 3-25, (*a*) at 8 o'clock; (*b*) at 4:30 o'clock.

Solutions

(*a*) At 8 o'clock, $m\angle a = \frac{1}{3}(360°) = 120°$.

(*b*) At 4:30 o'clock, $m\angle b = \frac{1}{2}(90°) = 45°$.

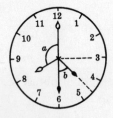

Fig. 3-25

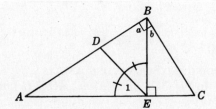

Fig. 3-26

3.10   APPLYING ANGLE FACTS

In Fig. 3-26, (*a*) name two pairs of perpendicular segments; (*b*) find $m\angle a$ if $m\angle b = 42°$; (*c*) find $m\angle AEB$ and $m\angle CED$.

Solutions

(*a*) Since $\angle ABC$ is a right angle, $\overline{AB} \perp \overline{BC}$. Since $\angle BEC$ is a right angle, $\overline{BE} \perp \overline{AC}$.

(*b*) $m\angle a = 90° - m\angle b = 90° - 42° = 48°$.

(*c*) $m\angle AEB = 180° - m\angle BEC = 180° - 90° = 90°$. $m\angle CED = 180° - m\angle 1 = 180° - 45° = 135°$.

3.6   TRIANGLES

A *polygon* is a closed plane figure bounded by straight line segments as sides. Thus Fig. 3-27 is a polygon of five sides, called a *pentagon*; it is named pentagon *ABCDE*, using its letters in order.

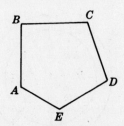

Fig. 3-27

A *triangle* is a polygon having three sides. A *vertex* of a triangle is a point at which two of the sides meet. (*Vertices* is the plural of vertex.) The symbol for triangle is △; for triangles, ⧍.

A triangle may be named with its three letters in any order or with a Roman numeral placed inside of it. Thus the triangle shown in Fig. 3-28 is △*ABC* or △I; its sides are $\overline{AB}$, $\overline{AC}$, and $\overline{BC}$; its vertices are A, B, and C; its angles are ∠A, ∠B, and ∠C.

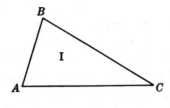

Fig. 3-28

3.6A   Classifying Triangles

Triangles are classified according to the equality of the lengths of their sides or according to the kind of angles they have.

Triangles According to the Equality of the Lengths of their Sides (Fig. 3-29)

1. *Scalene triangle*: A scalene triangle is a triangle having no congruent sides.

   Thus in scalene triangle ABC, $a \neq b \neq c$. The small letter used for the length of each side agrees with the capital letter of the angle opposite it. Also, $\neq$ means "is not equal to."

2. *Isosceles triangle*: An isosceles triangle is a triangle having at least two congruent sides.

   Thus in isosceles triangle ABC, $a = c$. These equal sides are called the *legs* of the isosceles triangle; the remaining side is the *base b*. The angles on either side of the base are the *base angles*; the angle opposite the base is the *vertex angle*.

3. *Equilateral triangle*: An equilateral triangle is a triangle having three congruent sides.

   Thus in equilateral triangle ABC, $a = b = c$. Note that an equilateral triangle is also an isosceles triangle.

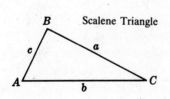
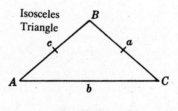
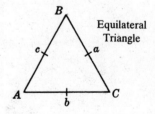

Fig. 3-29

Triangles According to the Kind of Angles (Fig. 3-30)

1. *Right triangle*: A right triangle is a triangle having a right angle.

   Thus in right triangle ABC, ∠C is the right angle. Side c opposite the right angle is the *hypotenuse*. The perpendicular sides, a and b, are the *legs* or *arms* of the right triangle.

2. *Obtuse triangle*: An obtuse triangle is a triangle having an obtuse angle.

    Thus in obtuse triangle *DEF*, ∠*D* is the obtuse angle.

3. *Acute triangle*: An acute triangle is a triangle having three acute angles.

    Thus in acute triangle *HJK*, ∠*H*, ∠*J*, and ∠*K* are acute angles.

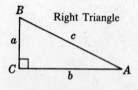 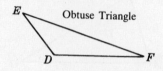 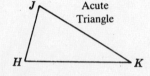

Fig. 3-30

3.6B   Special Lines in a Triangle

1. *Angle bisector of a triangle*: An angle bisector of a triangle is a segment or ray that bisects an angle and extends to the opposite side.

    Thus $\overline{BD}$, the angle bisector of ∠*B* in Fig. 3-31, bisects ∠*B*, making ∠1 = ∠2.

2. *Median of a triangle*: A median of a triangle is a segment from a vertex to the midpoint of the opposite side.

    Thus $\overline{BM}$, the median to $\overline{AC}$, in Fig. 3-32, bisects $\overline{AC}$, making *AM* = *MC*.

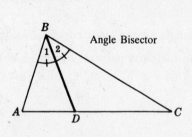 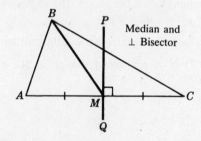

Fig. 3-31   **Fig. 3-32**

3. *Perpendicular bisector of a side*: A perpendicular bisector of a side of a triangle is a line that bisects and is perpendicular to a side.

    Thus $\overleftrightarrow{PQ}$, the perpendicular bisector of $\overline{AC}$ in Fig. 3-32, bisects $\overline{AC}$ and is perpendicular to it.

4. *Altitude to a side of a triangle*: An altitude of a triangle is a segment from a vertex perpendicular to the opposite side.

    Thus $\overline{BD}$, the altitude to $\overline{AC}$ in Fig. 3-33, is perpendicular to $\overline{AC}$ and forms right angles 1 and 2. Each angle bisector, median, and altitude of a triangle extends from a vertex to the opposite side.

5. *Altitudes of obtuse triangle*: In an obtuse triangle, the altitude drawn to either side of the obtuse angle falls outside the triangle.

Thus in obtuse triangle *ABC* (shaded) in Fig. 3-34, altitudes $\overline{BD}$ and $\overline{CE}$ fall outside the triangle. In each case, a side of the obtuse angle must be extended.

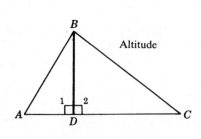

Fig. 3-33

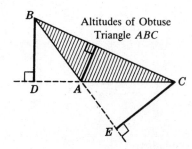

Fig. 3-34

SOLVED PROBLEMS

3.11 NAMING A TRIANGLE AND ITS PARTS

In Fig. 3-35, name (*a*) an obtuse triangle, and (*b*) two right triangles and the hypotenuse and legs of each. (*c*) In Fig. 3-36, name two isosceles triangles; also name the legs, base, and vertex angle of each.

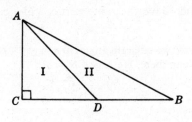

Fig. 3-35

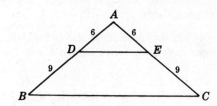

Fig. 3-36

Solutions

(*a*) Since ∠*ADB* is an obtuse angle, △*ADB* or △II is obtuse.

(*b*) Since ∠*C* is a right angle, △I and △*ABC* are right triangles. In △I, $\overline{AD}$ is the hypotenuse and $\overline{AC}$ and *CD* are the legs. In △*ABC*, *AB* is the hypotenuse and *AC* and *BC* are the legs.

(*c*) Since *AD* = *AE*, △*ADE* is an isosceles triangle. In △*ADE*, $\overline{AD}$ and $\overline{AE}$ are the legs, $\overline{DE}$ is the base, and ∠*A* is the vertex angle.

Since *AB* = *AC*, △*ABC* is an isosceles triangle. In △*ABC*, $\overline{AB}$ and $\overline{AC}$ are the legs, $\overline{BC}$ is the base, and ∠*A* is the vertex angle.

3.12 SPECIAL LINES IN A TRIANGLE

Name the equal segments and congruent angles in Fig. 3-37, (*a*) if $\overline{AE}$ is the altitude to $\overline{BC}$; (*b*) if $\overline{CG}$ bisects ∠*ACB*; (*c*) if $\overline{KL}$ is the perpendicular bisector of $\overline{AD}$; (*d*) if $\overline{DF}$ is the median to $\overline{AC}$.

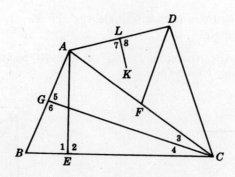

Fig. 3-37

Solutions

(a)   Since $\overline{AE} \perp \overline{BC}$, $\angle 1 \cong \angle 2$.

(b)   Since $\overline{CG}$ bisects $\angle ACB$, $\angle 3 \cong \angle 4$.

(c)   Since $\overline{LK}$ is the $\perp$ bisector of $\overline{AD}$, $AL = LD$ and $\angle 7 \cong \angle 8$.

(d)   Since $\overline{DF}$ is median to $\overline{AC}$, $AF = FC$.

3.7   PAIRS OF ANGLES

3.7A   Kinds of Pairs of Angles

1.   *Adjacent angles*: Adjacent angles are two angles which have the same vertex and a common side between them.

Thus, the entire angle of $c°$ in Fig. 3-38 has been cut into two adjacent angles of $a°$ and $b°$. These adjacent angles have same vertex A, and a common side $\overrightarrow{AD}$ between them. Here, $a° + b° = c°$.

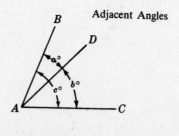

Fig. 3-38

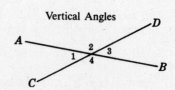

Fig. 3-39

2.   *Vertical angles*: Vertical angles are two nonadjacent angles formed by two intersecting lines.

Thus, $\angle 1$ and $\angle 3$ in Fig. 3-39 are vertical angles formed by intersecting lines $\overleftrightarrow{AB}$ and $\overleftrightarrow{CD}$. Also, $\angle 2$ and $\angle 4$ are another pair of vertical angles formed by the same lines.

3.   *Complementary angles*: Complementary angles are two angles whose measures total 90°.

Thus, in Fig. 3-40(a) the angles of $a°$ and $b°$ are adjacent complementary angles. However, in (b) the complementary angles are nonadjacent. In each case, $a° + b° = 90°$. Either of two complementary angles is said to be the *complement* of the other.

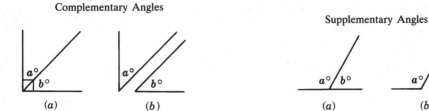

Fig. 3-40   Fig. 3-41

4. *Supplementary angles*: Supplementary angles are two angles whose measures total 180°.

Thus, in Fig. 3-41(a) the angles of $a°$ and $b°$ are adjacent supplementary angles. However, in (b) the supplementary angles are nonadjacent. In each case, $a° + b° = 180°$. Either of two supplementary angles is said to be the *supplement* of the other.

3.7B   Principles of Pairs of Angles

PRINCIPLE 1:   *If an angle of $c°$ is cut into two adjacent angles of $a°$ and $b°$, then $a° + b° = c°$.*

Thus if $a° = 25°$ and $b° = 35°$ in Fig. 3-42, then $c° = 25° + 35° = 60°$.

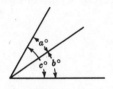

Fig. 3-42

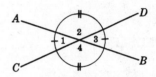

Fig. 3-43

PRINCIPLE 2:   *Vertical angles are congruent.*

Thus if AB and CD are straight lines in Fig. 3-43, then $\angle 1 \cong \angle 3$ and $\angle 2 \cong \angle 4$. Hence, if $m\angle 1 = 40°$, then $m\angle 3 = 40°$; in such a case, $m\angle 2 = m\angle 4 = 140°$.

PRINCIPLE 3:   *If two complementary angles contain $a°$ and $b°$, then $a° + b° = 90°$.*

Thus if angles of $a°$ and $b°$ are complementary and $a° = 40°$, then $b° = 50°$ [Fig. 3-44(a) or (b)].

PRINCIPLE 4:   *Adjacent angles are complementary if their exterior sides are perpendicular to each other.*

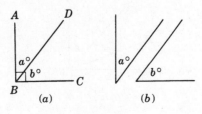

Fig. 3-44

Thus in Fig. 3-44(a), $a°$ and $b°$ are complementary since their exterior sides $\overline{AB}$ and $\overline{BC}$ are perpendicular to each other.

PRINCIPLE 5: *If two supplementary angles contain $a°$ and $b°$, then $a° + b° = 180°$.*

Thus if angles of $a°$ and $b°$ are supplementary and $a° = 140°$, then $b° = 40°$ [Fig. 3-45(a) or (b)].

PRINCIPLE 6: *Adjacent angles are supplementary if their exterior sides lie in the same straight line.*

Thus in Fig. 3-45(a) $a°$ and $b°$ are supplementary angles since their exterior sides $\vec{AB}$ and $\vec{BC}$ lie in the same straight line $\overleftrightarrow{AC}$.

Fig. 3-45   Fig. 3-46

PRINCIPLE 7: *If supplementary angles are congruent, each of them is a right angle. (Equal supplementary angles are right angles.)*

Thus if $\angle 1$ and $\angle 2$ in Fig. 3-46 are both congruent and supplementary, then each of them is a right angle.

SOLVED PROBLEMS

3.13   NAMING PAIRS OF ANGLES

(a) In Fig. 3-47(a), name two pairs of supplementary angles.

(b) In Fig. 3-47(b), name two pairs of complementary angles.

(c) In Fig. 3-47(c), name two pairs of vertical angles.

Solutions

(a) Since their sum is 180°, the supplementary angles are (1) $\angle 1$ and $\angle BED$; (2) $\angle 3$ and $\angle AEC$.

(b) Since their sum is 90°, the complementary angles are (1) $\angle 4$ and $\angle FJH$; (2) $\angle 6$ and $\angle EJG$.

(c) Since $\overleftrightarrow{KL}$ and $\overleftrightarrow{MN}$ are intersecting lines, the vertical angles are (1) $\angle 8$ and $\angle 10$; (2) $\angle 9$ and $\angle MOK$.

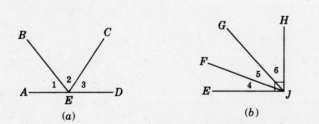   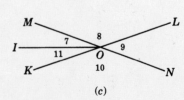

Fig. 3-47

3.14   FINDING PAIRS OF ANGLES

Find two angles such that:

(a) The angles are supplementary and the larger is twice the smaller.

(b) The angles are complementary and the larger is 20° more than the smaller.

(c) The angles are adjacent and form an angle of 120°. The larger is 20° less than three times the smaller.

(d) The angles are vertical and complementary.

Solutions

In each solution, x is a number only. This number indicates the number of degrees contained in the angle. Hence, if $x = 60$, the angle measures 60°.

(a) Let $x = m$ (smaller angle) and $2x = m$ (larger angle), as in Fig. 3-48(a).
     Principle 5: $x + 2x = 180$, so $3x = 180$; $x = 60$.
                                     $2x = 120$.   *Ans.* 60° and 120°

(b) Let $x = m$ (smaller angle) and $x + 20 = m$ (larger angle), as in Fig. 3-48(b).
     Principle 3: $x + (x + 20) = 90$, or $2x + 20 = 90$; $x = 35$.
                                     $x + 20 = 55$.   *Ans.* 35° and 55°

(c) Let $x = m$ (smaller angle) and $3x - 20 = m$ (larger angle) as in Fig. 3-48(c).
     Principle 1: $x + (3x - 20) = 120$, or $4x - 20 = 120$; $x = 35$.
                                     $3x - 20 = 85$.   *Ans.* 35° and 85°

(d) Let $x = m$ (each vertical angle), as in Fig. 3-48(d). They are congruent by Principle 2.
     Principle 3: $x + x = 90°$, or $2x = 90$; $x = 45$.   *Ans.* 45° each.

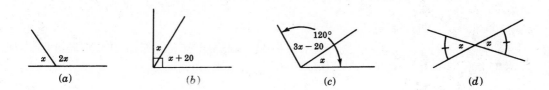

| (a) | (b) | (c) | (d) |

Fig. 3-48

3.15   FINDING A PAIR OF ANGLES USING TWO UNKNOWNS

For each of the following, let the two angles be represented by a and b. Obtain two equations for each case, and then find the angles.

(a) The angles are adjacent, forming an angle of 88°. One is 36° more than the other.

(b) The angles are complementary. One is twice as large as the other.

(c) The angles are supplementary. One is 60° less than twice the other.

(d) The angles are two angles of a triangle whose third angle measures 40°. The difference of the angles is 24°.

Solutions

(a) $a + b = 88$
     $a = b + 36$ *Ans.* 62° and 26°

(b) $a + b = 90$
     $a = 2b$ *Ans.* 60° and 30°

(c) $a + b = 180$
     $a = 2b - 60$ *Ans.* 100° and 80°

(d) $a + b = 140$
     $a - b = 24$ *Ans.* 82° and 58°

Supplementary Problems

1. Point, line, and plane are undefined terms. Which of these is illustrated by (*a*) the tip of a sharpened pencil; (*b*) the shaving edge of a blade; (*c*) a sheet of paper; (*d*) a side of a box; (*e*) the crease of a folded paper; (*f*) the junction of two roads on a map? (3.1)

2. (*a*) Name the line segments that intersect at *E* in Fig. 3-49. (3.2)

    (*b*) Name the line segments that intersect at *D*.

    (*c*) What other line segments can be drawn?

    (*d*) Name the point of intersection of $\overline{AC}$ and $\overline{BD}$.

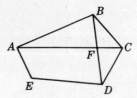

Fig. 3-49

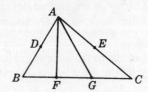

Fig. 3-50

3. (*a*) Find the length of $\overline{AB}$ in Fig. 3-50 if *AD* is 8 and *D* is the midpoint of $\overline{AB}$. (3.3)

    (*b*) Find the length of $\overline{AE}$ if *AC* is 21 and *E* is the midpoint of $\overline{AC}$.

4. (*a*) Find *OB* in Fig. 3-51 if diameter *AD* = 36. (3.4)

    (*b*) Find the number of degrees in $\overparen{AE}$ if *E* is the midpoint of semicircle $\overparen{AED}$. Find the number of degrees in (*c*) $\overparen{CD}$; (*d*) $\overparen{AC}$; (*e*) $\overparen{AEC}$.

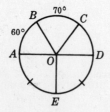

Fig. 3-51

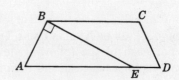

Fig. 3-52

5. Name the following angles in Fig. 3-52: (*a*) an acute angle at *B*; (*b*) an acute angle at *E*; (*c*) a right angle; (*d*) three obtuse angles; (*e*) a straight angle. (3.5)

6. (*a*) Find $m\angle ADC$ if $m\angle c = 45°$ and $m\angle d = 85°$ in Fig. 3-53. (3.6)

    (*b*) Find $m\angle AEB$ if $m\angle e = 60°$.

    (*c*) Find $m\angle EBD$ if $m\angle a = 15°$.

    (*d*) Find $m\angle ABC$ if $m\angle b = 42°$.

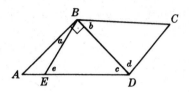

Fig. 3-53

7. Find (*a*) $\frac{5}{6}$ of a rt. $\angle$; (*b*) $\frac{2}{9}$ of a st. $\angle$; (*c*) $\frac{1}{3}$ of 31°; (*d*) $\frac{1}{3}$ of 45°55′.   (3.7)

8. What turn or rotation is made (*a*) by an hour hand in 3 hours; (*b*) by the minute hand in $\frac{1}{3}$ of an hour? What rotation is needed to turn from (*c*) west to northeast in a clockwise direction; (*d*) east to south in a counterclockwise direction; (*e*) southwest to northeast in either direction?   (3.8)

9. Find the angle formed by the hand of a clock (*a*) at 3 o'clock; (*b*) at 10 o'clock; (*c*) at 5:30 o'clock; (*d*) at 11:30 o'clock.   (3.9)

10. In Fig. 3-54:   (3.10)

    (*a*) Name two pairs of perpendicular lines.

    (*b*) Find $m\angle BCD$ if $m\angle 4$ is 39°.

    If $m\angle 1 = 78°$, find (*c*) $m\angle BAD$; (*d*) $m\angle 2$; (*e*) $m\angle CAE$.

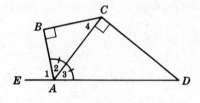

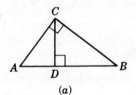

Fig. 3-54   **Fig. 3-55**

11. (*a*) In Fig. 3-55(*a*), name three right triangles and the hypotenuse and legs of each.   (3.11)

In Fig. 3-55(*b*), (*b*) name two obtuse triangles and (*c*) name two isosceles triangles, also naming the legs, base, and vertex angle of each.

12. In Fig. 3-56, name the congruent lines and angles (*a*) if $\overline{PR}$ is $\perp$ bisector of $\overline{AB}$; (*b*) if $\overline{BF}$ bisects $\angle ABC$; (*c*) if CG is an altitude to $\overline{AD}$; (*d*) if EM is a median to $\overline{AD}$.   (3.12)

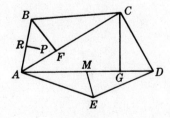

Fig. 3-56

13. In Fig. 3-57, state the relationship between: (3.13)

(a) $\angle 1$ and $\angle 4$ (d) $\angle 4$ and $\angle 5$

(b) $\angle 3$ and $\angle 4$ (e) $\angle 1$ and $\angle 3$

(c) $\angle 1$ and $\angle 2$ (f) $\angle AOD$ and $\angle 5$

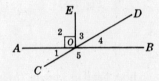

Fig. 3-57

14. Find two angles such that: (3.14)

(a)   The angles are complementary and the measure of the smaller is 40° less than the measure of the larger.

(b)   The angles are complementary and the measure of the larger is four times the measure of the smaller.

(c)   The angles are supplementary and the measure of the smaller is one-half the measure of the larger.

(d)   The angles are supplementary and the measure of the larger is 58° more than the measure of the smaller.

(e)   The angles are supplementary and the measure of the larger is 20° less than three times the measure of the smaller.

(f)   The angles are adjacent and form an angle measuring 140°. The measure of the smaller is 28° less than the measure of the larger.

(g)   The angles are vertical and supplementary.

15. For each of the following, let the two angles be represented by a and b. Obtain two equations for each case, and then find the angles.   (3.15)

(a)   The angles are adjacent and form an angle measuring 75°. Their difference is 21°.

(b)   The angles are complementary. One measures 10° less than three times the other.

(c)   The angles are supplementary. One measures 20° more than four times the other.

Methods of Proof

4.1 PROOF BY DEDUCTIVE REASONING

4.1A Deductive Reasoning is Proof

Deductive reasoning enables us to derive true or acceptably true conclusions from statements which are true or accepted as true. It consists of three steps as follows:

1. Making a *general statement* referring to a whole set or class of things, such as the class of dogs: *All dogs are quadrupeds (have four feet).*

2. Making a *particular statement* about one or some of the members of the set or class referred to in the general statement: *All greyhounds are dogs.*

3. Making a *deduction* that follows logically when the general statement is applied to the particular statement: *All greyhounds are quadrupeds.*

Deductive reasoning is called *syllogistic reasoning* because the three statements together constitute a syllogism. In a syllogism the general statement is called the major premise, the particular statement is the minor premise, and the deduction is the conclusion. Thus in the above syllogism:

1. The major premise is: *All dogs are quadrupeds.*

2. The minor premise is: *All greyhounds are dogs.*

3. The conclusion is: *All greyhounds are quadrupeds.*

Using a circle, as in Fig. 4-1, to represent each set or class will help you understand the relationships involved in deductive reasoning.

1. Since the major premise or general statement states that all dogs are quadrupeds, the circle representing dogs must be inside that for quadrupeds.

2. Since the minor premise or particular statement states that all greyhounds are dogs, the circle representing greyhounds must be inside that for dogs.

3. The conclusion is obvious. Since the circle of greyhounds must be inside the circle of quadrupeds, the only possible conclusion is that greyhounds are quadrupeds.

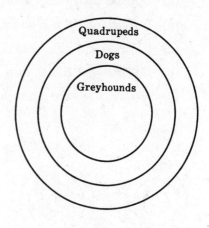

Fig. 4-1

4.1B   Observation, Measurement, and Experimentation are not Proof

Observation cannot serve as proof. Eyesight, as in the case of a color-blind person, may be defective. Appearances may be misleading. Thus in each part of Fig. 4-2, *AB* does not seem to equal *CD* although it actually does.

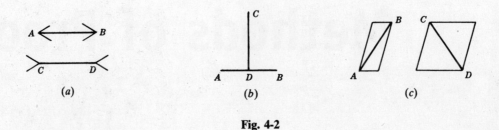

Fig. 4-2

Measurement cannot serve as proof. Measurement applies only to the limited number of cases involved. The conclusion it provides is not exact but approximate, depending on the precision of the measuring instrument and the care of the observer. In measurement, allowance should be made for possible error equal to half the smallest unit of measurement used. Thus if an angle is measured to the nearest degree, an allowance of half a degree of error should be made.

Experiment cannot serve as proof. Its conclusions are only probable ones. The degree of probability depends on the particular situations or instances examined in the process of experimentation. Thus it is probable that a pair of dice are loaded if ten successive 7s are rolled with the pair, and the probability is much greater if twenty successive 7s are rolled; however, neither probability is a certainty.

SOLVED PROBLEMS

4.1   **USING CIRCLES TO DETERMINE GROUP RELATIONSHIPS**

In (*a*) to (*e*) each letter, such as *A*, *B*, and *R*, represents a set or group. Complete each statement. Show how circles may be used to represent the sets or groups.

(*a*)   If *A* is *B* and *B* is *C*, then __?__ .

(*b*)   If *A* is *B* and *B* is *E* and *E* is *R*, then __?__ .

(*c*)   If *X* is *Y* and __?__ , then *X* is *M*.

(*d*)   If *C* is *D* and *E* is *C*, then __?__ .

(*e*)   If squares (*S*) are rectangles (*R*) and rectangles are parallelograms (*P*), then __?__ .

Solutions

(*a*)   *A* is *C*   (*b*)   *A* is *R*   (*c*)   *Y* is *M*   (*d*)   *E* is *D*   (*e*)   Squares are parallelograms

4.2   COMPLETING A SYLLOGISM

Write the statement needed to complete each syllogism:

Major Premise (General Statement)	**Minor Premise** (Particular Statement)	**Conclusion** (Deducted Statement)
(*a*) A cat is a domestic animal.	Fluffy is a cat.	___?___
(*b*) All people must die.	___?___	Jan must die.
(*c*) Vertical angles are congruent.	$\angle c$ and $\angle d$ are vertical angles.	___?___
(*d*) ___?___	A square is a rectangle.	A square has congruent diagonals.
(*e*) An obtuse triangle has only one obtuse angle.	___?___	$\triangle ABC$ has only one obtuse angle.

Solutions

(*a*) Fluffy is a domestic animal.   (*d*) A rectangle has congruent diagonals.

(*b*) Jan is a person.   (*e*) $\triangle ABC$ is an obtuse triangle.

(*c*) $\angle c \cong \angle d$.

4.2   POSTULATES (ASSUMPTIONS)

The entire structure of proof in geometry rests upon, or begins with, some unproved general statements called *postulates*. These are statements which we must willingly assume or accept as true so as to be able to deduce other statements.

4.2A   Algebraic Postulates

POSTULATE 1:   *Things equal to the same or equal things are equal to each other; if $a = b$ and $c = b$, then $a = c$.* (Transitive Postulate)

Thus the total value of a dime is equal to the value of two nickels, since each is equal to the value of ten pennies.

POSTULATE 2:   *A quantity may be substituted for its equal in any expression or equation.* (Substitution Postulate)

Thus if $x = 5$ and $y = x + 3$, we may substitute 5 for x and find $y = 5 + 3 = 8$.

POSTULATE 3:   *The whole equals the sum of its parts.* (Partition Postulate)

Thus the total value of a dime, a nickel, and a penny is 16 cents.

POSTULATE 4:   *Any quantity equals itself.* (Reflexive Postulate or Identity Postulate)

Thus $x = x$, $m\angle A = m\angle A$, and $AB = AB$.

POSTULATE 5:   *If equals are added to equals, the sums are equal; if $a = b$ and $c = d$, then $a + c = b + d$.* (Addition Postulate)

If	7 dimes = 70 cents	If	$x + y = 12$
and	2 dimes = 20 cents	and	$x - y = 8$
then	9 dimes = 90 cents	then	$2x = 20$

POSTULATE 6: *If equals are subtracted from equals, the differences are equal; if $a = b$ and $c = d$, then $a - c = b - d$.* (Subtraction Postulate)

If	7 dimes = 70 cents	If	$x + y = 12$
and	2 dimes = 20 cents	and	$x - y = 8$
then	5 dimes = 50 cents	then	$2y = 4$

POSTULATE 7: *If equals are multiplied by equals, the products are equal; if $a = b$ and $c = d$, then $ac = bd$.* (Multiplication Postulate)

Thus if the price of one book is $2, the price of three books is $6.

Special multiplication axiom: Doubles of equals are equal.

POSTULATE 8: *If equals are divided by equals, the quotients are equal; if $a = b$ and $c = d$, then $a/c = b/d$, where $c, d \neq 0$.* (Division Postulate)

Thus if the price of 1 lb of butter is 80 cents then, at the same rate, the price of $\frac{1}{4}$ lb is 20 cents.

POSTULATE 9: *Like powers of equals are equal; if $a = b$, then $a^n = b^n$.* (Powers Postulate)

Thus if $x = 5$, then $x^2 = 5^2$ or $x^2 = 25$.

POSTULATE 10: *Like roots of equals are equal; if $a = b$ then $\sqrt[n]{a} = \sqrt[n]{b}$.*

Thus if $y^3 = 27$, then $y = \sqrt[3]{27} = 3$.

4.2B   Geometric Postulates

POSTULATE 11: *One and only one straight line can be drawn through any two points.*

Thus $\overleftrightarrow{AB}$ is the only line that can be drawn between A and B in Fig. 4-3.

Fig. 4-3   Fig. 4-4

POSTULATE 12: *Two lines can intersect in one and only one point.*

Thus only P is the point of intersection of $\overleftrightarrow{AB}$ and $\overleftrightarrow{CD}$ in Fig. 4-4.

POSTULATE 13: *The length of a segment is the shortest distance between two points.*

Thus $\overline{AB}$ is shorter than the curved or broken line segment between A and B in Fig. 4-5.

Fig. 4-5   Fig. 4-6

POSTULATE 14: *One and only one circle can be drawn with any given point as center and a given line segment as a radius.*

Thus only circle A in Fig. 4-6 can be drawn with A as center and $\overline{AB}$ as a radius.

POSTULATE 15: *Any geometric figure can be moved without change in size or shape.*

Thus $\triangle I$ in Fig. 4-7 can be moved to a new position without changing its size or shape.

POSTULATE 16: *A segment has one and only one midpoint.*

Thus only *M* is the midpoint of $\overline{AB}$ in Fig. 4-8.

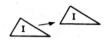

Fig. 4-7   **Fig. 4-8**

POSTULATE 17: *An angle has one and only one bisector.*

Thus only $\vec{AD}$ is the bisector of $\angle A$ in Fig. 4-9.

POSTULATE 18: *Through any point on a line, one and only one perpendicular can be drawn to the line.*

Thus only $\vec{PC} \perp \vec{AB}$ at point *P* on $\vec{AB}$ (Fig. 4-10).

Fig. 4-9   **Fig. 4-10**   **Fig. 4-11**

POSTULATE 19: *Through any point outside a line, one and only one perpendicular can be drawn to the given line.*

Thus only $\overline{PC}$ can be drawn $\perp \vec{AB}$ from point *P* outside $\vec{AB}$ in Fig. 4-11.

SOLVED PROBLEMS

4.3   APPLYING POSTULATE 1

In each part, what conclusion follows when Postulate 1 is applied to the given data from Figs. 4-12 and 4-13?

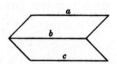

Fig. 4-12   **Fig. 4-13**

(*a*) Given: $a = 10, b = 10, c = 10$ (*d*) Given: $m\angle 1 = 40°, m\angle 2 = 40°, m\angle 3 = 40°$

(*b*) Given: $a = 25, a = c$ (*e*) Given: $m\angle 1 = m\angle 2, m\angle 3 = m\angle 1$

(*c*) Given: $a = b, c = b$ (*f*) Given: $m\angle 3 = m\angle 1, m\angle 2 = m\angle 3$

Solutions

(*a*) Since *a*, *b*, and *c* each equal 10, $a = b = c$.

(*b*) Since *c* and 25 each equal *a*, $c = 25$.

(*c*) Since *a* and *c* each equal *b*, $a = c$.

(d) Since $\angle 1$, $\angle 2$, and $\angle 3$ each measures $40°$, $\angle 1 \cong \angle 2 \cong \angle 3$.

(e) Since $\angle 2$ and $\angle 3$ each $\cong \angle 1$, $\angle 2 \cong \angle 3$.

(f) Since $\angle 1$ and $\angle 2$ each $\cong \angle 3$, $\angle 1 \cong \angle 2$.

4.4   APPLYING POSTULATE 2

In each part, what conclusion follows when Postulate 2 is applied to the given data?

(a) Evaluate $2a + 2b$ when $a = 4$ and $b = 8$.

(b) Find x if $3x + 4y = 35$ and $y = 5$.

(c) Given: $m\angle 1 + m\angle B + m\angle 2 = 180°$, $\angle 1 \cong \angle A$, and $\angle 2 \cong \angle C$ in Fig. 4-14.

Solutions

(a) Substitute 4 for a and 8 for b:
$$2a + 2b$$
$$2(4) + 2(8)$$
$$8 + 16 = 24 \; Ans.$$

(c) Substitute $\angle A$ for $\angle 1$ and $\angle C$ for $\angle 2$:
$$m\angle 1 + m\angle B + m\angle 2 \; = 180°$$
$$m\angle A + m\angle B + m\angle C = 180° \; Ans.$$

(b) Substitute 5 for y:
$$3x + 4y \quad = 35$$
$$3x + 4(5) = 35$$
$$3x + 20 \quad = 35$$
$$3x = 15, \quad x = 5 \; Ans.$$

Fig. 4-14

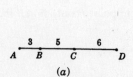

Fig. 4-15

4.5   APPLYING POSTULATE 3

State the conclusions that follow when Postulate 3 is applied to the data in (a) Fig. 4.15(a) and (b) Fig. 4-15(b).

Solutions

(a) $AC = 3 + 5 = 8$
    $BD = 5 + 6 = 11$
    $AD = 3 + 5 + 6 = 14$

(b) $m\angle AEC = 60° + 40° = 100°$
    $m\angle BED = 40° + 30° = 70°$
    $m\angle AED = 60° + 40° + 30° = 130°$

4.6   APPLYING POSTULATES 4, 5, AND 6

In each part, state a conclusion that follows when Postulates 4, 5, and 6 are applied to the given data.

(a) Given: $a = e$ (Fig. 4-16)

(b) Given: $a = c$, $b = d$ (Fig. 4-16)

(c) Given: $m\angle BAC = m\angle DAE$ (Fig. 4-17)

(d) Given: $m\angle BAC = m\angle BCA$, $m\angle 1 = m\angle 3$ (Fig. 4-17)

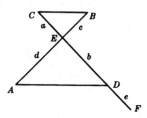

Fig. 4-16

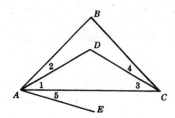

Fig. 4-17

Solutions

(a) $a = e$ Given
        $b = b$ Identity
        $\overline{a + b = b + e}$ Add. Post.
        $CD = EF$ Subst.

(b) $a = c$ Given
        $b = d$ Given
        $\overline{a + b = c + d}$ Add. Post.
        $CD = AB$ Subst.

(c) $m\angle BAC = m\angle DAE$ Given
             $m\angle 1 = m\angle 1$ Identity
        $\overline{m\angle BAC - m\angle 1 = m\angle DAE - m\angle 1}$ Subt. Post.
             $m\angle 2 = m\angle 5$ Subst.

(d) $m\angle BAC = m\angle BCA$ Given
             $m\angle 1 = m\angle 3$ Given
        $\overline{m\angle BAC - m\angle 1 = m\angle BCA - m\angle 3}$ Subt. Post.
             $m\angle 2 = m\angle 4$ Subst.

4.7   APPLYING POSTULATES 7 AND 8

State the conclusions that follow when the multiplication and division axioms are applied to the data in (a) Fig. 4-18 and (b) Fig. 4-19.

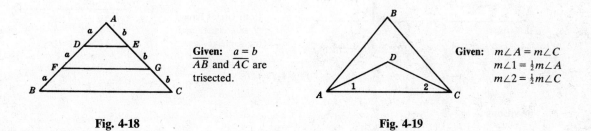

Given: $a = b$
$\overline{AB}$ and $\overline{AC}$ are trisected.

Fig. 4-18

Given: $m\angle A = m\angle C$
        $m\angle 1 = \frac{1}{2}m\angle A$
        $m\angle 2 = \frac{1}{2}m\angle C$

Fig. 4-19

Solutions

(a)   If $a = b$, then $2a = 2b$ since doubles of equals are equal. Hence, $AF = DB = AG = EC$. Also, $3a = 3b$, using the Multiplication Postulate. Hence, $AB = AC$.

(b)   If $m\angle A = m\angle C$, then $\frac{1}{2}m\angle A = \frac{1}{2}m\angle C$ since halves of equals are equal. Hence, $m\angle 1 = m\angle 2$.

4.8   APPLYING POSTULATES TO STATEMENTS
Complete each sentence and state the postulate that applies.

(*a*)   If Harry and Alice are the same age today, then in 10 years __?__ .

(*b*)   Since 32°F and 0°C both name the temperature at which water freezes, we know that __?__ .

(*c*)   If Henry and John are the same weight now and each reduces 20 lb, then __?__ .

(*d*)   If two stocks of equal value both triple in value, then __?__ .

(*e*)   If two ribbons of equal size are cut into five equal parts, then __?__ .

(*f*)   If Joan and Agnes are the same height as Anne, then __?__ .

(*g*)   If two air conditioners of the same price are each discounted 10 percent, then __?__ .

Solutions

(*a*)   They will be the same age. (Add. Post.)

(*b*)   32°F = 0°C. (Trans. Post.)

(*c*)   They will be the same weight. (Subt. Post.)

(*d*)   They will have the same value. (Mult. Post.)

(*e*)   Their parts will be of the same size. (Div. Post.)

(*f*)   Joan and Agnes are of the same height. (Trans. Post.)

(*g*)   They will have the same price. (Subt. Post.)

4.9   APPLYING GEOMETRIC POSTULATES
State the postulate needed to correct each diagram and accompanying statement in Fig. 4-20.

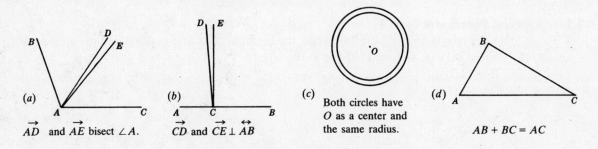

Fig. 4-20

Solutions

(*a*)   Postulate 17.   (*b*)   Postulate 18.   (*c*)   Postulate 14.   (*d*)   Postulate 13. (*AC* is less than the sum of *AB* and *BC*.)

4.3   BASIC ANGLE THEOREMS

A *theorem* is a statement (believed to be true), which, when proved, can be used to prove other statements or derive other results. Each of the following basic theorems requires the use of definitions and postulates for its proof.

Note: We shall use the term *principle* to include important geometric statements such as theorems, postulates, and definitions.

PRINCIPLE 1:   *All right angles are congruent.*

Thus $\angle A \cong \angle B$ in Fig. 4-21.

PRINCIPLE 2:   *All straight angles are congruent.*

Thus $\angle C \cong \angle D$ in Fig. 4-22.

Fig. 4-21   Fig. 4-22

PRINCIPLE 3:   *Complements of the same or of congruent angles are congruent.*

This is a combination of the following two principles:

1.   *Complements of the same angle are congruent.* Thus $\angle a \cong \angle b$ in Fig. 4.23; each is the complement of $\angle x$.

2.   *Complements of congruent angles are congruent.* Thus $\angle c \cong \angle d$ in Fig. 4-24; their complements are the congruent $\angle$s x and y.

Fig. 4-23   Fig. 4-24

PRINCIPLE 4:   *Supplements of the same or of congruent angles are congruent.*

This is a combination of the following two principles:

1.   *Supplements of the same angle are congruent.* Thus $\angle a \cong \angle b$ in Fig. 4-25; each is the supplement of $\angle x$.

2.   *Supplements of congruent angles are congruent.* Thus $\angle c \cong \angle d$ in Fig. 4-26; their supplements are the congruent angles x and y.

Fig. 4-25   Fig. 4-26

PRINCIPLE 5:   *Vertical angles are congruent.*

Thus in Fig. 4.27, $\angle a \cong \angle b$; this follows from Principle 4, since $\angle a$ and $\angle b$ are supplements of the same angle, $\angle c$.

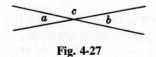

Fig. 4-27

SOLVED PROBLEMS

4.10 APPLYING BASIC THEOREMS: PRINCIPLES 1 TO 5

State the basic angle theorem needed to prove $\angle a \cong \angle b$ in each part of Fig. 4-28.

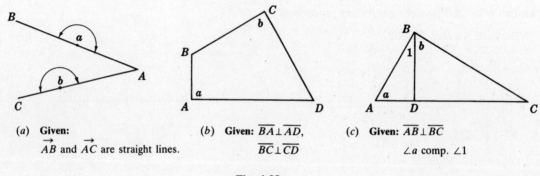

(a) **Given:**
$\overrightarrow{AB}$ and $\overrightarrow{AC}$ are straight lines.

(b) **Given:** $\overline{BA} \perp \overline{AD}$,
$\overline{BC} \perp \overline{CD}$

(c) **Given:** $\overline{AB} \perp \overline{BC}$
$\angle a$ comp. $\angle 1$

Fig. 4-28

Solutions

(a) Since $\overleftrightarrow{AB}$ and $\overleftrightarrow{AC}$ are straight lines, $\angle a$ and $\angle b$ are straight $\angle$s. Hence, $\angle a \cong \angle b$.   *Ans.* All straight angles are congruent.

(b) Since $\overline{BA} \perp \overline{AD}$ and $\overline{BC} \perp \overline{CD}$, $\angle a$ and $\angle b$ are rt. $\angle$s. Hence, $\angle a \cong \angle b$.   *Ans.* All right angles are congruent.

(c) Since $\overline{AB} \perp \overline{BC}$, $\angle B$ is a rt. $\angle$, making $\angle b$ the complement of $\angle 1$. Since $\angle a$ is the complement of $\angle 1$, $\angle a \cong \angle b$.   *Ans.* Complements of the same angle are congruent.

4.4 DETERMINING THE HYPOTHESIS AND CONCLUSION

4.4A Statement Forms: Subject-Predicate Form and If-Then Form

The statements "A heated metal expands" and "If a metal is heated, then it expands" are two forms of the same idea. The following table shows how each form may be divided into its two important parts, the *hypothesis*, which tells *what is given*, and the *conclusion*, which tells *what is to be proved*. Note that in the if-then form, the word *then* may be omitted.

Form	Hypothesis (What is given)	Conclusion (What is to be proved)
Subject-predicate form: *A heated metal expands.*	**Hypothesis is subject:** *A heated metal*	**Conclusion is predicate:** *expands*
If-then form: *If a metal is heated, then it expands.*	**Hypothesis is if clause:** *If a metal is heated*	**Conclusion is then clause:** *then it expands*

4.4B Converse of a Statement

The converse of a statement is formed by interchanging the hypothesis and conclusion. Hence to form the converse of an if-then statement, interchange the if and then clauses. In the case of the subject-predicate form, interchange the subject and predicate.

Thus the converse of "triangles are polygons" is "polygons are triangles." Also, the converse of "if a metal is heated, then it expands" is "if a metal expands, then it is heated." Note in each of these cases that the statement is true but its converse need not necessarily be true.

PRINCIPLE 1: *The converse of a true statement is not necessarily true.*

Thus the statement "triangles are polygons" is true. Its converse need not be true.

PRINCIPLE 2: *The converse of a definition is always true.*

Thus the converse of the definition "a triangle is a polygon of three sides" is "a polygon of three sides is a triangle." Both the definition and its converse are true.

SOLVED PROBLEMS

4.11 DETERMINING THE HYPOTHESIS AND CONCLUSION IN SUBJECT-PREDICATE FORM
Determine the hypothesis and conclusion of each statement.

	Solutions	
Statements	**Hypothesis** (subject)	**Conclusion** (predicate)
(a) Perpendiculars form right angles.	Perpendiculars	form right angles
(b) Complements of the same angle are congruent.	Complements of the same angle	are congruent
(c) An equilateral triangle is equiangular.	An equilateral triangle	is equiangular
(d) A right triangle has only one right angle.	A right triangle	has only one right angle
(e) A triangle is not a quadrilateral.	A triangle	is not a quadrilateral

4.12 DETERMINING THE HYPOTHESIS AND CONCLUSION IN IF-THEN FORM
Determine the hypothesis and conclusion of each statement.

	Solutions	
Statements	**Hypothesis** (if clause)	**Conclusion** (then clause)
(a) If a line bisects an angle, then it divides the angle into two congruent parts.	If a line bisects an angle	then it divides the angle into two congruent parts
(b) A triangle has an obtuse angle if it is an obtuse triangle.	If it is an obtuse triangle	(then) a triangle has an obtuse angle
(c) If a student is sick, she should not go to school.	If a student is sick	(then) she should not go to school
(d) A student, if he wishes to pass, must study regularly.	If he wishes to pass	(then) a student must study regularly

4.13   FORMING CONVERSES AND DETERMINING THEIR TRUTH

State whether the given statement is true. Then form its converse and state whether this is necessarily true.

(a) A quadrilateral is a polygon.

(b) An obtuse angle has greater measure than a right angle.

(c) Florida is a state of the United States.

(d) If you are my pupil, then I am your teacher.

(e) An equilateral triangle is a triangle that has all congruent sides.

Solutions

(a) Statement is true. Its converse, "a polygon is a quadrilateral," is not necessarily true; it might be a triangle.

(b) Statement is true. Its converse, "an angle with greater measure than a right angle is an obtuse angle," is not necessarily true; it might be a straight angle.

(c) Statement is true. Its converse, "a state of the United States is Florida," is not necessarily true; it might be any one of the other 49 states.

(d) Statement is true. Its converse, "if I am your teacher, then you are my pupil," is also true.

(e) The statement, a definition, is true. Its converse, "a triangle that has all congruent sides is an equilateral triangle," is also true.

4.5   PROVING A THEOREM

Theorems should be proved using the following step-by-step procedure. The form of the proof is shown in the example that follows the procedure. Note that accepted symbols and abbreviations may be used.

1. Divide the theorem into its hypothesis (what is given) and its conclusion (what is to be proved). Underline the hypothesis with a single line, and the conclusion with a double line.

2. On one side, make a marked diagram. Markings on the diagram should include such helpful symbols as square corners for right angles, cross marks for equal parts, and question marks for parts to be proved equal.

3. On the other side, next to the diagram, state what is given and what is to be proved. The "Given" and "To Prove" must refer to the parts of the diagram.

4. Present a plan. Although not essential, a plan is very advisable. It should state the major methods of proof to be used.

5. On the left, present statements in successively numbered steps. The last statement must be the one to be proved. All the statements must refer to parts of the diagram.

6. On the right, next to the statements, provide a reason for each statement. Acceptable reasons in the proof of a theorem are given facts, definitions, postulates, assumed theorems, and previously proven theorems.

Step 1: **Prove:** <u>All right angles are equal</u>
                              <u>in measure.</u>

Steps 2 **Given:** $\angle A$ and $\angle B$ are rt. $\angle$s

and 3:   **To Prove:** $m\angle A = m\angle B$

Step 4: **Plan:** Since each angle equals 90°,
                              the angles are equal in measure,
                              using Post. 1: Things equal to the
                              same thing are equal to each other.

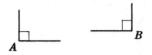

Steps 5
and 6:

Statements	Reasons
1. $\angle A$ and $\angle B$ are rt. $\angle$s.	1. Given
2. $m\angle A$ and $m\angle B$ each = 90°.	2. m(rt. $\angle$) = 90°
3. $m\angle A = m\angle B$	3. Things = to same thing = each other.

SOLVED PROBLEM

4.14 PROVING A THEOREM

Use the proof procedure to prove that supplements of angles of equal measure have equal measure.

Step 1: **Prove:** <u>Supplements of angles of</u>
                                <u>equal measure have equal measure.</u>

Steps 2 **Given:** $\angle a$ sup. $\angle 1$, $\angle b$ sup. $\angle 2$
and 3: $m\angle 1 = m\angle 2$

            To Prove: $m\angle a = m\angle b$

Step 4: **Plan:** Using the subtraction postulate,
                              the equal angle measures may be
                              subtracted from the equal sums of
                              measures of pairs of supplementary
                              angles. The equal remainders are the
                              measures of the supplements.

Steps 5
and 6:

Statements	Reasons
1. $\angle a$ sup. $\angle 1$, $\angle b$ sup. $\angle 2$	1. Given

2. $m\angle a + m\angle 1 = 180°$
     $m\angle b + m\angle 2 = 180°$
	2. Sup. $\angle$s are $\angle$s the sum of
	     whose measures = 180°.

3. $m\angle a + m\angle 1 = m\angle b + m\angle 2$	3. Things = to the same thing = each other.
4. $m\angle 1 = m\angle 2$	4. Given
5. $m\angle a = m\angle b$	5. If =s are subtracted from =s, the differences are =.

Supplementary Problems

1. Complete each statement. In (a) to (e), each letter, such as C, D, or R, represents a set or group.   (4.1)

    (a)   If A is B and B is H, then __?__ .

    (b)   If C is D and P is C, then __?__ .

    (c)   If __?__ and B is R, then B is S.

    (d)   If E is F, F is G, and G is K, then __?__ .

    (e)   If G is H, H is R, and __?__ , then A is R.

    (f)   If triangles are polygons and polygons are geometric figures, then __?__ .

    (g)   If a rectangle is a parallelogram and a parallelogram is a quadrilateral, then __?__ .

2. State the conclusions which follow when Postulate 1 is applied to the given data, which refer to Fig. 4.29.   (4.3)

    (a) $a = 7, c = 7, f = 7$

    (b) $b = 15, b = g$

    (c) $f = h, h = a$

    (d) $a = c, c = f, f = h$

    (e) $b = d, d = g, g = e$

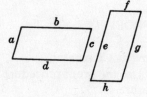

Fig. 4-29

3. State the conclusions which follow when Postulate 2 is applied in each case.   (4.4)

    (a)   Evaluate $a^2 + 3a$ when $a = 10$.

    (b)   Evaluate $x^2 - 4y$ when $x = 4$ and $y = 3$.

    (c)   Does $b^2 - 8 = 17$ when $b = 5$?

    (d)   Find x if $x + y = 20$ and $y = x + 3$.

    (e)   Find y if $x + y = 20$ and $y = 3x$.

    (f)   Find x if $5x - 2y = 24$ and $y = 3$.

    (g)   Find x if $x^2 + 3y = 45$ and $y = 3$.

4. State the conclusions that follow when Postulate 3 is applied to the data in Fig. 4-30(a) and (b).   (4.5)

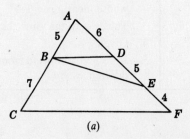
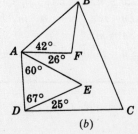

(a)   (b)

Fig. 4-30

5. State a conclusion involving two new equals that follows when Postulate 4, 5, or 6 is applied to the given data.   (4.6)

   (*a*) Given: $b = e$ (Fig. 4-31).

   (*b*) Given: $b = c$, $a = d$ (Fig. 4-31).

   (*c*) Given: $\angle 4 \cong \angle 5$ (Fig. 4-32).

   (*d*) Given: $\angle 1 \cong \angle 3$, $\angle 2 \cong \angle 4$ (Fig. 4-32).

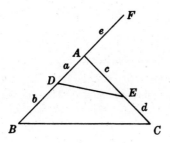

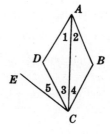

Fig. 4-31   Fig. 4-32

6. In Fig. 4-33 $\overline{AD}$ and $\overline{BC}$ are trisected.   (4.7)

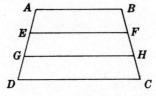

Fig. 4-33

   (*a*) If $\overline{AD} \cong \overline{BC}$, why is $\overline{AE} \cong \overline{BF}$?

   (*b*) If $\overline{EG} \cong \overline{FH}$, why is $\overline{AG} \cong \overline{BH}$?

   (*c*) If $\overline{GD} \cong \overline{HC}$, why is $\overline{AD} \cong \overline{BC}$?

   (*d*) If $\overline{ED} \cong \overline{FC}$, why is $\overline{EG} \cong \overline{FH}$?

7. In Fig. 4-34, $\angle BCD$ and $\angle ADC$ are trisected.

   (*a*) If $m\angle BCD = m\angle ADC$, why does $m\angle FCD = m\angle FDC$?

   (*b*) If $m\angle 1 = m\angle 2$, why does $m\angle BCD = m\angle ADC$?

   (*c*) If $m\angle 1 = m\angle 2$, why does $m\angle ADF = m\angle BCF$?

   (*d*) If $m\angle EDC = m\angle ECD$, why does $m\angle 1 = m\angle 2$?

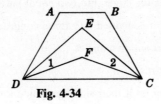

Fig. 4-34

8. Complete each statement, and name the postulate that applies.   (4.8)

(a) If Bill and Helen earn the same amount of money each hour and their rate of pay is increased by the same amount, then ___?___ .

(b) In the past year, those stocks have tripled in value. If they had the same value last year, then ___?___ .

(c) A week ago, there were two classes that had the same register. If the same number of pupils were dropped in each, then ___?___ .

(d) Since 100°C and 212°F are the boiling temperatures of water, then ___?___ .

(e) If two boards have the same length and each is cut into four equal parts, then ___?___ .

(f) Since he has $2,000 in Bank A, $3,000 in Bank B and $5,000 in Bank C, then ___?___ .

(g) If three quarters and four nickels are compared with three quarters and two dimes, ___?___ .

9. Answer each of the following by stating the basic angle theorem needed. The questions refer to Fig. 4-35.   (4.10)

(a) Why does $m\angle 1 = m\angle 2$?

(b) Why does $m\angle DBC = m\angle ECB$?

(c) If $m\angle 3 = m\angle 4$, why does $m\angle 5 = m\angle 6$?

(d) If $\overrightarrow{AF} \perp \overline{DE}$ and $\overrightarrow{GC} \perp \overline{DE}$, why does $m\angle 7 = m\angle 8$?

(e) If $\overrightarrow{AF} \perp \overline{DE}$, $\overrightarrow{GC} \perp \overline{DE}$, and $m\angle 11 = m\angle 12$, why does $m\angle 9 = m\angle 10$?

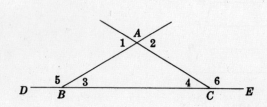

Fig. 4-35

10. Determine the hypothesis and conclusion of each statement.   (4.11 and 4.12)

(a) Stars twinkle.

(b) Jet planes are the speediest.

(c) Water boils at 212° Fahrenheit.

(d) If it is the American flag, its colors are red, white, and blue.

(e) You cannot learn geometry if you fail to do homework in the subject.

(f) A batter goes to first base if the umpire calls a fourth ball.

(g) If A is B's brother and C is B's son, then A is C's uncle.

(h) An angle bisector divides the angle into two equal parts.

(i) A segment is trisected if it is divided into three congruent parts.

(j) A pentagon has five sides and five angles.

(k) Some rectangles are squares.

(l) Angles do not become larger if their sides are made longer.

(*m*) Angles, if they are congruent and supplementary, are right angles.

(*n*) The figure cannot be a polygon if one of its sides is not a straight line segment.

11. State the converse of each of the following true statements. State whether the converse is necessarily true. (4.13)

(*a*) Half a right angle is an acute angle.

(*b*) An obtuse triangle is a triangle having one obtuse angle.

(*c*) If the umpire called a third strike, then the batter is out.

(*d*) If I am taller than you, then you are shorter than I.

(*e*) If I am heavier than you, then our weights are unequal.

12. Prove each of the following. (4.14)

(*a*) Straight angles are congruent.

(*b*) Complements of congruent angles are congruent.

(*c*) Vertical angles are congruent.

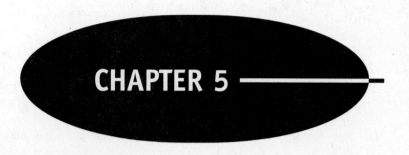

CHAPTER 5

Congruent Triangles

5.1 CONGRUENT TRIANGLES

Congruent figures are figures that have the same size and the same shape; they are the exact duplicates of each other. Such figures can be made to coincide so that their corresponding parts will fit together. Two circles having the same radius are congruent circles.

Congruent triangles are triangles that have the same size and the same shape.

If two triangles are congruent, their corresponding sides and angles must be congruent. Thus congruent triangles ABC and $A'B'C'$ in Fig. 5-1 have congruent corresponding sides ($AB \cong A'B'$, $BC \cong B'C'$, and $AC \cong A'C'$) and congruent corresponding angles ($\angle A \cong \angle A'$, $\angle B \cong \angle B'$, and $\angle C \cong \angle C'$).

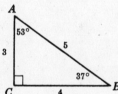

Fig. 5-1

(Read $\triangle ABC \cong \triangle A'B'C'$ as "Triangle ABC is congruent to triangle A-prime, B-prime, C-prime.")

Note in the congruent triangles how corresponding equal parts may be located. Corresponding sides lie opposite congruent angles, and corresponding angles lie opposite congruent sides.

5.1A Basic Principles of Congruent Triangles

PRINCIPLE 1: *If two triangles are congruent, then their corresponding parts are congruent.* (Corresponding parts of congruent triangles are congruent.)

Thus if $\triangle ABC \cong \triangle A'B'C'$ in Fig. 5-2, then $\angle A \cong \angle A'$, $\angle B \cong \angle B'$, $\angle C \cong \angle C'$, $a \cong a'$, $b \cong b'$, and $c \cong c'$.

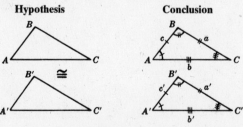

Fig. 5-2

Methods of Proving that Triangles are Congruent

PRINCIPLE 2: (s.a.s. ≅ s.a.s.) *If two sides and the included angle of one triangle are congruent to the corresponding parts of another, then the triangles are congruent.*

Thus if $b \cong b'$, $c \cong c'$, and $\angle A \cong \angle A'$ in Fig. 5-3 then $\triangle ABC \cong \triangle A'B'C'$.

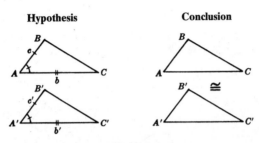

Fig. 5-3

PRINCIPLE 3: (a.s.a. ≅ a.s.a.) *If two angles and the included side of one triangle are congruent to the corresponding parts of another, then the triangles are congruent.*

Thus if $\angle A \cong \angle A'$, $\angle C \cong \angle C'$, and $b \cong b'$ in Fig. 5-4, then $\triangle ABC \cong \triangle A'B'C'$.

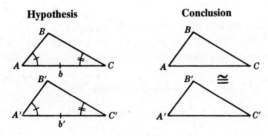

Fig. 5-4

PRINCIPLE 4: (s.s.s. ≅ s.s.s.) *If three sides of one triangle are congruent to three sides of another, then the triangles are congruent.*

Thus if $a \cong a'$, $b \cong b'$, and $c \cong c'$ in Fig. 5-5, then $\triangle ABC \cong \triangle A'B'C'$.

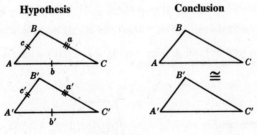

Fig. 5-5

SOLVED PROBLEMS

5.1   SELECTING CONGRUENT TRIANGLES

From each set of three triangles in Fig. 5-6, select the congruent triangles and state the congruency principle that is involved.

Solutions

(a)   △I ≅ △II, by s.a.s. ≅ s.a.s. In △III, the right angle is not between 3 and 4.

(b)   △II ≅ △III, by a.s.a. ≅ a.s.a. In △I, side 10 is not between 70° and 30°.

(c)   △I ≅ △II ≅ △III by s.s.s. ≅ s.s.s.

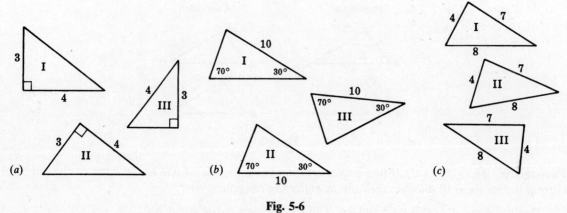

Fig. 5-6

5.2   DETERMINING THE REASON FOR CONGRUENCY OF TRIANGLES

In each part of Fig. 5-7, △I can be proved congruent of △II. Make a diagram showing the equal parts of both triangles and state the congruency principle that is involved.

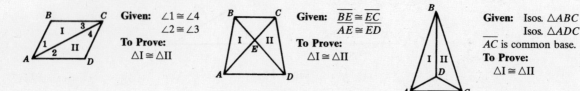

Fig. 5-7

Solutions

(a)   *AC* is a common side of both ⧌ [Fig. 5-8(a)]. △I ≅ △II by a.s.a. ≅ a.s.a.

(b)   ∠1 and ∠2 are vertical angles [Fig. 5-8(b)]. △I ≅ △II by s.a.s. ≅ s.a.s.

(c)   *BD* is a common side of both ⧌ [Fig. 5-8(c)]. △I ≅ △II by s.s.s. ≅ s.s.s.

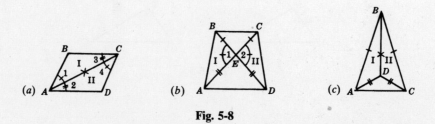

Fig. 5-8

5.3   FINDING PARTS NEEDED TO PROVE TRIANGLES CONGRUENT

State the additional parts needed to prove △I ≅ △II in the given figure by the given congruency principle.

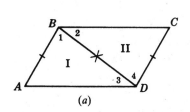 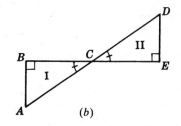 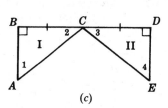

Fig. 5-9

(*a*) In Fig. 5-9(*a*) by s.s.s. ≅ s.s.s.

(*b*) In Fig. 5-9(*a*) by s.a.s. ≅ s.a.s.

(*c*) In Fig. 5-9(*b*) by a.s.a. ≅ a.s.a.

(*d*) In Fig. 5-9(*c*) by a.s.a. ≅ a.s.a.

(*e*) In Fig. 5-9(*c*) by s.a.s. ≅ s.a.s.

Solutions

(*a*) If $\overline{AD} \cong \overline{BC}$, then △I ≅ △II by s.s.s. ≅ s.s.s.

(*b*) If ∠1 ≅ ∠4, then △I ≅ △II by s.a.s. ≅ s.a.s.

(*c*) If $\overline{BC} \cong \overline{CE}$, then △I ≅ △II by a.s.a. ≅ a.s.a.

(*d*) If ∠2 ≅ ∠3, then △I ≅ △II by a.s.a. ≅ a.s.a.

(*e*) If $\overline{AB} \cong \overline{DE}$, then △I ≅ △II by s.a.s. ≅ s.a.s.

5.4   SELECTING CORRESPONDING PARTS OF CONGRUENT TRIANGLES

In each part of Fig. 5-10, the equal parts needed to prove △I ≅ △II are marked. List the remaining parts that are congruent.

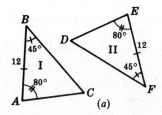 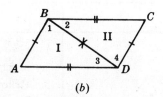 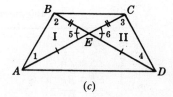

Fig. 5-10

Solutions

Congruent corresponding sides lie opposite congruent angles. Congruent corresponding angles lie opposite congruent sides.

(*a*) Opposite 45°, $\overline{AC} \cong \overline{DE}$. Opposite 80°, $\overline{BC} \cong \overline{DF}$. Opposite side 12; ∠C ≅ ∠D.

(*b*) Opposite $\overline{AB}$ and $\overline{CD}$, ∠3 ≅ ∠2. Opposite $\overline{BC}$ and $\overline{AD}$, ∠1 ≅ ∠4. Opposite common side $\overline{BD}$, ∠A ≅ ∠C.

(*c*) Opposite $\overline{AE}$ and $\overline{ED}$, ∠2 ≅ ∠3. Opposite $\overline{BE}$ and $\overline{EC}$, ∠1 ≅ ∠4. Opposite ∠5 and ∠6, $\overline{AB} \cong \overline{CD}$.

5.5   APPLYING ALGEBRA TO CONGRUENT TRIANGLES

In each part of Fig. 5-11, find x and y.

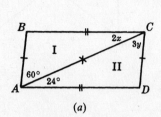

(a)

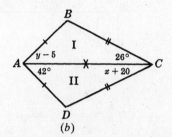

(b)

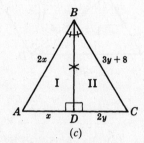

(c)

Fig. 5-11

Solutions

(a)   Since $\triangle I \cong \triangle II$, corresponding angles are congruent. Hence, $2x = 24$ or $x = 12$, and $3y = 60$ or $y = 20$.

(b)   Since $\triangle I \cong \triangle II$, corresponding angles are congruent. Hence, $x + 20 = 26$ or $x = 6$, and $y - 5 = 42$ or $y = 47$.

(c)   Since $\triangle I \cong \triangle II$, corresponding sides are congruent. Then $2x = 3y + 8$ and $x = 2y$. Substituting $2y$ for x in the first of these equations, we obtain $2(2y) = 3y + 8$ or $y = 8$. Then $x = 2y = 16$.

5.6   PROVING A CONGRUENCY PROBLEM

Given: $\overline{BF} \perp \overline{DE}$
$\overline{BF} \perp \overline{AC}$
$\angle 3 \cong \angle 4$

To Prove: $\overline{AF} \cong \overline{FC}$

Prove: Prove $\triangle I \cong \triangle II$

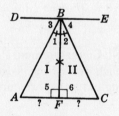

PROOF:

Statements	Reasons
1. $\overline{BF} \perp \overline{AC}$	1. Given
2. $\angle 5 \cong \angle 6$	2. ⊥s form rt. ∡s; rt. ∡s are ≅
3. $\overline{BF} \cong \overline{BF}$	3. Reflexive property
4. $\overline{BF} \perp \overline{DE}$	4. Given
5. $\angle 1$ is the complement of $\angle 3$. $\angle 2$ is the complement of $\angle 4$.	5. Adjacent angles are complementary if exterior sides are ⊥ to each other.
6. $\angle 3 \cong \angle 4$	6. Given
7 $\angle 1 \cong \angle 2$	7. Complements of ≅ ∡s are =
8. $\triangle I \cong \triangle II$	8. a.s.a. ≅ a.s.a.
9. $\overline{AF} \cong \overline{FC}$	9. Corresponding parts of congruent ▲s are ≅

5.7   PROVING A CONGRUENCY PROBLEM STATED IN WORDS

Prove that if the opposite sides of a quadrilateral are equal and a diagonal is drawn, equal angles are formed between the diagonal and the sides.

Solution

If the opposite sides of a quadrilateral are congruent and a diagonal is drawn, congruent angles are formed between the diagonal and the sides.

Given: Quadrilateral *ABCD*
$\overline{AB} \cong \overline{CD}$, $\overline{BC} \cong \overline{AD}$
$\overline{AC}$ is a diagonal.

To Prove: $\angle 1 \cong \angle 4$, $\angle 2 \cong \angle 3$

Plan: Prove $\triangle I \cong \triangle II$

PROOF:

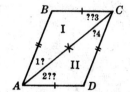

Statements	Reasons
1. $\overline{AB} \cong \overline{CD}$, $\overline{BC} \cong \overline{AD}$	1. Given
2. $\overline{AC} \cong \overline{AC}$	2. Reflexive property
3. $\triangle I \cong \triangle II$	3. s.s.s. $\cong$ s.s.s.
4. $\angle 1 \cong \angle 4$, $\angle 2 \cong \angle 3$	4. Corresponding parts of $\cong$ $\triangle$ are $\cong$.

5.2 ISOSCELES AND EQUILATERAL TRIANGLES

5.2A Principles of Isosceles and Equilateral Triangles

PRINCIPLE 1: *If two sides of a triangle are congruent, the angles opposite these sides are congruent.* (Base angles of an isosceles triangle are congruent.)

Thus in $\triangle ABC$ in Fig. 5-12, if $\overline{AB} \cong \overline{BC}$, then $\angle A \cong \angle C$.

A proof of Principle 1 is given in Chapter 16.

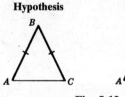

Fig. 5-12

Fig. 5-13

PRINCIPLE 2: *If two angles of a triangle are congruent, the sides opposite these angles are congruent.*

Thus in $\triangle ABC$ in Fig. 5-13, if $\angle A \cong \angle C$, then $\overline{AB} \cong \overline{BC}$.

Principle 2 is the converse of Principle 1. A proof of Principle 2 is given in Chapter 16.

PRINCIPLE 3: *An equilateral triangle is equiangular.*

Thus in $\triangle ABC$ in Fig. 5-14, if $\overline{AB} \cong \overline{BC} \cong \overline{CA}$, then $\angle A \cong \angle B \cong \angle C$.

Principle 3 is a corollary of Principle 1. A *corollary* of a theorem is another theorem whose statement and proof follow readily from the theorem.

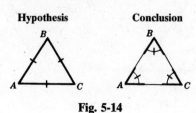

Fig. 5-14

PRINCIPLE 4:   *An equiangular triangle is equilateral.*

Thus in $\triangle ABC$ in Fig. 5-15, if $\angle A \cong \angle B \cong \angle C$, then $\overline{AB} \cong \overline{BC} \cong \overline{CA}$.

Principle 4 is the converse of Principle 3 and a corollary of Principle 2.

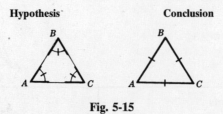

Fig. 5-15

SOLVED PROBLEMS

5.8   APPLYING PRINCIPLES 1 AND 3

In each part of Fig. 5-16, name the congruent angles that are opposite congruent sides of a triangle.

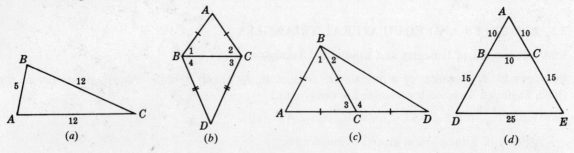

Fig. 5-16

Solutions

(a)   Since $\overline{AC} \cong \overline{BC}$, $\angle A \cong \angle B$.

(b)   Since $\overline{AB} \cong \overline{AC}$, $\angle 1 \cong \angle 2$. Since $\overline{BD} \cong \overline{CD}$, $\angle 3 \cong \angle 4$.

(c)   Since $\overline{AB} \cong \overline{AC} \cong \overline{BC}$, $\angle A \cong \angle 1 \cong \angle 3$. Since $\overline{BC} \cong \overline{CD}$, $\angle 2 \cong \angle D$.

(d)   Since $\overline{AB} \cong \overline{BC} \cong \overline{AC}$, $\angle A \cong \angle ACB \cong \angle ABC$. Since $\overline{AE} \cong \overline{AD} \cong \overline{DE}$, $\angle A \cong \angle D \cong \angle E$.

5.9   APPLYING PRINCIPLES 2 AND 4

In each part of Fig. 5-17, name the congruent sides that are opposite congruent angles of a triangle.

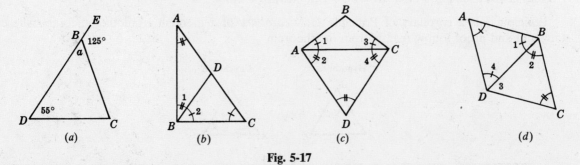

Fig. 5-17

Solutions

(a) Since $m\angle a = 55°$, $\angle a \cong \angle D$. Hence, $\overline{BC} \cong \overline{CD}$.

(b) Since $\angle A \cong \angle 1$, $\overline{AD} \cong \overline{BD}$. Since $\angle 2 \cong \angle C$, $\overline{BD} \cong \overline{CD}$.

(c) Since $\angle 1 \cong \angle 3$, $\overline{AB} \cong \overline{BC}$. Since $\angle 2 \cong \angle 4 \cong \angle D$, $\overline{CD} \cong \overline{AD} \cong \overline{AC}$.

(d) Since $\angle A \cong \angle 1 \cong \angle 4$, $\overline{AB} \cong \overline{BD} \cong \overline{AD}$. Since $\angle 2 \cong \angle C$, $\overline{BD} \cong \overline{CD}$.

5.10 APPLYING ISOSCELES TRIANGLE PRINCIPLES

In each of Fig. 5-18(*a*) and (*b*), $\triangle$I can be proved congruent to $\triangle$II. Make a diagram showing the congruent parts of both triangles and state the congruency principle involved.

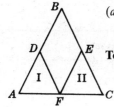 (*a*) **Given:** $\overline{AB} \cong \overline{BC}$
$\overline{AD} \cong \overline{EC}$
F is midpoint of $\overline{AC}$.
To Prove: $\triangle$I $\cong$ $\triangle$II

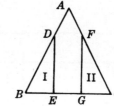 (*b*) **Given:** $\overline{AB} \cong \overline{AC}$
BC is trisected at *E* and *G*.
$\overline{DE} \perp \overline{BC}$
$\overline{FG} \perp \overline{BC}$
To Prove: $\triangle$I $\cong$ $\triangle$II

Fig. 5-18

Solutions

(a) Since $\overline{AB} \cong \overline{BC}$, $\angle A \cong \angle C$. $\triangle$I $\cong$ $\triangle$II by s.a.s. $\cong$ s.a.s. [see Fig. 5-19(*a*)].

(b) Since $\overline{AB} \cong \overline{AC}$, $\angle B \cong \angle C$. $\triangle$I $\cong$ $\triangle$II by a.s.a. $\cong$ a.s.a. [see Fig. 5-19(*b*)].

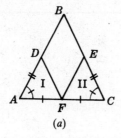

Fig. 5-19

5.11 PROVING AN ISOSCELES TRIANGLE PROBLEM

Given: $\overline{AB} \cong \overline{BC}$
AC is trisected at *D* and *E*
To Prove: $\angle 1 \cong \angle 2$
Plan: Prove $\triangle$I $\cong$ $\triangle$II to obtain $\overline{BD} \cong \overline{BE}$.

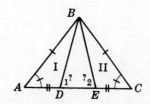

PROOF:

Statements	Reasons
1. *AC* is trisected at *D* and *E*.	1. Given
2. $\overline{AD} \cong \overline{EC}$	2. To trisect is to divide into three congruent parts.
3. $\overline{AB} \cong \overline{BC}$	3. Given
4. $\angle A \cong \angle C$	4. In a $\triangle$, $\angle\!s$ opposite $\cong$ sides are $\cong$.
5. $\triangle$I $\cong$ $\triangle$II	5. s.a.s. $\cong$ s.a.s.
6. $\overline{BD} \cong \overline{BE}$	6. Corresponding parts of $\cong$ $\triangle\!s$ are $\cong$.
7. $\angle 1 \cong \angle 2$	7. Same as 4.

5.12 PROVING AN ISOSCELES TRIANGLE PROBLEM STATED IN WORDS

Prove that the bisector of the vertex angle of an isosceles triangle is a median to the base.

Solution

<u>The bisector of the vertex angle of an isosceles triangle is a median to the base.</u>

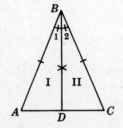

Given: Isosceles $\triangle ABC$ ($\overline{AB} \cong \overline{BC}$)
$\overline{BD}$ bisects $\angle B$

To Prove: $\overline{BD}$ is a median to $\overline{AC}$

Plan: Prove $\triangle I \cong \triangle II$ to obtain $\overline{AD} \cong \overline{DC}$.

PROOF:

Statements	Reasons
1. $\overline{AB} \cong \overline{BC}$	1. Given
2. $\overline{BD}$ bisects $\angle B$.	2. Given
3. $\angle 1 \cong \angle 2$	3. To bisect is to divide into two congruent parts.
4. $\overline{BD} \cong \overline{BD}$	4. Reflexive property
5. $\triangle I \cong \triangle II$	5. s.a.s. $\cong$ s.a.s.
6. $\overline{AD} \cong \overline{DC}$	6. Corresponding parts of $\cong$ ⧍ are $\cong$.
7. $\overline{BD}$ is a median to $\overline{AC}$.	7. A line from a vertex of a $\triangle$ bisecting opposite side is a median.

Supplementary Problems

1. Select the congruent triangles in (*a*) Fig. 5-20, (*b*) Fig. 5-21, and (*c*) Fig. 5-22, and state the congruency principle in each case.   (5.1)

Fig. 5-20

Fig. 5-21

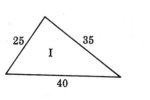 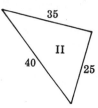 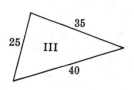

Fig. 5-22

2. In each figure below, △I can be proved congruent to △II. State the congruency principle involved.   (5.2)

(a)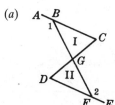

Given: ∠1 ≅ ∠2
          G is midpoint of $\overline{BF}$.
To Prove: △I ≅ △II

(e)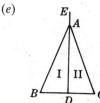

Given: ∠EAB ≅ ∠EAC
          $\overline{AD} \perp \overline{BC}$
To Prove: △I ≅ △II

(b)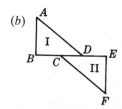

Given: $\overline{AB} \perp \overline{BE}$
          $\overline{EF} \perp \overline{BE}$
          $\overline{BC} \cong \overline{DE}$
          $\overline{AB} \cong \overline{EF}$
To Prove: △I ≅ △II

(f)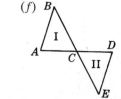

Given: $\overline{AD}$ and $\overline{BE}$
          bisect each
          other.
To Prove: △I ≅ △II

(c)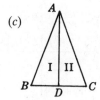

Given: $\overline{AB} \cong \overline{AC}$
          $\overline{AD}$ is median to $\overline{BC}$.
To Prove: △I ≅ △II

(g)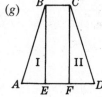

Given: $\overline{BE} \perp \overline{AD}$
          $\overline{CF} \perp \overline{AD}$
          $\overline{BE} \cong \overline{CF}$
          $\overline{AD}$ is trisected.
To Prove: △I ≅ △II

(d)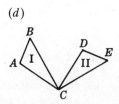

Given: $\overline{BC} \perp \overline{CE}$
          $\overline{AC} \perp \overline{CD}$
          $\overline{AC} \cong \overline{CD}$
          $\overline{BC} \cong \overline{CE}$
To Prove: △I ≅ △II

(h)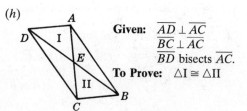

Given: $\overline{AD} \perp \overline{AC}$
          $\overline{BC} \perp \overline{AC}$
          $\overline{BD}$ bisects $\overline{AC}$.
To Prove: △I ≅ △II

3. State the additional parts needed to prove △I ≅ △II in the given figure by the given congruency
   principle.   (5.3)

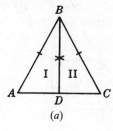

(a)

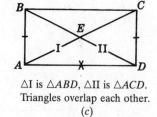

(b)

△I is △ABD, △II is △ACD.
Triangles overlap each other.

(c)

Fig. 5-23

(a)   In Fig. 5-23(a) by s.s.s. ≅ s.s.s.

(b)   In Fig. 5-23(a) by s.a.s. ≅ s.a.s.

(c)   In Fig. 5-23(b) by a.s.a. ≅ a.s.a.

(d)   In Fig. 5-23(b) by s.a.s. ≅ s.a.s.

(e)   In Fig. 5-23(c) by s.s.s. ≅ s.s.s.

(f)   In Fig. 5-23(c) by s.a.s. ≅ s.a.s.

4. In each part of Fig. 5-24, the congruent parts needed to prove △I ≅ △II are marked. Name the remaining parts that are congruent.   (5.4)

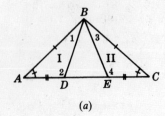
(a)

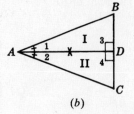
(b)

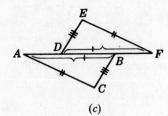
(c)

Fig. 5-24

5. In each part of Fig. 5-25, find x and y.   (5.5)

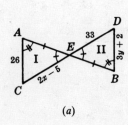
(a)

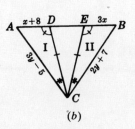
(b)

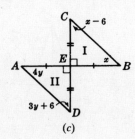
(c)

Fig. 5-25

6. Prove each of the following.   (5.6)

(a) In Fig. 5-26: **Given:** $\overline{BD} \perp \overline{AC}$
                           D is midpoint of $\overline{AC}$.
            To Prove: $\overline{AB} \cong \overline{BC}$

(c) In Fig. 5-27: **Given:** $\angle 1 \cong \angle 2$, $\overline{BF} \cong \overline{DE}$
                           $\overline{BF}$ bisects $\angle B$.
                           $\overline{DE}$ bisects $\angle D$.
                           $\angle B$ and $\angle D$ are rt. ⦞.
            To Prove: $\overline{AB} \cong \overline{CD}$

(b) In Fig. 5-26: $\overline{BD}$ is altitude to $\overline{AC}$.
                           $\overline{BD}$ bisects $\angle B$.
            To Prove: $\angle A \cong \angle C$

(d) In Fig. 5-27: **Given:** $\overline{BC} \cong \overline{AD}$
                           E is midpoint of $\overline{BC}$.
                           F is midpoint of $\overline{AD}$.
                           $\overline{AB} \cong \overline{CD}$, $\overline{BF} \cong \overline{DE}$
            To Prove: $\angle A \cong \angle C$

Fig. 5-26

Fig. 5-27

(e) In Fig. 5-28: **Given:** $\angle 1 \cong \angle 2$
                              $\overline{CE}$ bisects $\overline{BF}$.
             To Prove: $\angle C \cong \angle E$

(g) In Fig. 5-29: **Given:** $\overline{CD} \cong \overline{C'D'}$, $\overline{AD} \cong \overline{A'D'}$
                              $\overline{CD}$ is altitude to $\overline{AB}$.
                              $\overline{C'D'}$ is altitude to $\overline{A'B'}$.
             To Prove: $\angle A \cong \angle A'$

(f) In Fig. 5-28: **Given:** $\overline{BF}$ and $\overline{CE}$ bisect
                              each other.
             To Prove: $\overline{BC} \cong \overline{EF}$

(h) In Fig. 5-29: **Given:** $\overline{CD}$ bisects $\angle C$.
                              $\overline{C'D'}$ bisects $\angle C'$.
                              $\angle C \cong \angle C'$,
                              $\angle B \cong \angle B \cong \angle B'$,
                              $\overline{BC} \cong \overline{B'C'}$.
             To Prove: $\overline{CD} \cong \overline{C'D'}$

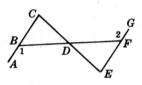

Fig. 5-28

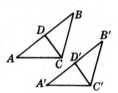

Fig. 5-29

7. Prove each of the following.   (5.7)

(a) If a line bisects an angle of a triangle and is perpendicular to the opposite side, then it bisects that side.

(b) If the diagonals of a quadrilateral bisect each other, then its opposite sides are congruent.

(c) If the base and a leg of one isosceles triangle are congruent to the base and a leg of another isosceles triangle, then their vertex angles are congruent.

(d) Lines drawn from a point on the perpendicular bisector of a given line to the ends of the given line are congruent.

(e) If the legs of one right triangle are congruent respectively to the legs of another, their hypotenuses are congruent.

8. In each part of Fig. 5-30, name the congruent angles that are opposite sides of a triangle.   (5.8)

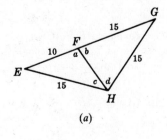

(a)

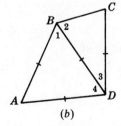

(b)

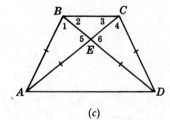

(c)

Fig. 5-30

9. In each part of Fig. 5-31, name the congruent sides that are opposite congruent angles of a triangle.   (5.9)

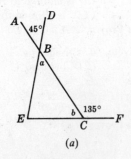
(a)

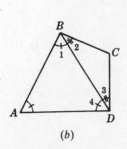
(b)

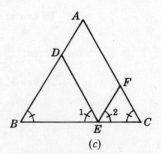
(c)

Fig. 5-31

10. In each part of Fig. 5-32, two triangles are to be proved congruent. Make a diagram showing the congruent parts of both triangles and state the reason for congruency.   (5.10)

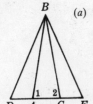

(a) **Given:**
$\overline{AD} \cong \overline{CE}$
$\angle 1 \cong \angle 2$
To Prove:
$\triangle ABD \cong \triangle CBE$

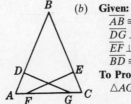

(b) **Given:**
$\overline{AB} \cong \overline{BC}$
$\overline{DG} \perp \overline{AB}$
$\overline{EF} \perp \overline{BC}$
$\overline{BD} \cong \overline{BE}$
To Prove:
$\triangle AGD \cong \triangle CFE$

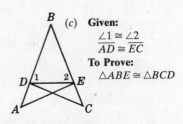

(c) **Given:**
$\angle 1 \cong \angle 2$
$\overline{AD} \cong \overline{EC}$
To Prove:
$\triangle ABE \cong \triangle BCD$

Fig. 5-32

11. In each part of Fig. 5-33, $\triangle$I, $\triangle$II, and $\triangle$III can be proved congruent. Make a diagram showing the congruent parts and state the reason for congruency.   (5.10)

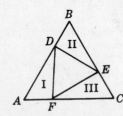

(a) **Given:**
$\triangle ABC$ is equilateral.
$\overline{AF} \cong \overline{BD} \cong \overline{CE}$
To Prove:
$\triangle I \cong \triangle II \cong \triangle III$

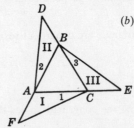

(b) **Given:**
$\triangle ABC$ is equilateral.
$\overline{AF}$, $\overline{BD}$, and $\overline{CE}$ are extensions of the sides of $\triangle ABC$.
$\angle 1 \cong \angle 2 \cong \angle 3$
To Prove:
$\triangle I \cong \triangle II \cong \triangle III$

Fig. 5-33

12. Prove each of the following: (5.11)

(a) In Fig. 5-34: **Given:** $\overline{AB} \cong \overline{AC}$
F is midpoint of $\overline{BC}$.
$\angle 1 \cong \angle 2$
To Prove: $\overline{FD} \cong \overline{FE}$

(b) In Fig. 5-34: **Given:** $\overline{AB} \cong \overline{AC}$
$\overline{AD} \cong \overline{AE}$
$\overline{FD} \perp \overline{AB}$, $\overline{FE} \perp \overline{AC}$
To Prove: $\overline{BF} \cong \overline{FC}$

(c) In Fig. 5-35: **Given:** $\overline{AB} \cong \overline{AC}$
$\angle A$ is trisected.
To Prove: $\overline{AD} \cong \overline{AE}$

(d) In Fig. 5-35: **Given:** $\overline{AB} \cong \overline{AC}$
$\overline{DB} \cong \overline{BC}$
$\overline{CE} \cong \overline{BC}$
To Prove: $\overline{AD} \cong \overline{AE}$

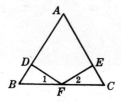

Fig. 5-34

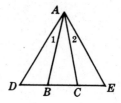

Fig. 5-35

13. Prove each of the following: (5.12)

   (*a*)   The median to the base of an isosceles triangle bisects the vertex angle.

   (*b*)   If the bisector of an angle of a triangle is also an altitude to the opposite side, then the other two sides of the triangle are congruent.

   (*c*)   If a median to a side of a triangle is also an altitude to that side, then the triangle is isosceles.

   (*d*)   In an isosceles triangle, the medians to the legs are congruent.

   (*e*)   In an isosceles triangle, the bisectors of the base angles are congruent.

CHAPTER 6

Parallel Lines, Distances, and Angle Sums

6.1 PARALLEL LINES

Parallel lines are straight lines which lie in the same plane and do not intersect however far they are extended. The symbol for parallel is $\parallel$; thus $\overleftrightarrow{AB} \parallel \overleftrightarrow{CD}$ is read "line $\overleftrightarrow{AB}$ is parallel to line $\overleftrightarrow{CD}$." In diagrams, arrows are used to indicate that lines are parallel (see Fig. 6-1).

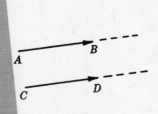

Fig. 6-1

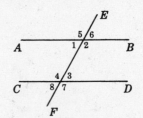

Fig. 6-2

A *transversal* of two or more lines is a line that cuts across these lines. Thus $\overleftrightarrow{EF}$ is a transversal of $\overleftrightarrow{AB}$ and $\overleftrightarrow{CD}$, in Fig. 6-2.

The *interior angles* formed by two lines cut by a transversal are the angles between the two lines, while the *exterior angles* are those outside the lines. Thus, of the eight angles formed by $\overleftrightarrow{AB}$ and $\overleftrightarrow{CD}$ cut by $\overleftrightarrow{EF}$ in Fig. 6-2, the interior angles are $\angle 1$, $\angle 2$, $\angle 3$, and $\angle 4$; the exterior angles are $\angle 5$, $\angle 6$, $\angle 7$, and $\angle 8$.

6.1A Pairs of Angles Formed by Two Lines Cut by a Transversal

Corresponding angles of two lines cut by a transversal are angles on the same side of the transversal and on the same side of the lines. Thus $\angle 1$ and $\angle 2$ in Fig. 6-3 are corresponding angles of

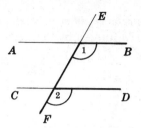

Fig. 6-3

$\overleftrightarrow{AB}$ and $\overleftrightarrow{CD}$ cut by transversal $\overleftrightarrow{EF}$. Note that in this case the two angles are both to the right of the transversal and both below the lines.

When two parallel lines are cut by a transversal, the sides of two corresponding angles form a capital F in varying positions, as shown in Fig. 6-4.

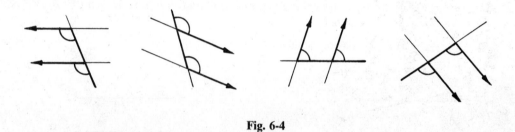

Fig. 6-4

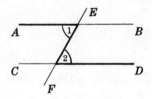

Fig. 6-5

Alternate interior angles of two lines cut by a transversal are nonadjacent angles between the two lines and on opposite sides of the transversal. Thus $\angle 1$ and $\angle 2$ in Fig. 6-5 are alternate interior angles of $\overleftrightarrow{AB}$ and $\overleftrightarrow{CD}$ cut by $\overleftrightarrow{EF}$. When parallel lines are cut by a transversal, the sides of two alternate interior angles form a capital Z or N in varying positions, as shown in Fig. 6-6.

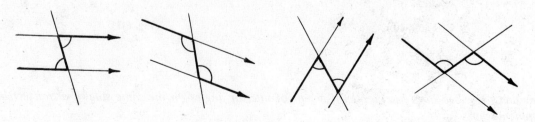

Fig. 6-6

When parallel lines are cut by a transversal, *interior angles on the same side of the transversal* can be readily located by noting the capital U formed by their sides (Fig. 6-7).

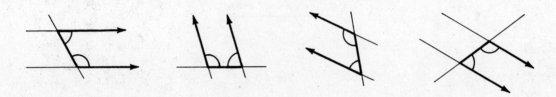

Fig. 6-7

6.1B Principles of Parallel Lines

PRINCIPLE 1:   *Through a given point not on a given line, one and only one line can be drawn parallel to a given line.* (Parallel-Line Postulate)

Thus, either l_1 or l_2 but not both may be parallel to l_3 in Fig. 6-8.

Fig. 6-8   **Fig. 6-9**

Proving that Lines are Parallel

PRINCIPLE 2:   *Two lines are parallel if a pair of corresponding angles are congruent.*

Thus, $l_1 \parallel l_2$ if $\angle a \cong \angle b$ in Fig. 6-9.

PRINCIPLE 3:   *Two lines are parallel if a pair of alternate interior angles are congruent.*

Thus, $l_1 \parallel l_2$ if $\angle c \cong \angle d$ in Fig. 6-10.

Fig. 6-10   **Fig. 6-11**

PRINCIPLE 4:   *Two lines are parallel if a pair of interior angles on the same side of a transversal are supplementary.*

Thus, $l_1 \parallel l_2$ if $\angle e$ and $\angle f$ are supplementary in Fig. 6-11.

PRINCIPLE 5: *Lines are parallel if they are perpendicular to the same line. (Perpendiculars to the same line are parallel.)*

Thus, $l_1 \parallel l_2$ if l_1 and l_2 are each perpendicular to l_3 in Fig. 6-12.

Fig. 6-12   **Fig. 6-13**

PRINCIPLE 6: *Lines are parallel if they are parallel to the same line. (Parallels to the same line are parallel.)*

Thus, $l_1 \parallel l_2$ if l_1 and l_2 are each parallel to l_3 in Fig. 6-13.

Properties of Parallel Lines

PRINCIPLE 7: *If two lines are parallel, each pair of corresponding angles are congruent. (Corresponding angles of parallel lines are congruent.)*

Thus, if $l_1 \parallel l_2$, then $\angle a \cong \angle b$ in Fig. 6-14.

Fig. 6-14   **Fig. 6-15**

PRINCIPLE 8: *If two lines are parallel, each pair of alternate interior angles are congruent. (Alternate interior angles of parallel lines are congruent.)*

Thus, if $l_1 \parallel l_2$, then $\angle c \cong \angle d$ in Fig. 6-15.

PRINCIPLE 9: *If two lines are parallel, each pair of interior angles on the same side of the transversal are supplementary.*

Thus, if $l_1 \parallel l_2$, $\angle e$ and $\angle f$ are supplementary in Fig. 6-16.

PRINCIPLE 10: *If lines are parallel, a line perpendicular to one of them is perpendicular to the others also.*

Thus, if $l_1 \parallel l_2$ and $l_3 \perp l_1$, then $l_3 \perp l_2$ in Fig. 6-17.

Fig. 6-16   **Fig. 6-17**

PRINCIPLE 11:   *If lines are parallel, a line parallel to one of them is parallel to the others also.*

Thus, if $l_1 \| l_2$ and $l_3 \| l_1$, then $l_3 \| l_2$ in Fig. 6-18.

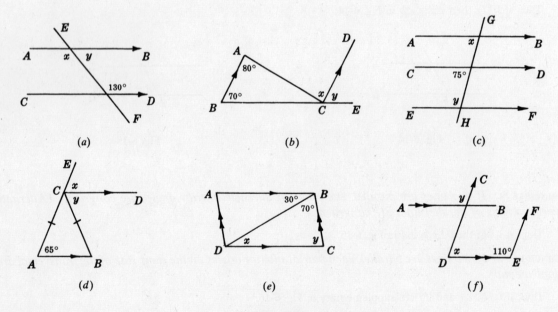

Fig. 6-18   **Fig. 6-19**

PRINCIPLE 12:   *If the sides of two angles are respectively parallel to each other, the angles are either congruent or supplementary.*

Thus, if $l_1 \| l_3$ and $l_2 \| l_4$ in Fig. 6-19, then $\angle a \cong \angle b$ and $\angle a$ and $\angle c$ are supplementary.

SOLVED PROBLEMS

6.1   NUMERICAL APPLICATIONS OF PARALLEL LINES

In each part of Fig. 6-20, find the measure x and the measure y of the indicated angles.

Fig. 6-20

Solutions

(a) $x = 130°$ (Principle 8). $y = 180° - 130° = 50°$ (Principle 9).

(b) $x = 80°$ (Principle 8). $y = 70°$ (Principle 7).

(c) $x = 75°$ (Principle 7). $y = 180° - 75° = 105°$ (Principle 9).

(d) $x = 65°$ (Principle 7). Since $m\angle B = m\angle A$, $m\angle B = 65°$. Hence, $y = 65°$ (Principle 8).

(e) $x = 30°$ (Principle 8). $y = 180° + (30° + 70°) = 80°$ (Principle 9).

(f) $x = 180° - 110° = 70°$ (Principle 9). $y = 110°$ (Principle 12).

6.2   APPLYING PARALLEL LINE PRINCIPLES AND THEIR CONVERSES

The following short proofs refer to Fig. 6-21. In each, the first statement is given. State the parallel-line principle needed as the reason for each of the remaining statements.

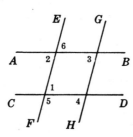

Fig. 6-21

(a) 1. $\angle 1 \cong \angle 2$ 1. Given
    2. $\overline{AB} \| \overline{CD}$ 2. ?
    3. $\angle 3 \cong \angle 4$ 3. ?

(b) 1. $\angle 2 \cong \angle 3$ 1. Given
    2. $\overline{EF} \| \overline{GH}$ 2. ?
    3. $\angle 4$ sup. $\angle 5$ 3. ?

(c) 1. $\angle 5$ sup. $\angle 4$ 1. Given
    2. $\overline{EF} \| \overline{GH}$ 2. ?
    3. $\angle 3 \cong \angle 6$ 3. ?

(d) 1. $\overline{EF} \perp \overline{AB}, \overline{GH} \perp \overline{AB},$ 1. Given
       $\overline{EF} \perp \overline{CD}$
    2. $\overline{EF} \| \overline{GH}$ 2. ?
    3. $\overline{CD} \perp \overline{GH}$ 3. ?

Solutions

(a) 2: Principle 3; 3: Principle 7.

(b) 2: Principle 2; 3: Principle 9.

(c) 2: Principle 4; 3: Principle 8.

(d) 2: Principle 5; 3: Principle 10.

6.3   ALGEBRAIC APPLICATIONS OF PARALLEL LINES

In each part of Fig. 6-22, find x and y. Provide the reason for each equation obtained from the diagram.

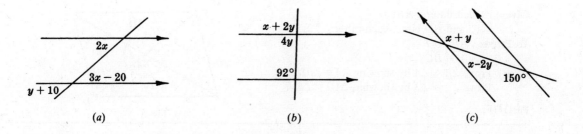

Fig. 6-22

Solutions

(a) $3x - 20 = 2x$ (Principle 8)
        $x = 20°$
    $y + 10 = 2x$ (Principle 7)
        $y + 10 = 40$
          $y = 30°$

(b) $4y = 180 - 92 = 88$ (Principle 9)
         $y = 22°$
    $x + 2y = 92$ (Principle 7)
    $x + 44 = 92$
        $x = 48°$

(c) (1) $x + y = 150$ (Principle 8)
    (2) $x - 2y = 30$ (Principle 9)
          $3y = 120$ (Subt. Postulate)
           $y = 40°$
     $x + 40 = 150$
          $x = 110°$

6.4 PROVING A PARALLEL-LINE PROBLEM

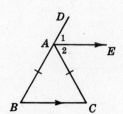

Given: $\overline{AB} \cong \overline{AC}$
        $\overrightarrow{AE} \parallel \overline{BC}$
To Prove: $\overrightarrow{AE}$ bisects $\angle DAC$
Plan: Show that $\angle 1$ and $\angle 2$ are congruent
      to the congruent angles B and C.

PROOF:

Statements	Reasons
1. $\overrightarrow{AE} \parallel \overline{BC}$	1. Given
2. $\angle 1 \cong \angle B$	2. Corresponding ⧦ of ∥ lines are ≅.
3. $\angle 2 \cong \angle C$	3. Alternate interior ⧦ of ∥ lines are ≅.
4. $\overline{AB} \cong \overline{AC}$	4. Given
5. $\angle B \cong \angle C$	5. In a △, ⧦ opposite ≅ sides are ≅.
6. $\angle 1 \cong \angle 2$	6. Things ≅ to ≅ things are ≅ to each other.
7. $\overrightarrow{AE}$ bisects $\angle DAC$.	7. To divide into two congruent parts is to bisect.

6.5 PROVING A PARALLEL-LINE PROBLEM STATED IN WORDS

    Prove that if the diagonals of a quadrilateral bisect each other, the opposite sides are parallel.

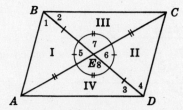

Given: Quadrilateral $ABCD$
      $\overline{AC}$ and $\overline{BD}$ bisect each other.
To Prove: $\overline{AB} \parallel \overline{CD}$
          $\overline{AD} \parallel \overline{BC}$
Plan: Prove $\angle 1 \cong \angle 4$ by showing $\triangle I \cong \triangle II$.
      Prove $\angle 2 \cong \angle 3$ by showing $\triangle III \cong \triangle IV$.

PROOF:

Statements	Reasons
1. $\overline{AC}$ and $\overline{BD}$ bisect each other.	1. Given
2. $\overline{BE} \cong \overline{ED}, \overline{AE} \cong \overline{EC}$	2. To bisect is to divide into two congruent parts.
3. $\angle 5 \cong \angle 6, \angle 7 \cong \angle 8$	3. Vertical ⧦ are ≅.
4. $\triangle I \cong \triangle II, \triangle III \cong \triangle IV$	4. s.a.s. ≅ s.a.s.
5. $\angle 1 \cong \angle 4, \angle 2 \cong \angle 3$	5. Corresponding parts of congruent △ are ≅.
6. $\overline{AB} \parallel \overline{CD}, \overline{BC} \parallel \overline{AD}$	6. Lines cut by a transversal are ∥ if alternate interior ⧦ are ≅.

6.2 DISTANCES

6.2A Distances between Two Geometric Figures

The distance between two geometric figures is the straight line segment which is the *shortest segment between the figures*.

1. The distance *between two points*, such as P and Q in Fig. 6-23(a), is the line segment $\overline{PQ}$ between them.

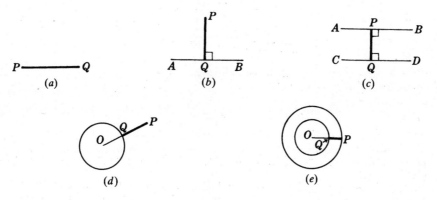

Fig. 6-23

2. The distance *between a point and a line*, such as P and $\overleftrightarrow{AB}$ in (b), is the line segment PQ, the perpendicular from the point to the line.

3. The distance *between two parallels*, such as $\overleftrightarrow{AB}$ and $\overleftrightarrow{CD}$ in (c), is the segment PQ, a perpendicular between the two parallels.

4. The distance *between a point and a circle*, such as P and circle O in (d), is $\overline{PQ}$, the segment of $\overline{OP}$ between the point and the circle.

5. The distance *between two concentric circles*, such as two circles whose center is O, is $\overline{PQ}$, the segment of the larger radius that lies between the two circles, as shown in (e).

6.2B Distance Principles

PRINCIPLE 1: *If a point is on the perpendicular bisector of a line segment, then it is equidistant from the ends of the line segment.*

Thus if P is on $\overleftrightarrow{CD}$, the $\perp$ bisector of $\overline{AB}$ in Fig. 6-24, then $\overline{PA} \cong \overline{PB}$.

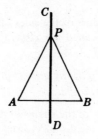

Fig. 6-24

PRINCIPLE 2:   *If a point is equidistant from the ends of a line segment, then it is on the perpendicular bisector of the line segment.* (Principle 2 is the converse of Principle 1.)

Thus if $\overline{PA} \cong \overline{PB}$ in Fig. 6-24, then P is on $\overleftrightarrow{CD}$, the $\perp$ bisector of $\overline{AB}$.

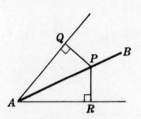

Fig. 6-25

PRINCIPLE 3:   *If a point is on the bisector of an angle, then it is equidistant from the sides of the angle.*

Thus if P is on $\overrightarrow{AB}$, the bisector of $\angle A$ in Fig. 6-25, then $\overline{PQ} \cong \overline{PR}$, where PQ and PR are the distances of P from the sides of the angle.

PRINCIPLE 4:   *If a point is equidistant from the sides of an angle, then it is on the bisector of the angle.* (Principle 4 is the converse of Principle 3.)

Thus if $PQ = PR$, where PQ and PR are the distances of P from the sides of $\angle A$ in Fig. 6-25, then P is on $\overrightarrow{AB}$, the bisector of $\angle A$.

PRINCIPLE 5:   *Two points each equidistant from the ends of a line segment determine the perpendicular bisector of the line segment.* (The line joining the vertices of two isosceles triangles having a common base is the perpendicular bisector of the base.)

Thus if $\overline{PA} \cong \overline{PB}$ and $\overline{QA} \cong \overline{QB}$ in Fig. 6-26, then P and Q determine $\overleftrightarrow{CD}$, the $\perp$ bisector of $\overline{AB}$.

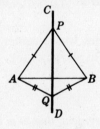

Fig. 6-26

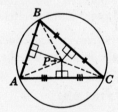

Fig. 6-27

PRINCIPLE 6:   *The perpendicular bisectors of the sides of a triangle meet in a point which is equidistant from the vertices of the triangle.*

Thus if P is the intersection of the $\perp$ bisectors of the sides of $\triangle ABC$ in Fig. 6-27, then $\overline{PA} \cong \overline{PB} \cong \overline{PC}$. P is the center of the circumscribed circle and is called the *circumcenter* of $\triangle ABC$.

PRINCIPLE 7:   *The bisectors of the angles of a triangle meet in a point which is equidistant from the sides of the triangle.*

Thus if Q is the intersection of the bisectors of the angles of $\triangle ABC$ in Fig. 6-28, then $\overline{QR} \cong \overline{QS} \cong \overline{QT}$, where these are the distances from Q to the sides of $\triangle ABC$. Q is the center of the inscribed circle and is called the *incenter* of $\triangle ABC$.

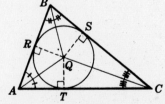

Fig. 6-28

SOLVED PROBLEMS

6.6   FINDING DISTANCES
In each of the following, find the distance and indicate the kind of distance involved. In Fig. 6-29(a), find the distance (a) from P to A; (b) from P to $\overleftrightarrow{CD}$; (c) from A to $\overrightarrow{BC}$; (d) from $\overrightarrow{AB}$ to $\overleftrightarrow{CD}$. In Fig. 6-29(b), find the distance (e) from P to inner circle O; (f) from P to outer circle O; (g) between the concentric circles.

Solutions

(a) $PA = 7$, distance between two points

(b) $PG = 4$, distance from a point to a line

(c) $AE = 10$, distance from a point to a line

(d) $FG = 6$, distance between two parallel lines

(e) $PQ = 12 - 3 = 9$, distance from a point to a circle

(f) $PR = 12 - 5 = 7$, distance from a point to a circle

(g) $QR = 5 - 3 = 2$, distance between two concentric circles

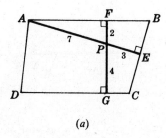

(a)   (b)

Fig. 6-29

6.7   LOCATING A POINT SATISFYING GIVEN CONDITIONS
In Fig. 6-30,

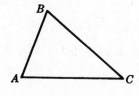

Fig. 6-30

(a)   Locate P, a point on $\overrightarrow{BC}$ and equidistant from A and C.

(b)   Locate Q, a point on $\overrightarrow{AB}$ and equidistant from $\overline{BC}$ and $\overline{AC}$.

(c)   Locate R, the center of the circle circumscribed about $\triangle ABC$.

(d)   Locate S, the center of the circle inscribed in $\triangle ABC$.

Solutions

See Fig. 6-31.

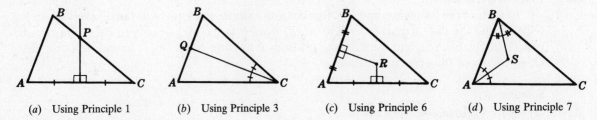

(a) Using Principle 1   (b) Using Principle 3   (c) Using Principle 6   (d) Using Principle 7

Fig. 6-31

6.8   APPLYING PRINCIPLES 2 AND 4

For each $\triangle ABC$ in Fig. 6-32, describe P, Q, and R as equidistant points, and locate each on a bisector.

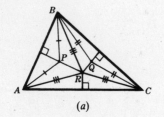   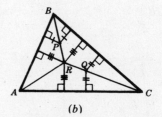

(a)   (b)

Fig. 6-32

Solutions

(a) Since P is equidistant from A and B, it is on the $\perp$ bisector of $\overline{AB}$. Since Q is equidistant from B and C, it is on the $\perp$ bisector of $\overline{BC}$. Since R is equidistant from A, B, and C, it is on the $\perp$ bisectors of $\overline{AB}$, $\overline{BC}$, and $\overline{AC}$.

(b) Since P is equidistant from $\overleftrightarrow{AB}$ and $\overleftrightarrow{BC}$, it is on the bisector of $\angle B$. Since Q is equidistant from $\overleftrightarrow{AC}$ and $\overleftrightarrow{BC}$, it is on the bisector of $\angle C$. Since R is equidistant from $\overleftrightarrow{AB}$, $\overleftrightarrow{BC}$, and $\overleftrightarrow{AC}$, it is on the bisectors of $\angle A$, $\angle B$, and $\angle C$.

6.9   APPLYING PRINCIPLES 1, 3, 6, AND 7

For each $\triangle ABC$ in Fig. 6-33, describe P, Q, and R as equidistant points. Also, describe R as the center of a circle.

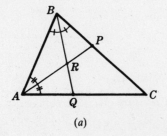   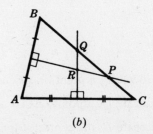

(a)   (b)

Fig. 6-33

Solutions

(a) Since P is on the bisector of $\angle A$, it is equidistant from $\overrightarrow{AB}$ and $\overrightarrow{AC}$. Since Q is on the bisector of $\angle B$, it is equidistant from $\overrightarrow{AB}$ and $\overrightarrow{BC}$. Since R is on the bisectors of $\angle A$ and $\angle B$, it is equidistant from $\overrightarrow{AB}$, $\overrightarrow{BC}$, and $\overrightarrow{AC}$. R is the incenter of $\triangle ABC$, that is, the center of its inscribed circle.

(b) Since P is on the $\perp$ bisector of $\overline{AB}$, it is equidistant from A and B. Since Q is on the $\perp$ bisector of $\overline{AC}$, it is equidistant from A and C. Since R is on the $\perp$ bisectors of $\overline{AB}$ and $\overline{AC}$, it is equidistant from A, B, and C. R is the circumcenter of $\triangle ABC$, that is, the center of its circumscribed circle.

6.10 APPLYING PRINCIPLES 1, 3, 6, AND 7

In each part of Fig. 6-34, find two points equidistant from the ends of a line segment, and find the perpendicular bisector determined by the two points.

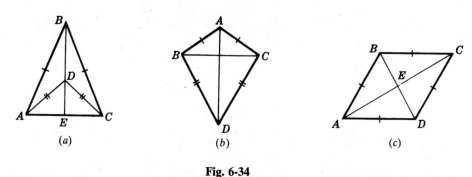

Fig. 6-34

Solutions

(a) B and D are equidistant from A and C. Hence $\overline{BE}$ is the $\perp$ bisector of $\overline{AC}$.

(b) A and D are equidistant from B and C. Hence $\overline{AD}$ is the $\perp$ bisector of $\overline{BC}$.

(c) B and D are equidistant from A and C; hence $\overline{BD}$ is the $\perp$ bisector of $\overline{AC}$. A and C are equidistant from B and D; hence $\overline{AC}$ is the $\perp$ bisector of $\overline{BD}$.

6.3 SUM OF THE MEASURES OF THE ANGLES OF A TRIANGLE

The angles of any triangle may be torn off, as in Fig. 6-35(a), and then fitted together as shown in (b). The three angles will form a straight angle.

We can prove that the sum of the measures of the angles of a triangle equals 180° by drawing a line through one vertex of the triangle parallel to the side opposite the vertex. In Fig. 6-35(c), $\overleftrightarrow{MN}$ is drawn through B parallel to AC. Note that the measure of the straight angle at B equals the sum of the measures of the angles of $\triangle ABC$; that is, $a° + b° + c° = 180°$. Each pair of congruent angles is a pair of alternate interior angles of parallel lines.

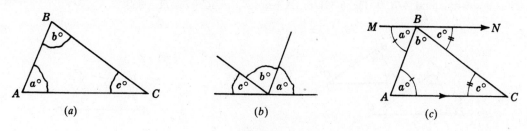

Fig. 6-35

6.3A Interior and Exterior Angles of a Polygon

An exterior angle of a polygon is formed whenever one of its sides is extended through a vertex. If each of the sides of a polygon is extended, as shown in Fig. 6-36, an exterior angle will be formed at each vertex. Each of these exterior angles is the supplement of its adjacent interior angle.

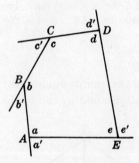

Fig. 6-36

Thus in the case of pentagon *ABCDE*, there will be five exterior angles, one at each vertex. Note that each exterior angle is the supplement of an adjacent interior angle. For example, $m\angle a + m\angle a' = 180°$.

6.3B Angle-Measure-Sum Principles

PRINCIPLE 1: *The sum of the measures of the angles of a triangle equals the measure of a straight angle or 180°.*

Thus in △*ABC* of Fig. 6-37, $m\angle A + m\angle B + m\angle C = 180°$.

PRINCIPLE 2: *If two angles of one triangle are congruent respectively to two angles of another triangle, the remaining angles are congruent.*

Thus in △*ABC* and △*A'B'C'* in Fig. 6-38, if $\angle A \cong \angle A'$ and $\angle B \cong \angle B'$, then $\angle C \cong \angle C'$.

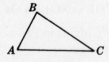

Fig. 6-37

PRINCIPLE 3: *The sum of the measures of the angles of a quadrilateral equals 360°.*

Thus in quadrilateral *ABCD* (Fig. 6-39), $m\angle A + m\angle B + m\angle C + m\angle D = 360°$.

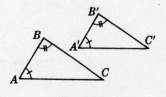

Fig. 6-38

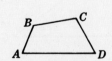

Fig. 6-39

PRINCIPLE 4:   *The measure of each exterior angle of a triangle equals the sum of the measures of its two nonadjacent interior angles.*

Thus in $\triangle ABC$ in Fig. 6-40, $m\angle ECB = m\angle A + m\angle B$.

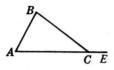

Fig. 6-40

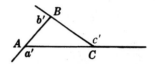

Fig. 6-41

PRINCIPLE 5:   *The sum of the measures of the exterior angles of a triangle equals 360°.*

Thus in $\triangle ABC$ in Fig. 6-41, $m\angle a' + m\angle b' + m\angle c' = 360°$.

PRINCIPLE 6:   *The measure of each angle of an equilateral triangle equals 60°.*

Thus if $\triangle ABC$ in Fig. 6-42 is equilateral, then $m\angle A = 60°$, $m\angle B = 60°$, and $m\angle C = 60°$.

Fig. 6-42

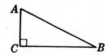

Fig. 6-43

PRINCIPLE 7:   *The acute angles of a right triangle are complementary.*

Thus in rt. $\triangle ABC$ in Fig. 6-43, if $m\angle C = 90°$, then $m\angle A + m\angle B = 90°$.

PRINCIPLE 8:   *The measure of each acute angle of an isosceles right triangle equals 45°.*

Thus in isos. rt. $\triangle ABC$ in Fig. 6-44, if $m\angle C = 90°$, then $m\angle A = 45°$ and $m\angle B = 45°$.

PRINCIPLE 9:   *A triangle can have no more than one right angle.*

Thus in rt. $\triangle ABC$ in Fig. 6-43, if $m\angle C = 90°$, then $\angle A$ and $\angle B$ cannot be rt. $\angle$s.

PRINCIPLE 10:   *A triangle can have no more than one obtuse angle.*

Thus in obtuse $\triangle ABC$ in Fig. 6-45, if $\angle C$ is obtuse, then $\angle A$ and $\angle B$ cannot be obtuse angles.

Fig. 6-44

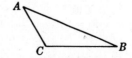

Fig. 6-45

PRINCIPLE 11: *Two angles are congruent or supplementary if their sides are respectively perpendicular to each other.*

Thus if $l_1 \perp l_3$ and $l_2 \perp l_4$ in Fig. 6-46, then $\angle a \cong \angle b$, and a and $\angle c$ are supplementary.

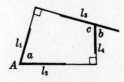

Fig. 6-46

SOLVED PROBLEMS

6.11 NUMERICAL APPLICATIONS OF ANGLE-MEASURE-SUM PRINCIPLES
In each part of Fig. 6-47, find x and y.

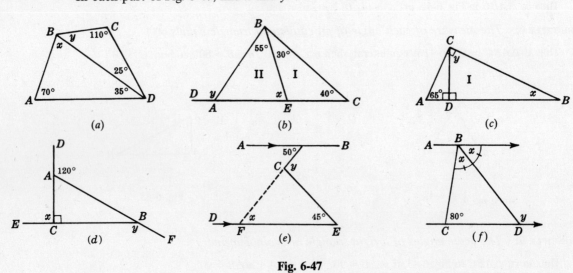

Fig. 6-47

Solutions

(a) $x + 35 + 70 = 180$ (Pr. 1)
$x = 75°$
$y + 110 + 25 = 180$ (Pr. 1)
$y = 45°$

Check: The sum of the measures of the angles of quad. *ABCD* should equal 360°.
$70 + 120 + 110 + 60 \stackrel{?}{=} 360$
$360 = 360$

(b) x is ext. $\angle$ of $\triangle$I.
$x = 30 + 40$ (Pr. 4)
$x = 70°$
y is an ext. $\angle$ of $\triangle ABC$.
$y = m\angle B + 40$ (Pr. 4)
$y = 85 + 40 = 125°$

(c) In $\triangle ABC$, $x + 65 = 90$ (Pr. 7)
$x = 25°$
In $\triangle$I, $x + y = 90$ (Pr. 7)
$25 + y = 90$
$y = 65°$

(d) Since $\overrightarrow{DC} \perp \overrightarrow{EB}$, $x = 90°$
$x + y + 120 = 360$ (Pr. 5)
$90 + y + 120 = 360$
$y = 150°$

(e) Since $\overrightarrow{AB} \parallel \overrightarrow{DE}$, $x = 50°$
$y = x + 45$ (Pr. 4)
$y = 50 + 45 = 95°$

(f) Since $\overrightarrow{AB} \parallel \overrightarrow{CD}$, $2x + 80 = 180$
$2x = 100$
$x = 50°$
$y = x + 80°$ (Pr. 4)
$y = 50 + 80 = 130°$

6.12   Applying Angle-Measure-Sum Principles to Isosceles and Equilateral Triangles

Find x and y in each part of Fig. 6-48.

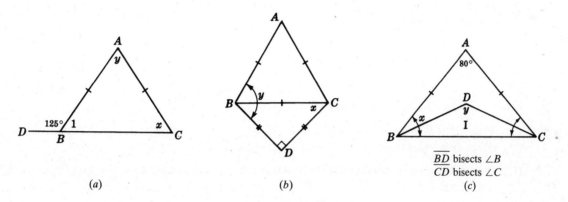

Fig. 6-48

Solutions

(a) Since $\overline{AB} \cong \overline{AC}$, we have $\angle 1 \cong \angle x$

$x = 180 - 125 = 55°$

$2x + y = 180$ (Pr. 1)

$110 + y = 180$

$y = 70°$

(b) By Pr. 8, $x = 45°$

Since $m\angle ABC = 60°$ (Pr. 6)

and $m\angle CBD = 45°$ (Pr. 8)

$y = 60 + 45 = 105°$

(c) Since $\overline{AB} \cong \overline{AC}$, $\angle ABC \cong \angle ACB$

$2x + 80 = 180$ (Pr. 1)

$x = 50°$

In $\triangle I$, $\frac{1}{2}x + \frac{1}{2}x + y = 180$ (Pr. 1)

$x + y = 180$

$50 + y = 180$

$y = 130°$

6.13   Applying Ratios to Angle-Measure Sums

Find the measure of each angle

(a) Of a triangle if its angle measures are in the ratio of 3:4:5 [Fig. 6-49(a)]

(b) Of a quadrilateral if its angle measures are in the ratio of 3:4:5:6 [(b)]

(c) Of a right triangle if the ratio of the measures of its acute angles is 2:3 [(c)]

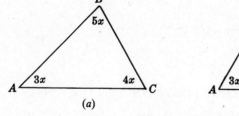
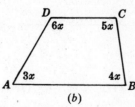
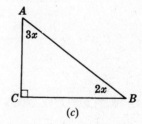

Fig. 6-49

Solutions

(a) Let $3x$, $4x$, and $5x$ represent the measures of the angles. Then $12x = 180$ by Principle 1, so that $x = 15$. Now $3x = 45$, $4x = 60$, and $5x = 75$.   *Ans.* 45°, 60°, 75°

(b) Let $3x$, $4x$, $5x$, and $6x$ represent the measures of the angles. Then $18x = 360$ by Principle 3, so that $x = 20$. Now $3x = 60$, $4x = 80$, and so forth.   *Ans.* 60°, 80°, 100°, 120°

(c) Let $2x$ and $3x$ represent the measures of the acute angles. Then $5x = 90$ by Principle 7 so that $x = 18$. Now $2x = 36$ and $3x = 54$.   *Ans.* 36°, 54°, 90°

6.14   Using Algebra to Prove Angle-Measure-Sum Problems

(a) Prove that if the measure of one angle of a triangle equals the sum of the measures of the other two, then the triangle is a right triangle.

(b) Prove that if the opposite angles of a quadrilateral are congruent, then its opposite sides are parallel.

Solutions

(a) **Given:** $\triangle ABC$, $m\angle C = m\angle A + m\angle B$

   To Prove: $\triangle ABC$ is a right triangle.

   Plan: Prove $m\angle C = 90°$

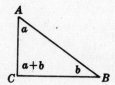

   ALGEBRAIC PROOF:

   Let a = number of degrees in $\angle A$
   b = number of degrees in $\angle B$
   Then $a + b$ = number of degrees in $\angle C$

   $$a + b + (a + b) = 180 \qquad \text{(Pr. 1)}$$
   $$2a + 2b = 180$$
   $$a + b = 90$$

   Since $m\angle C = 90°$, $\triangle ABC$ is a rt. $\triangle$.

(b) **Given:**   Quadrilateral $ABCD$
   $\angle A \cong \angle C$, $\angle B \cong \angle D$

   To Prove: $\overline{AB} \parallel \overline{CD}$, $\overline{BC} \parallel \overline{AD}$

   Plan: Prove int. $\angle$s on same side of transversal are supplementary.

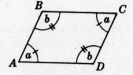

   ALGEBRAIC PROOF:

   Let a = number of degrees in $\angle A$ and $\angle C$,
   b = number of degrees in $\angle B$ and $\angle D$.

   $$2a + 2b = 360 \qquad \text{(Pr. 3)}$$
   $$a + b = 180$$

   Since $\angle A$ and $\angle B$ are supplementary, $\overline{BC} \parallel \overline{AD}$.

   Since $\angle A$ and $\angle D$ are supplementary, $\overline{AB} \parallel \overline{CD}$.

6.4   SUM OF THE MEASURES OF THE ANGLES OF A POLYGON

A *polygon* is a closed plane figure bounded by straight line segments as sides.

An *n-gon* is a polygon of n sides. Thus a polygon of 20 sides is a 20-gon.

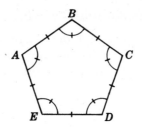

Regular Pentagon

Fig. 6-50

A *regular polygon* is an equilateral and equiangular polygon. Thus a regular pentagon is a polygon having 5 congruent angles and 5 congruent sides (Fig. 6-50). A square is a regular polygon of 4 sides.

Names of Polygons According to the Number of Sides

Number of Sides	Polygon	Number of Sides	Polygon
3	Triangle	8	Octagon
4	Quadrilateral	9	Nonagon
5	Pentagon	10	Decagon
6	Hexagon	12	Dodecagon
7	Heptagon	n	n-gon

6.4A   Sum of the Measures of the Interior Angles of a Polygon

By drawing diagonals from any vertex to each of the other vertices, as in Fig. 6-51, a polygon of 7 sides is divisible into 5 triangles. Note that each triangle has one side of the polygon, except the first and last triangles which have two such sides.

In general, this process will divide a polygon of n sides into $n - 2$ triangles; that is, the number of such triangles is always two less than the number of sides of the polygon.

The sum of the measures of the interior angles of the polygon equals the sum of the measures of the interior angles of the triangles. Hence:

Sum of measures of interior angles of a polygon of n sides = $(n - 2)180°$

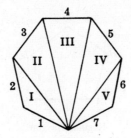

Fig. 6-51

6.4B Sum of the Measures of the Exterior Angles of a Polygon

The exterior angles of a polygon can be reproduced together, so that they have the same vertex. To do this, draw lines parallel to the sides of the polygon from a point, as shown in Fig. 6-52. If this is done, it can be seen that regardless of the number of sides, the sum of the measures of the exterior angle equals 360°. Hence:

Sum of measures of exterior angles of a polygon of n sides = 360°

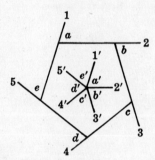

Fig. 6-52

6.4C Polygon-Angle Principles

For any polygon

PRINCIPLE 1: *If S is the sum of the measures of the interior angles of a polygon of n sides, then*

$$S = n - 2 \text{ straight angles} = (n-2)180°$$

The sum of the measures of the interior angles of a polygon of 10 sides (decagon) equals 1440°, since $S = 8(180) = 1440$.

PRINCIPLE 2: *The sum of the measures of the exterior angles of any polygon equals 360°.*

Thus the sum of the measures of the exterior angles of a polygon of 23 sides equals 360°.

For a regular polygon

PRINCIPLE 3: *If a regular polygon of n sides (Fig. 6-53) has an interior angle of measure i and an exterior angle of measure e (in degrees), then*

$$i = \frac{180(n-2)}{n} \qquad e = \frac{360}{n} \qquad \text{and} \qquad i + e = 180$$

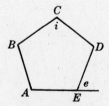

Regular Polygon

Fig. 6-53

Thus for a regular polygon of 20 sides,

$$i = \frac{180(20-2)}{20} = 162 \qquad e = \frac{360}{20} = 18 \qquad i + e = 162 + 18 = 180$$

SOLVED PROBLEMS

6.15 APPLYING ANGLE-MEASURE FORMULAS TO A POLYGON

(a) Find the sum of the measures of the interior angles of a polygon of 9 sides (express your answer in straight angles and in degrees).

(b) Find the number of sides a polygon has if the sum of the measures of the interior angles is 3600°.

(c) Is it possible to have a polygon the sum of whose angle measures is 1890°?

Solutions

(a) S (in straight angles) $= n - 2 = 9 - 2 = 7$ straight angles; $m\angle S = (n-2)180 = 7(180) = 1260°$

(b) S (in degrees) $= (n-2)180$. Then $3600 = (n-2)180$, from which $n = 22$.

(c) Since $1890 = (n-2)180$, then $n = 12\frac{1}{2}$. A polygon cannot have $12\frac{1}{2}$ sides.

6.16 APPLYING ANGLE-MEASURE FORMULAS TO A REGULAR POLYGON

(a) Find each exterior angle measure of a regular polygon having 9 sides.

(b) Find each interior angle measure of a regular polygon having 9 sides.

(c) Find the number of sides a regular polygon has if each exterior angle measure is 5°.

(d) Find the number of sides a regular polygon has if each interior angle measure is 165°.

Solutions

(a) Since $n = 9$, $m\angle e = \dfrac{360}{n} = \dfrac{360}{9} = 40$.   *Ans.* 40°

(b) Since $n = 9$, $m\angle i = \dfrac{(n-2)180}{n} = \dfrac{(9-2)180}{9} = 140$.   *Ans.* 140°

    Another method: Since $i + e = 180$, $i = 180 - e = 180 - 40 = 140$.

(c) Substituting $e = 5$ in $e = \dfrac{360}{n}$, we have $5 = \dfrac{360}{n}$. Then $5n = 360$, so $n = 72$.   *Ans.* 72 sides

(d) Substituting $i = 165$ in $i + e = 180$, we have $165 + e = 180$ or $e = 15$.

    Then, using $e = \dfrac{360}{n}$ with $e = 15$, we have $15 = \dfrac{360}{n}$, or $n = 24$.   *Ans.* 24 sides

6.17 APPLYING ALGEBRA TO THE ANGLE-MEASURE SUMS OF A POLYGON

Find each interior angle measure of a quadrilateral (a) if its interior angles are represented by $x + 10$, $2x + 20$, $3x - 50$, and $2x - 20$; (b) if its exterior angles are in the ratio $2:3:4:6$.

Solutions

(a)   Since the sum of the measures of the interior $\angle$s is 360°, we add

$$(x + 10) + (2x + 20) + (3x - 50) + (2x - 20) = 360$$
$$8x - 40 = 360$$
$$x = 50$$

Then $x + 10 = 60$; $2x + 20 = 120$; $3x - 50 = 100$; $2x - 20 = 80$.   *Ans.* 60°, 120°, 100°, 80°

(b)   Let the exterior angles be represented respectively by $2x$, $3x$, $4x$, and $6x$. Then $2x + 3x + 4x + 6x = 360$. Solving gives us $15x = 360$ and $x = 24$. Hence the exterior angles measure 48°, 72°, 96°, and 144°.

The interior angles are their supplements.   *Ans.* 132°, 108°, 84°, 36°

6.5   TWO NEW CONGRUENCY THEOREMS

Three methods of proving triangles congruent have already been introduced here. These are

1.   s.a.s. $\cong$ s.a.s.

2.   a.s.a. $\cong$ a.s.a.

3.   s.s.s. $\cong$ s.s.s.

Two additional methods of proving that triangles are congruent are

4.   s.a.a. $\cong$ s.a.a.

5.   hy. leg $\cong$ hy. leg

6.5A   Two New Congruency Principles

PRINCIPLE 1:   (s.a.a. $\cong$ s.a.a.) *If two angles and a side opposite one of them of one triangle are congruent to the corresponding parts of another, the triangles are congruent.*

Thus if $\angle A \cong \angle A'$, $\angle B \cong \angle B'$, and $\overline{BC} \cong \overline{B'C'}$ in Fig. 6-54, then $\triangle ABC \cong \triangle A'B'C'$.

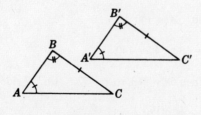

Fig. 6-54

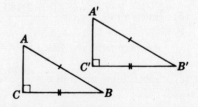

Fig. 6-55

PRINCIPLE 2:   (hy. leg $\cong$ hy. leg) *If the hypotenuse and a leg of one right triangle are congruent to the corresponding parts of another right triangle, the triangles are congruent.*

Thus if hy. $\overline{AB} \cong$ hy. $\overline{A'B'}$ and leg $\overline{BC} \cong$ leg $\overline{B'C'}$ in Fig. 6-55, then rt. $\triangle ABC \cong$ rt. $\triangle A'B'C'$.

A proof of this principle is given in Chapter 16.

SOLVED PROBLEMS

6.18   SELECTING CONGRUENT TRIANGLES USING s.a.a. ≅ s.a.a. OR hy. leg ≅ hy. leg
   In (*a*) Fig. 6-56 and (*b*) Fig. 6-57, select congruent triangles and state the reason for the congruency.

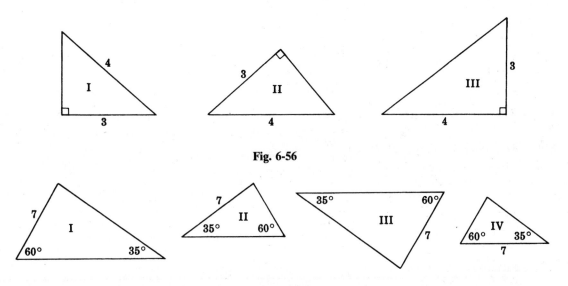

Fig. 6-56

Fig. 6-57

Solutions

(*a*)   △I ≅ △II hy. leg ≅ hy. leg. In △III, 4 is not a hypotenuse.

(*b*)   △I ≅ △III by s.a.a. ≅ s.a.a. In △II, 7 is opposite 60° instead of 35°. In △IV, 7 is included between 60° and 35°.

6.19   DETERMINING THE REASON FOR CONGRUENCY OF TRIANGLES
   In each part of Fig. 6-58, △I can be proved congruent to △II. Make a diagram showing the congruent parts of both triangles and state the reason for the congruency.

Solutions

(*a*)   See Fig. 6-59(*a*). △I ≅ △II by hy. leg ≅ hy. leg.

(*b*)   See Fig. 6-59(*b*). △I ≅ △II by s.a.a. ≅ s.a.a.

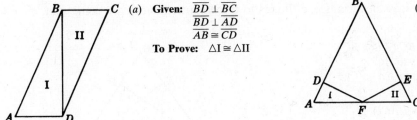

(*a*)   **Given:** $\overline{BD} \perp \overline{BC}$
                            $\overline{BD} \perp \overline{AD}$
                            $\overline{AB} \cong \overline{CD}$
       To Prove: △I ≅ △II

(*b*)   **Given:** $\overline{AB} \cong \overline{BC}$
                            $\overline{FD} \perp \overline{AB}$
                            $\overline{FE} \perp \overline{BC}$
       F is midpoint of $\overline{AC}$.
       To Prove: △I ≅ △II

Fig. 6-58

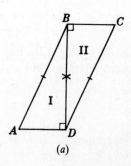

Fig. 6-59

6.20 **PROVING A CONGRUENCY PROBLEM**

Given: Quadrilateral *ABCD*
        $\overline{DF} \perp \overline{AC}$, $\overline{BE} \perp \overline{AC}$
        $\overline{AE} \cong \overline{FC}$, $\overline{BC} \cong \overline{AD}$

To Prove: $\overline{BE} \cong \overline{FD}$

Plan: Prove $\triangle I \cong \triangle II$

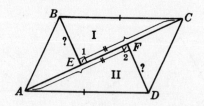

PROOF:

Statements	Reasons
1. $\overline{BC} \cong \overline{AD}$	1. Given
2. $\overline{DF} \perp \overline{AC}$, $\overline{BE} \perp \overline{AC}$	2. Given
3. $\angle 1 \cong \angle 2$	3. Perpendiculars form rt. ∕s, and all rt. ∕s are congruent.
4. $\overline{AE} \cong \overline{FC}$	4. Given
5. $\overline{EF} \cong \overline{EF}$	5. Identity
6. $\overline{AF} \cong \overline{EC}$	6. If equals are added to equals, the sums are equal. Definition of congruent segments.
7. $\triangle I \cong \triangle II$	7. Hy. leg ≅ hy. leg
8. $\overline{BE} \cong \overline{FD}$	8. Corresponding parts of congruent ∕s are congruent.

6.21 **PROVING A CONGRUENCY PROBLEM STATED IN WORDS**

Prove that in an isosceles triangle, altitudes to the congruent sides are congruent.

Given: Isosceles $\triangle ABC$ $(\overline{AB} \cong \overline{BC})$
        $\overline{AD}$ is altitude to $\overline{BC}$
        $\overline{CE}$ is altitude to $\overline{AB}$

To Prove: $\overline{AD} \cong \overline{CE}$

Plan: Prove $\triangle ACE \cong \triangle ACD$
      or $\triangle I \cong \triangle II$

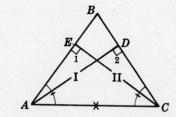

PROOF:

Statements	Reasons
1. $\overline{AB} \cong \overline{BC}$	1. Given
2. $\angle A \cong \angle C$	2. In a $\triangle$, angles opposite equal sides are equal.
3. $\overline{AD}$ is altitude to $\overline{BC}$, $\overline{CE}$ is altitude to $\overline{AB}$.	3. Given
4. $\angle 1 \cong \angle 2$	4. Altitudes form rt. $\angle$s and rt. $\angle$s are congruent.
5. $\overline{AC} \cong \overline{AC}$	5. Identity
6. $\triangle I \cong \triangle II$	6. s.a.a. $\cong$ s.a.a.
7. $\overline{AD} \cong \overline{CE}$	7. Corresponding parts of congruent $\triangle$s are congruent.

Supplementary Problems

1. In each part of Fig. 6-60, find x and y.   (6.1)

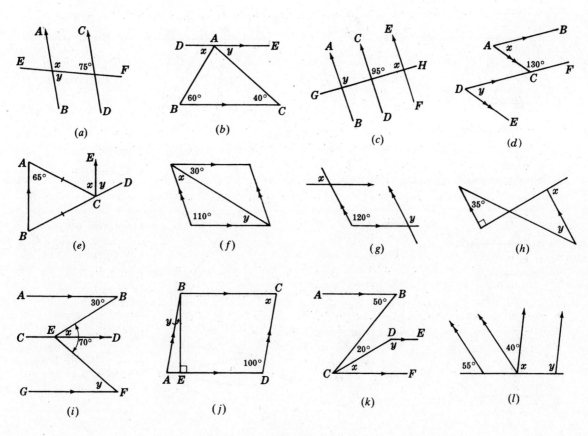

Fig. 6-60

2. In each part of Fig. 6-61, find x and y.   (6.3)

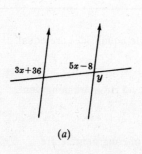

(a)

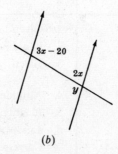

(b)

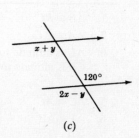

(c)

Fig. 6-61

3. If two parallel lines are cut by a transversal, find   (6.3)

(a) Two alternate interior angles represented by $3x$ and $5x - 70$

(b) Two corresponding angles represented by $2x + 10$ and $4x - 50$

(c) Two interior angles on the same side of the transversal represented by $2x$ and $3x$.

4. Provide the proofs requested in Fig. 6-62.   (6.4)

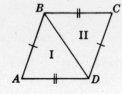

(a) **Given:** Quad. $ABCD$
$\overline{AB} \cong \overline{CD}$
$\overline{BC} \cong \overline{AD}$

To Prove: $\overline{AB} \parallel \overline{CD}$
$\overline{BC} \parallel \overline{AD}$

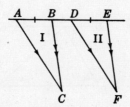

(b) **Given:** $\overline{AB} \cong \overline{DE}$
$\overline{AC} \parallel \overline{DF}$
$\overline{BC} \parallel \overline{EF}$

To Prove: $\overline{AC} \cong \overline{DF}$

Fig. 6-62

5. Provide the proofs requested in Fig. 6-63.   (6.4)

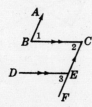

(a) **Given:** $\overline{AB} \parallel \overline{CF}$
$\overline{BC} \parallel \overline{DE}$

To Prove: $\angle 1 \cong \angle 3$

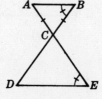

(b) **Given:** $\overline{AC} \cong \overline{BC}$
$\angle B \cong \angle E$

To Prove: $\overline{AB} \parallel \overline{DE}$

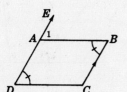

(c) **Given:** Quad. $ABCD$
$\overline{DE} \parallel \overline{BC}$
$\angle B \cong \angle D$

To Prove: $\overline{AB} \parallel \overline{CD}$

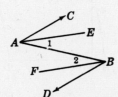

(d) **Given:** $\overline{AC} \parallel \overline{BD}$
$\overline{AE}$ bisects $\angle A$
$\overline{BF}$ bisects $\angle B$

To Prove: $\overline{BF} \parallel \overline{AE}$

Fig. 6-63

6. Prove each of the following: (6.5)

    (*a*) If the opposite sides of a quadrilateral are parallel, then they are also congruent.

    (*b*) If $\overline{AB}$ and $\overline{CD}$ bisect each other at E, then $\overline{AC} \parallel \overline{BD}$.

    (*c*) In quadrilateral $ABCD$, let $\overline{BC} \parallel \overline{AD}$. If the diagonals $\overline{AC}$ and $\overline{BD}$ intersect at E and $\overline{AE} \cong \overline{DE}$, then $\overline{BE} \cong \overline{CE}$.

    (*d*) $\overset{\leftrightarrow}{AB}$ and $\overset{\leftrightarrow}{CD}$ are parallel lines cut by a transversal at E and F. If $\overset{\rightarrow}{EG}$ and $\overset{\rightarrow}{FH}$ bisect a pair of corresponding angles, then $\overset{\rightarrow}{EG} \parallel \overset{\rightarrow}{FH}$.

    (*e*) If a line through vertex B of $\triangle ABC$ is parallel to $\overline{AC}$ and bisects the angle formed by extending $\overline{AB}$ through B, then $\triangle ABC$ is isosceles.

7. In Fig. 6-64, find the distance from (*a*) A to B; (*b*) E to $\overline{AC}$; (*c*) A to $\overline{BC}$; (*d*) $\overline{ED}$ to $\overline{BC}$. (6.6)

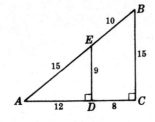

Fig. 6-64

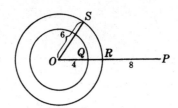

Fig. 6-65

8. In Fig. 6-65, find the distance (*a*) from P to the outer circle; (*b*) from P to the inner circle; (*c*) between the concentric circles; (*d*) from P to O. (6.6)

9. In Fig. 6-66: (6.6)

    (*a*) Locate P, a point on $\overline{AD}$, equidistant from B and C. Then locate Q, a point on $\overline{AD}$, equidistant from $\overline{AB}$ and $\overline{BC}$.

    (*b*) Locate R, a point equidistant from A, B, and C. Then locate S, a point equidistant from B, C, and D.

    (*c*) Locate T, a point equidistant from $\overline{BC}$, $\overline{CD}$, and $\overline{AD}$. Then locate U, a point equidistant from $\overline{AB}$, $\overline{BC}$, and $\overline{CD}$.

10. In each part of Fig. 6-67, describe P, Q, and R as equidistant points and locate them on a bisector.

Fig. 6-66

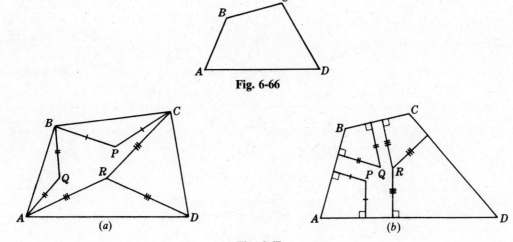

Fig. 6-67

11. In each part of Fig. 6-68, describe P, Q, and R as equidistant points.   (6.9)

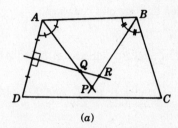

(a)   (b)

Fig. 6-68

12. Find x and y in each part of Fig. 6-69.   (6.11)

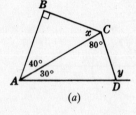   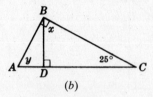   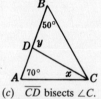

(a)   (b)   (c) $\overline{CD}$ bisects $\angle C$.

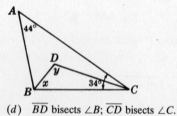   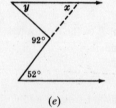   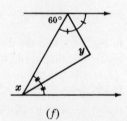

(d) $\overline{BD}$ bisects $\angle B$; $\overline{CD}$ bisects $\angle C$.   (e)   (f)

Fig. 6-69

13. Find x and y in each part of Fig. 6-70.   (6.12)

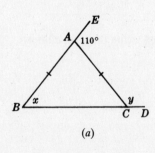   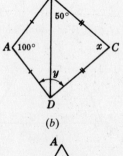   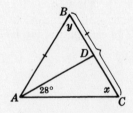

(a)   (b)   (c) $\overline{AD}$ bisects $\angle A$.

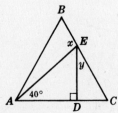   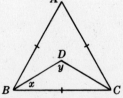   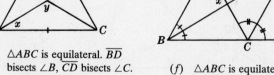

(d) $\triangle ABC$ is equilateral.   (e) $\triangle ABC$ is equilateral. $\overline{BD}$ bisects $\angle B$, $\overline{CD}$ bisects $\angle C$.   (f) $\triangle ABC$ is equilateral.

Fig. 6-70

14. Find the measure of each angle   (6.13)

    (*a*)   Of a triangle if its angle measures are in the ratio 1:3:6

    (*b*)   Of a right triangle if its acute angle measures are in the ratio 4:5

    (*c*)   Of an isosceles triangle if the ratio of the measures of its base angle to a vertex angle is 1:3

    (*d*)   Of a quadrilateral if its angle measures are in the ratio 1:2:3:4

    (*e*)   Of a triangle, one of whose angles measures 55° and whose other two angle measures are in the ratio 2:3

    (*f*)   Of a triangle if the ratio of the measures of its exterior angles is 2:3:4

15. Prove each of the following: (6.14)

    (*a*)   In quadrilateral $ABCD$ if $\angle A \cong \angle D$ and $\angle B \cong \angle C$, then $\overline{BC} \parallel \overline{AD}$.

    (*b*)   Two parallel lines are cut by a transversal. Prove that the bisectors of two interior angles on the same side of the transversal are perpendicular to each other.

16. Show that a triangle is   (6.14)

    (*a*)   Equilateral if its angles are represented by $x + 15$, $3x - 75$, and $2x - 30$

    (*b*)   Isosceles if its angles are represented by $x + 15$, $3x - 35$, and $4x$

    (*c*)   A right triangle if its angle measures are in the ratio 2:3:5

    (*d*)   An obtuse triangle if one angle measures 64° and the larger of the other two measures 10° less than five times the measure of the smaller.

17. (*a*)   Find the sum of the measures of the interior angles (in straight angles) of a polygon of 9 sides; of 32 sides.   (6.15)

    (*b*)   Find the sum of the measures of the interior angles (in degrees) of a polygon of 11 sides; of 32 sides; of 1002 sides.

    (*c*)   Find the number of sides a polygon has if the sum of the measures of the interior angles is 28 straight angles; 20 right angles; 4500°; 36,000°.

18. (*a*)   Find the measure of each exterior angle of a regular polygon having 18 sides; 20 sides; 40 sides.   (6.16)

    (*b*)   Find the measure of each interior angle of a regular polygon having 18 sides; 20 sides; 40 sides.

    (*c*)   Find the number of sides a regular polygon has if each exterior angle measures 120°; 40°; 18°; 2°.

    (*d*)   Find the number of sides a regular polygon has if each interior angle measures 60°, 150°; 170°, 175°; 179°.

19. (*a*)   Find each interior angle of a quadrilateral if its interior angles are represented by $x - 5$, $x + 20$, $2x - 45$, and $2x - 30$.   (6.17)

    (*b*)   Find the measure of each interior angle of a quadrilateral if the measures of its exterior angles are in the ratio of 1:2:3:3.

20. In each part of Fig. 6-71, select congruent triangles and state the reason for the congruency.   (6.18)

21. In each part of Fig. 6-72, two triangles can be proved congruent. Make a diagram showing the congruent parts of both triangles, and state the reason for the congruency.   (6.19)

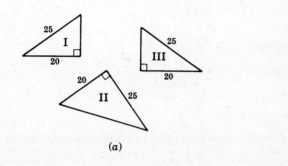

(a)   (b)

Fig. 6-71

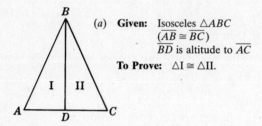

(a) **Given:** Isosceles $\triangle ABC$
       $(\overline{AB} \cong \overline{BC})$
       $\overline{BD}$ is altitude to $\overline{AC}$

To Prove: $\triangle I \cong \triangle II$.

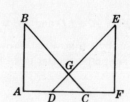

(b) **Given:** $\overline{AB} \cong \overline{EF}$
       $\overline{AB} \perp \overline{AF}, \overline{EF} \perp \overline{AF}$
       $\overline{DG} \cong \overline{GC}$

To Prove: $\triangle ABC \cong \triangle DEF$

Fig. 6-72

22. Provide the proofs requested in Fig. 6-73.   (6.20)

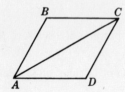

(a) **Given:** $\angle B \cong \angle D$
       $\overline{BC} \parallel \overline{AD}$

To Prove: $\overline{BC} \cong \overline{AD}$

(b) **Given:** $\overline{AE} \cong \overline{EC}$
       $\overline{BD} \perp \overline{BC}$
       $\overline{BD} \perp \overline{AD}$

To Prove: $\overline{BE} \cong \overline{ED}$

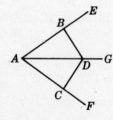

(c) **Given:** $\overline{BD} \cong \overline{DC}$
       $\overline{BD} \perp \overline{AE}$
       $\overline{DC} \perp \overline{AF}$

To Prove: $\overline{AG}$ bisects $\angle A$.

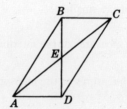

(d) **Given:** $\overline{AC} \cong \overline{BD}$
       $\overline{AB} \perp \overline{AD}$
       $\overline{CD} \perp \overline{AD}$

To Prove: $\overline{AB} \cong \overline{CD}$

Fig. 6-73

23. Prove each of the following: (6.21)

(a) If the perpendiculars to two sides of a triangle from the midpoint of the third side are congruent, then the triangle is isosceles.

(b) Perpendiculars from a point in the bisector of an angle to the sides of the angle are congruent.

(c) If the altitudes to two sides of a triangle are congruent, then the triangle is isosceles.

(d) Two right triangles are congruent if the hypotenuse and an acute angle of one are congruent to the corresponding parts of the other.

CHAPTER 7

Parallelograms, Trapezoids, Medians, and Midpoints

7.1 TRAPEZOIDS

A *trapezoid* is a quadrilateral having two and only two parallel sides. The *bases* of the trapezoid are its parallel sides; the *legs* are its nonparallel sides. The *median* of the trapezoid is the segment joining the midpoints of its legs.

Thus in trapezoid $ABCD$ in Fig. 7-1, the bases are $\overline{AD}$ and $\overline{BC}$, and the legs are $\overline{AB}$ and $\overline{CD}$. If M and N are midpoints, then $\overline{MN}$ is the median of the trapezoid.

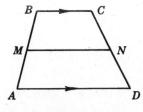

Fig. 7-1

An *isosceles trapezoid* is a trapezoid whose legs are congruent. Thus in isosceles trapezoid $ABCD$ in Fig. 7-2, $\overline{AB} \cong \overline{CD}$.

The *base angles* of a trapezoid are the angles at the ends of its longer base: $\angle A$ and $\angle D$ are the base angles of isosceles trapezoid $ABCD$.

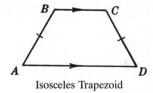

Isosceles Trapezoid

Fig. 7-2

7.1A Trapezoid Principles

PRINCIPLE 1: *The base angles of an isosceles trapezoid are congruent.*

Thus in trapezoid $ABCD$ of Fig. 7.3, if $\overline{AB} \cong \overline{CD}$, then $\angle A \cong \angle D$.

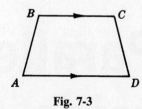

Fig. 7-3

PRINCIPLE 2: *If the base angles of a trapezoid are congruent, the trapezoid is isosceles.*

Thus in Fig. 7-3, if $\angle A \cong \angle D$, then $\overline{AB} \cong \overline{CD}$.

SOLVED PROBLEMS

7.1 APPLYING ALGEBRA TO THE TRAPEZOID

In each trapezoid in Fig. 7-4, find x and y.

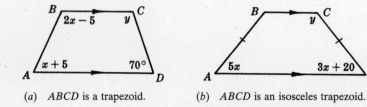

(a) $ABCD$ is a trapezoid.   (b) $ABCD$ is an isosceles trapezoid.

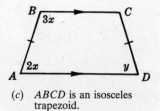

(c) $ABCD$ is an isosceles trapezoid.

Fig. 7-4

Solutions

(a) Since $\overline{AD} \| \overline{BC}$, $(2x - 5) + (x + 5) = 180$; then $3x = 180$ and $x = 60$.
Also, $y + 70 = 180$ or $y = 110$.

(b) Since $\angle A \cong \angle D$, $5x = 3x + 20$, so that $2x = 20$ or $x = 10$.
Since $\overline{BC} \| \overline{AD}$, $y + (3x + 20) = 180$, so $y + 50 = 180$ or $y = 130$.

(c) Since $\overline{BC} \| \overline{AD}$, $3x + 2x = 180$ or $x = 36$.
Since $\angle D \cong \angle A$, $y = 2x$ or $y = 72$.

7.2 PROOF OF A TRAPEZOID PRINCIPLE STATED IN WORDS

Prove that the base angles of an isosceles trapezoid are congruent.

Given: Isosceles trapezoid $ABCD$
$(\overline{BC} \| \overline{AD}, \overline{AB} \cong \overline{CD})$
To Prove: $\angle A \cong \angle D$
Plan: Draw $\perp$s to base from B and C.
Prove $\triangle \text{I} \cong \text{II}$.

PROOF:

Statements	Reasons
1. Draw $\overline{BE} \perp \overline{AD}$ and $\overline{CF} \perp \overline{AD}$.	1. A $\perp$ may be drawn to a line from an outside point.
2. $\overline{BC} \parallel \overline{AD}$, $\overline{AB} \cong \overline{CD}$	2. Given
3. $\overline{BE} \cong \overline{CF}$	3. Parallel lines are everywhere equidistant. Definition of congruent segments.
4. $\angle 1 \cong \angle 2$	4. Perpendiculars form rt. $\angle$. All rt. $\angle$ are congruent.
5. $\triangle I \cong \triangle II$	5. hy. leg $\cong$ hy. leg
6. $\angle A \cong \angle D$	6. Corresponding parts of congruent $\triangle$ are congruent.

7.2 PARALLELOGRAMS

A parallelogram is a quadrilateral whose opposite sides are parallel. The symbol for parallelogram is $\square$. Thus in $\square ABCD$ in Fig. 7-5, $\overline{AB} \parallel \overline{CD}$ in $\overline{AD} \parallel \overline{BC}$.

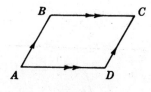

Fig. 7-5

If the opposite sides of a quadrilateral are parallel, then it is a parallelogram. (This is the converse of the above definition.) Thus if $\overline{AB} \parallel \overline{CD}$ and $\overline{AD} \parallel \overline{BC}$, then $ABCD$ is a $\square$.

7.2A Principles Involving Properties of Parallelograms

PRINCIPLE 1: *The opposite sides of a parallelogram are parallel. (This is the definition.)*

PRINCIPLE 2: *A diagonal of a parallelogram divides it into two congruent triangles.*

$\overline{BD}$ is a diagonal of $\square ABCD$ in Fig. 7-6, so $\triangle I \cong \triangle II$.

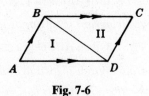

Fig. 7-6

PRINCIPLE 3: *The opposite sides of a parallelogram are congruent.*

Thus in $\square ABCD$ in Fig. 7-5, $\overline{AB} \cong \overline{CD}$ and $\overline{AD} \cong \overline{BC}$.

PRINCIPLE 4: *The opposite angles of a parallelogram are congruent.*

Thus in $\square ABCD$, $\angle A \cong \angle C$ and $\angle B \cong \angle D$.

PRINCIPLE 5:   *The consecutive angles of a parallelogram are supplementary.*

Thus in $\square ABCD$, $\angle A$ is the supplement of both $\angle B$ and $\angle D$.

PRINCIPLE 6:   *The diagonals of a parallelogram bisect each other.*

Thus in $\square ABCD$ in Fig. 7-7, $\overline{AE} \cong \overline{EC}$ and $\overline{BE} \cong \overline{ED}$.

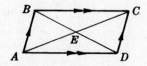

Fig. 7-7

7.2B   Proving a Quadrilateral is a Parallelogram

PRINCIPLE 7:   *A quadrilateral is a parallelogram if its opposite sides are parallel.*

Thus if $\overline{AB} \parallel \overline{CD}$ and $\overline{AD} \parallel \overline{BC}$, then $ABCD$ is a $\square$.

PRINCIPLE 8:   *A quadrilateral is a parallelogram if its opposite sides are congruent.*

Thus if $\overline{AB} \cong \overline{CD}$ and $\overline{AD} \cong \overline{BC}$ in Fig. 7-8, then $ABCD$ is a $\square$.

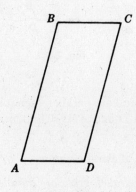

Fig. 7-8

PRINCIPLE 9:   *A quadrilateral is a parallelogram if two sides are congruent and parallel.*

Thus if $\overline{BC} \cong \overline{AD}$ and $\overline{BC} \parallel \overline{AD}$ in Fig. 7-8, then $ABCD$ is a $\square$.

PRINCIPLE 10:   *A quadrilateral is a parallelogram if its opposite angles are congruent.*

Thus if $\angle A \cong \angle C$ and $\angle B \cong \angle D$ in Fig. 7-8, then $ABCD$ is a $\square$.

PRINCIPLE 11:   *A quadrilateral is a parallelogram if its diagonals bisect each other.*

Thus if $\overline{AE} \cong \overline{EC}$ and $\overline{BE} \cong \overline{ED}$ in Fig. 7-9, then $ABCD$ is a $\square$.

Fig. 7-9

SOLVED PROBLEMS

7.3   APPLYING PROPERTIES OF PARALLELOGRAMS

Assuming $ABCD$ is a parallelogram, find x and y in each part of Fig. 7.10.

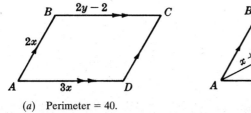

(a)   Perimeter = 40.   (b)   (c)

Fig. 7-10

Solutions

(a)   By Principle 3, $BC = AD = 3x$ and $CD = AB = 2x$; then $2(2x + 3x) = 40$, so that $10x = 40$ or $x = 4$.
By Principle 3, $2y - 2 = 3x$; then $2y - 2 = 3(4)$, so $2y = 14$ or $y = 7$.

(b)   By Principle 6, $x + 2y = 15$ and $x = 3y$.
Substituting $3y$ for x in the first equation yields $3y + 2y = 15$ or $y = 3$. Then $x = 3y = 9$.

(c)   By Principle 4, $3x - 20 = x + 40$, so $2x = 60$ for $x = 30$.
By Principle 5, $y + (x + 40) = 180$. Then $y + (30 + 40) = 180$ or $y = 110$.

7.4   APPLYING PRINCIPLE 7 TO DETERMINE PARALLELOGRAMS

Name the parallelograms in each part of Fig. 7-11.

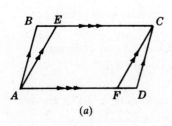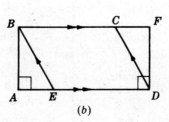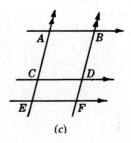

(a)   (b)   (c)

Fig. 7-11

Solutions

(a) $ABCD$, $AECF$; (b) $ABFD$, $BCDE$; (c) $ABDC$, $CDFE$, $ABFE$.

7.5   APPLYING PRINCIPLES 9, 10, AND 11

State why $ABCD$ is a parallelogram in each part of Fig. 7-12.

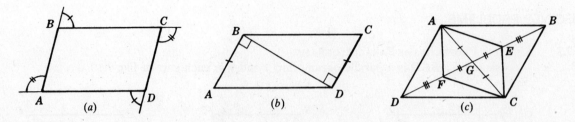

Fig. 7-12

Solutions

(a) Since supplements of congruent angles are congruent, opposite angles of *ABCD* are congruent. Thus by Principle 10, *ABCD* is a parallelogram.

(b) Since perpendiculars to the same line are parallel, $\overline{AB} \| \overline{CD}$. Hence by Principle 9, *ABCD* is a parallelogram.

(c) By the addition axiom, $\overline{DG} \cong \overline{GB}$. Hence by Principle 11, *ABCD* is a parallelogram.

7.6 PROVING A PARALLELOGRAM PROBLEM

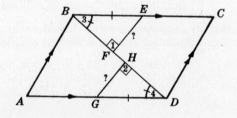

Given: $\square ABCD$
        E is midpoint of $\overline{BC}$.
        G is midpoint of $\overline{AD}$.
        $\overline{EF} \perp \overline{BD}, \overline{GH} \perp \overline{BD}$

To Prove: $\overline{EF} \cong \overline{GH}$

Plan: Prove $\triangle BFE \cong \triangle GHD$

PROOF:

Statements	Reasons
1. *E* is midpoint of $\overline{BC}$. *G* is midpoint of $\overline{AD}$.	1. Given
2. $BE = \frac{1}{2}BC, GC = \frac{1}{2}AD$	2. A midpoint cuts a segment in half.
3. *ABCD* is a $\square$.	3. Given
4. $\overline{BC} \cong \overline{AD}$	4. Opposite sides of a $\square$ are congruent.
5. $\overline{BE} \cong \overline{GD}$	5. Halves of equals are equal.
6. $\overline{EF} \perp \overline{BD}, \overline{GH} \perp \overline{BD}$	6. Given
7. $\angle 1 \cong \angle 2$	7. Perpendiculars form rt. $\angle$s. Rt. $\angle$s are $\cong$.
8. $\overline{BC} \| \overline{AD}$	8. Opposite sides of a $\square$ are $\|$.
9. $\angle 3 \cong \angle 4$	9. Alternate interior $\angle$s of $\|$ lines are $\cong$.
10. $\triangle BFE \cong \triangle GHD$	10. s.a.a. $\cong$ s.a.a.
11. $\overline{EF} \cong \overline{GH}$	11. Corresponding parts of congruent $\triangle$ are $\cong$.

7.3 SPECIAL PARALLELOGRAMS: RECTANGLE, RHOMBUS, SQUARE

7.3A Definitions and Relationships among the Special Parallelograms

Rectangles, rhombuses, and squares belong to the set of parallelograms. Each of these may be defined as a parallelogram, as follows:

1. A *rectangle* is an equiangular parallelogram.

2. A *rhombus* is an equilateral parallelogram.

3. A *square* is an equilateral and equiangular parallelogram. Thus a square is both a rectangle and a rhombus.

The relations among the special parallelograms can be pictured by using a circle to represent each set. Note the following in Fig. 7-13.

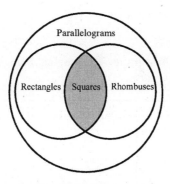

Fig. 7-13

1. Since every rectangle and every rhombus must be a parallelogram, the circle for the set of rectangles and the circle for the set of rhombuses must be inside the circle for the set of parallelograms.

2. Since every square is both a rectangle and a rhombus, the overlapping shaded section must represent the set of squares.

7.3B   Principles Involving Properties of the Special Parallelograms

PRINCIPLE 1:   *A rectangle, rhombus, or square has all the properties of a parallelogram.*

PRINCIPLE 2:   *Each angle of a rectangle is a right angle.*

PRINCIPLE 3:   *The diagonals of a rectangle are congruent.*

Thus in rectangle $ABCD$ in Fig. 7-14, $\overline{AC} \cong \overline{BD}$.

PRINCIPLE 4:   *All sides of a rhombus are congruent.*

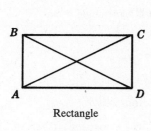

Rectangle

Fig. 7-14

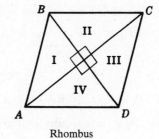

Rhombus

Fig. 7-15

PRINCIPLE 5: *The diagonals of a rhombus are perpendicular bisectors of each other.*

Thus in rhombus *ABCD* in Fig. 7-15, $\overline{AC}$ and $\overline{BD}$ are ⊥ bisectors of each other.

PRINCIPLE 6: *The diagonals of a rhombus bisect the vertex angles.*

Thus in rhombus *ABCD*, $\overline{AC}$ bisects $\angle A$ and $\angle C$.

PRINCIPLE 7: *The diagonals of a rhombus form four congruent triangles.*

Thus in rhombus *ABCD*, $\triangle \text{I} \cong \triangle \text{II} \cong \triangle \text{III} \cong \triangle \text{IV}$.

PRINCIPLE 8: *A square has all the properties of both the rhombus and the rectangle.*

By definition, a square is both a rectangle and a rhombus.

7.3C Diagonal Properties of Parallelograms, Rectangles, Rhombuses, and Squares

Each check in the following table indicates a diagonal property of the figure.

Diagonal Properties	Parallelogram	Rectangle	Rhombus	Square
Diagonals bisect each other.	✓	✓	✓	✓
Diagonals are congruent.		✓		✓
Diagonals are perpendicular.			✓	✓
Diagonals bisect vertex angles.			✓	✓
Diagonals form 2 pairs of congruent triangles.	✓	✓	✓	✓
Diagonals form 4 congruent triangles			✓	✓

7.3D Proving that a Parallelogram is a Rectangle, Rhombus, or a Square

Proving that a Parallelogram is a Rectangle

The basic or minimum definition of a rectangle is this: *A rectangle is a parallelogram having one right angle.* Since the consecutive angles of a parallelogram are supplementary, if one angle is a right angle, the remaining angles must be right angles.

The converse of this basic definition provides a useful method of proving that a parallelogram is a rectangle, as follows:

PRINCIPLE 9: *If a parallelogram has one right angle, then it is a rectangle.*

Thus if *ABCD* in Fig. 7-16 is a ▱ and $m\angle A = 90°$, then *ABCD* is a rectangle.

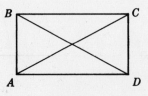

Fig. 7-16

PRINCIPLE 10:   *If a parallelogram has congruent diagonals, then it is a rectangle.*

Thus if $ABCD$ is a $\square$ and $\overline{AC} \cong \overline{BD}$, then $ABCD$ is a rectangle.

Proving that a Parallelogram is a Rhombus

The basic or minimum definition of a rhombus is this: *A rhombus is a parallelogram having two congruent adjacent sides.*

The converse of this basic definition provides a useful method of proving that a parallelogram is a rhombus, as follows:

PRINCIPLE 11:   *If a parallelogram has congruent adjacent sides, then it is a rhombus.*

Thus if $ABCD$ in Fig. 7-17 is a $\square$ and $\overline{AB} \cong \overline{BC}$, then $ABCD$ is a rhombus.

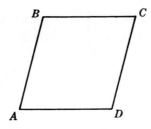

Fig. 7-17

Proving that a Parallelogram is a Square

PRINCIPLE 12:   *If a parallelogram has a right angle and two congruent adjacent sides, then it is a square.*

This follows from the fact that a square is both a rectangle and a rhombus.

SOLVED PROBLEMS

7.7   **APPLYING ALGEBRA TO THE RHOMBUS**

Assuming $ABCD$ is a rhombus, find x and y in each part of Fig. 7-18.

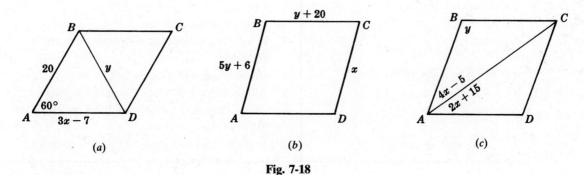

Fig. 7-18

Solutions

(a)   Since $\overline{AB} \cong \overline{AD}$, $3x - 7 = 20$ or $x = 9$. Since $\triangle ABD$ is equiangular it is equilateral, and so $y = 20$.

(b)   Since $\overline{BC} \cong \overline{AB}$, $5y + 6 = y + 20$ or $y = 3\frac{1}{2}$. Since $\overline{CD} \cong \overline{BC}$, $x = y + 20$ or $x = 23\frac{1}{2}$.

(c)   Since $\overline{AC}$ bisects $\angle A$, $4x - 5 = 2x + 15$ or $x = 10$. Hence, $2x + 15 = 35$ and $m\angle A = 2(35°) = 70°$. Since $\angle B$ and $\angle A$ are supplementary, $y + 70 = 180$ or $y = 110$.

7.8   PROVING A SPECIAL PARALLELOGRAM PROBLEM

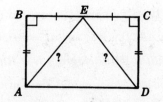

Given: Rectangle $ABCD$
E is midpoint of $\overline{BC}$.

To Prove: $\overline{AE} \cong \overline{ED}$

Plan: Prove $\triangle AEB \cong \triangle CED$.

PROOF:

Statements	Reasons
1. $ABCD$ is a rectangle.	1. Given
2. E is midpoint of $\overline{BC}$.	2. Given
3. $\overline{BE} \cong \overline{EC}$	3. A midpoint divides a line into two congruent parts.
4. $\angle B \cong \angle C$	4. A rectangle is equiangular.
5. $\overline{AB} \cong \overline{CD}$	5. Opposite sides of a $\square$ are congruent.
6. $\triangle AEB \cong \triangle CED$	6. s.a.s. $\cong$ s.a.s.
7. $\overline{AE} \cong \overline{ED}$	7. Corresponding parts of congruent $\triangle$ are congruent.

7.9   PROVING A SPECIAL PARALLELOGRAM PROBLEM STATED IN WORDS
Prove that a diagonal of a rhombus bisects each vertex angle through which it passes.

Solution

Given: Rhombus $ABCD$
$\overline{AC}$ is a diagonal.

To Prove: $\overline{AC}$ bisects $\angle A$ and $\angle C$.

Plan: Prove (1) $\angle 1$ and $\angle 2$ are congruent
to $\angle 3$.
(2) $\angle 3$ and $\angle 4$ are congruent
to $\angle 1$.

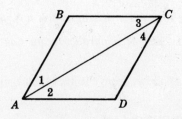

PROOF:

Statements	Reasons
1. $ABCD$ is a rhombus.	1. Given
2. $\overline{AB} \cong \overline{BC}$	2. A rhombus is equilateral.
3. $\angle 1 \cong \angle 3$	3. In a $\triangle$, angles opposite congruent sides are congruent.
4. $\overline{BC} \| \overline{AD}, \overline{AB} \| \overline{CD}$	4. Opposite sides of a $\square$ are $\|$.
5. $\angle 2 \cong \angle 3, \angle 1 \cong \angle 4$	5. Alternate interior $\triangle$ of $\|$ lines are congruent.
6. $\angle 1 \cong \angle 2, \angle 3 \cong \angle 4$	6. Things congruent to the same thing are congruent to each other.
7. $\overline{AC}$ bisects $\angle A$ and $\angle C$.	7. To divide into two congruent parts is to bisect.

7.4   THREE OR MORE PARALLELS; MEDIANS AND MIDPOINTS

7.4A   Three or More Parallels

PRINCIPLE 1:   *If three or more parallels cut off congruent segments on one transversal, then they cut off congruent segments on any other transversal.*

Thus if $l_1 \| l_2 \| l_3$ in Fig. 7-19, and segments a and b of transversal $\overset{\leftrightarrow}{AB}$ are congruent, then segments c and d of transversal $\overset{\leftrightarrow}{CD}$ are congruent.

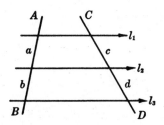

Fig. 7-19

7.4B   Midpoint and Median Principles of Triangles and Trapezoids

PRINCIPLE 2:   *If a line is drawn from the midpoint of one side of a triangle and parallel to a second side, then it passes through the midpoint of the third side.*

Thus in $\triangle ABC$ in Fig. 7-20, if M is the midpoint of $\overline{AB}$ and $\overline{MN} \| \overline{AC}$, then N is the midpoint of $\overline{BC}$.

PRINCIPLE 3:   *If a line joins the midpoints of two sides of a triangle, then it is parallel to the third side and its length is one-half the length of the third side.*

Thus in $\triangle ABC$, if M and N are the midpoints of $\overline{AB}$ and $\overline{BC}$, then $\overline{MN} \| \overline{AC}$ and $MN = \frac{1}{2}AC$.

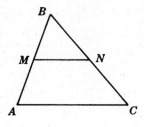

Fig. 7-20

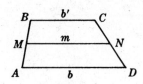

Fig. 7-21

PRINCIPLE 4:   *The median of a trapezoid is parallel to its bases, and its length is equal to one half of the sum of their lengths.*

Thus if m is the median of trapezoid $ABCD$ in Fig. 7-21, then $m \| \overline{AD}$, $m \| \overline{BC}$, and $m = \frac{1}{2}(b + b')$.

PRINCIPLE 5:   *The length of the median to the hypotenuse of a right triangle equals one-half the length of the hypotenuse.*

Thus in rt. $\triangle ABC$ in Fig. 7-22, if $\overline{CM}$ is the median to hypotenuse $\overline{AB}$, then $CM = \frac{1}{2}AB$; that is, $\overline{CM} \cong \overline{AM} \cong \overline{MB}$.

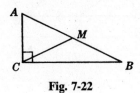

Fig. 7-22

PRINCIPLE 6:   *The medians of a triangle meet in a point which is two-thirds of the distance from any vertex to the midpoint of the opposite side.*

Thus if $\overline{AN}$, $\overline{BP}$, and $\overline{CM}$ are medians of $\triangle ABC$ in Fig. 7-23, then they meet in a point G which is two-thirds of the distance from A to N, B to P, and C to M.

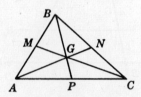

Fig. 7-23

SOLVED PROBLEMS

7.10   APPLYING PRINCIPLE 1 TO THREE OR MORE PARALLELS
Find x and y in each part of Fig. 7-24.

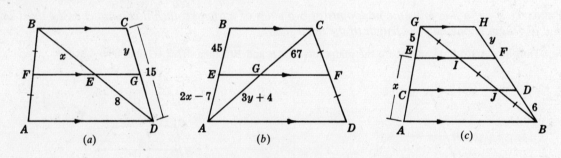

Fig. 7-24

Solutions

(a)   Since $BE = ED$ and $GC = \frac{1}{2}CD$, $x = 8$ and $y = 7\frac{1}{2}$.

(b)   Since $BE = EA$ and $CG = AG$, $2x - 7 = 45$ and $3y + 4 = 67$. Hence $x = 26$ and $y = 21$.

(c)   Since $AC = CE = EG$ and $HF = FD = DB$, $x = 10$ and $y = 6$.

7.11   APPLYING PRINCIPLES 2 AND 3
Find x and y in each part of Fig. 7-25.

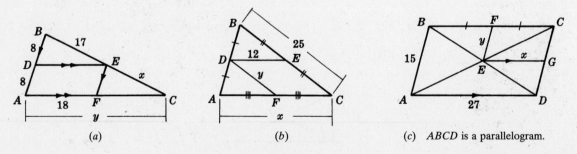

(c)   *ABCD* is a parallelogram.

Fig. 7-25

Solutions

(a) By Principle 2, E is the midpoint of $\overline{BC}$ and F is the midpoint of $\overline{AC}$. Hence $x = 17$ and $y = 36$.

(b) By Principle 3, $DE = \frac{1}{2}AC$ and $DF = \frac{1}{2}BC$. Hence $x = 24$ and $y = 12\frac{1}{2}$.

(c) Since $ABCD$ is a parallelogram, E is the midpoint of $\overline{AC}$. Then by Principle 2, G is the midpoint of $\overline{CD}$.
   By Principle 3, $x = \frac{1}{2}(27) = 13\frac{1}{2}$ and $y = \frac{1}{2}(15) = 7\frac{1}{2}$.

7.12   APPLYING PRINCIPLE 4 TO THE MEDIAN OF A TRAPEZOID
   If $\overline{MP}$ is the median of trapezoid $ABCD$ in Fig. 7-26,

(a) Find m if $b = 20$ and $b' = 28$.

(b) Find b' if $b = 30$ and $m = 26$.

(c) Find b if $b' = 35$ and $m = 40$.

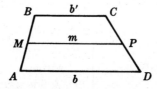

Fig. 7-26

Solutions

   In each case, we apply the formula $m = \frac{1}{2}(b + b')$. The results are:

(a) $m = \frac{1}{2}(20 + 28)$ or $m = 24$

(b) $26 = \frac{1}{2}(30 + b')$ or $b' = 22$

(c) $40 = \frac{1}{2}(b + 35)$ or $b = 45$

7.13   APPLYING PRINCIPLES 5 AND 6 TO THE MEDIANS OF A TRIANGLE
   Find x and y in each part of Fig. 7-27.

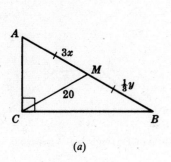

(a)

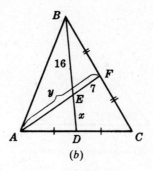

(b)

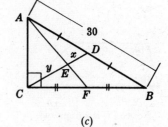

(c)

Fig. 7-27

Solutions

(a) Since $AM = MB$, $\overline{CM}$ is the median to hypotenuse $\overline{AB}$. Hence by Principle 5, $3x = 20$ and $\frac{1}{3}y = 20$. Thus $x = 6\frac{2}{3}$ and $y = 60$.

(b) $\overline{BD}$ and $\overline{AF}$ are medians of $\triangle ABC$. Hence by Principle 6, $x = \frac{1}{2}(16) = 8$ and $y = 3(7) = 21$.

(c) $\overline{CD}$ is the median to hypotenuse $\overline{AB}$; hence by Principle 5, $CD = 15$.
$\overline{CD}$ and $\overline{AF}$ are medians of $\triangle ABC$; hence by Principle 6, $x = \frac{1}{3}(15) = 5$ and $y = \frac{2}{3}(15) = 10$.

7.14   PROVING A MIDPOINT PROBLEM

Given: Quadrilateral $ABCD$
E, F, G, and H are midpoints of $\overline{AD}$, $\overline{AB}$, $\overline{BC}$, and $\overline{CD}$, respectively.

To Prove: $EFGH$ is a $\square$.

Plan: Prove $\overline{EF}$ and $\overline{GH}$ are congruent and parallel.

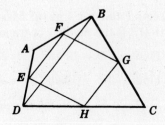

PROOF:

Statements	Reasons
1. Draw $\overline{BD}$.	1. A segment may be drawn between any two points.
2. E, F, G, and H are midpoints.	2. Given
3. $\overline{EF} \parallel \overline{BD}$ and $\overline{GH} \parallel \overline{BD}$ $EF = \frac{1}{2}BD$, $GH = \frac{1}{2}BD$	3. A line segment joining the midpoints of two sides of a $\triangle$ is parallel to the third side and equal in length to half the third side.
4. $\overline{EF} \parallel \overline{GH}$	4. Two lines parallel to a third line are parallel to each other.
5. $\overline{EF} \cong \overline{GH}$	5. Segments of the same length are congruent.
6. $EFGH$ is a $\square$.	6. If two sides of a quadrilateral are $\cong$ and $\parallel$, the quadrilateral is a $\square$.

Supplementary Problems

1. Find x and y in each part of Fig. 7-28.   (7.1)

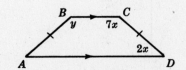

(a) $ABCD$ is a trapezoid.   (b) $ABCD$ is an isosceles trapezoid.   (b) $ABCD$ is an isosceles trapezoid.

Fig. 7-28

2. Prove that if the base angles of a trapezoid are congruent, the trapezoid is isosceles.   (7.2)

3. Prove that (*a*) the diagonals of an isosceles trapezoid are congruent; (*b*) if the nonparallel sides $\overline{AB}$ and $\overline{CD}$ of an isosceles trapezoid are extended until they meet at *E*, triangle *ADE* thus formed is isosceles.   (7.2)

4. Name the parallelograms in each part of Fig. 7-29.   (7.4)

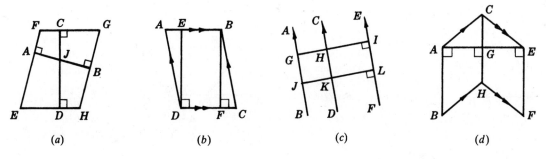

(*a*)   (*b*)   (*c*)   (*d*)

Fig. 7-29

5. State why *ABCD* in each part of Fig. 7-30 is a parallelogram.   (7.5)

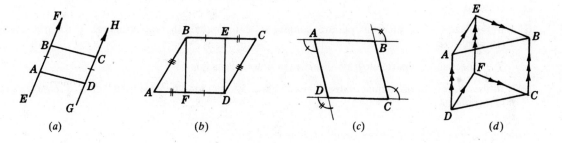

(*a*)   (*b*)   (*c*)   (*d*)

Fig. 7-30

6. Assuming *ABCD* in Fig. 7-31 is a parallelogram, find *x* and *y* if: (7.3)

(*a*) *AD* = 5*x*, *AB* = 2*x*, *CD* = *y*, perimeter = 84

(*b*) *AB* = 2*x*, *BC* = 3*y* + 8, *CD* = 7*x* − 25, *AD* = 5*y* − 10

(*c*) *m*∠*A* = 4*y* − 60, *m*∠*C* = 2*y*, *m*∠*D* = *x*

(*d*) *m*∠*A* = 3*x*, *m*∠*B* = 10*x* − 15, *m*∠*C* = *y*

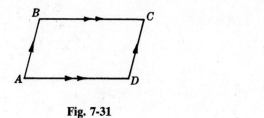

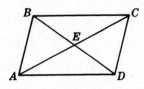

Fig. 7-31   **Fig. 7-32**

7. Assuming $ABCD$ in Fig. 7-32 is a parallelogram, find x and y if: (7.3)

(a) $AE = x + y$, $EC = 20$, $BE = x - y$, $ED = 8$

(b) $AE = x$, $EC = 4y$, $BE = x - 2y$, $ED = 9$

(c) $AE = 3x - 4$, $EC = x + 12$, $BE = 2y - 7$, $ED = x - y$

(d) $AE = 2x + y$, $AC = 30$, $BE = x + y$, $BD = 24$

8. Provide the proofs requested in Fig. 7-33 (7.6)

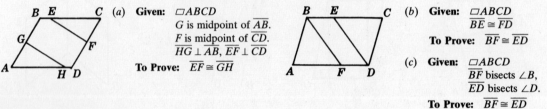

(a) **Given:** $\square ABCD$
G is midpoint of $\overline{AB}$.
F is midpoint of $\overline{CD}$.
$\overline{HG} \perp \overline{AB}$, $\overline{EF} \perp \overline{CD}$

To Prove: $\overline{EF} \cong \overline{GH}$

(b) **Given:** $\square ABCD$
$\overline{BE} \cong \overline{FD}$

To Prove: $\overline{BF} \cong \overline{ED}$

(c) **Given:** $\square ABCD$
$\overline{BF}$ bisects $\angle B$,
$\overline{ED}$ bisects $\angle D$.

To Prove: $\overline{BF} \cong \overline{ED}$

Fig. 7-33

9. Prove each of the following:

(a) The opposite sides of a parallelogram are congruent. (Principle 3)

(b) If the opposite sides of a quadrilateral are congruent, then the quadrilateral is a parallelogram. (Principle 8)

(c) If two sides of a quadrilateral are congruent and parallel, the quadrilateral is a parallelogram. (Principle 9)

(d) The diagonals of a parallelogram bisect each other. (Principle 6)

(e) If the diagonals of a quadrilateral bisect each other, then the quadrilateral is a parallelogram. (Principle 11)

10. Assuming $ABCD$ in Fig. 7-34 is a rhombus, find x and y if: (7.7)

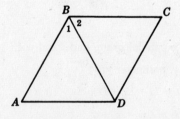

Fig. 7-34

(a) $BC = 35$, $CD = 8x - 5$, $BD = 5y$, $m\angle C = 60°$

(b) $AB = 43$, $AD = 4x + 3$, $BD = y + 8$, $m\angle B = 120°$

(c) $AB = 7x$, $AD = 3x + 10$, $BC = y$

(d) $AB = x + y$, $D = 2x - y$, $BC = 12$

(e) $m\angle B = 130°$, $m\angle 1 = 3x - 10$, $m\angle A = 2y$

(f) $m\angle 1 = 8x - 29$, $m\angle 2 = 5x + 4$, $m\angle D = y$

11. Provide the proofs requested in Fig. 7-35.   (7.8)

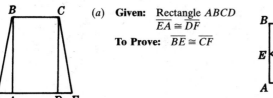

(a) **Given:** Rectangle $ABCD$
$\overline{EA} \cong \overline{DF}$
   To Prove: $\overline{BE} \cong \overline{CF}$

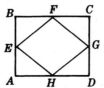

(b) **Given:** Rectangle $ABCD$
E, F, G, and H are
the midpoints of the
sides of the rectangle.
   To Prove: $EFGH$ is a rhombus.

Fig. 7-35

12. Prove each of the following: (7.9)

    (a) If the diagonals of a parallelogram are congruent, the parallelogram is a rectangle.

    (b) If the diagonals of a parallelogram are perpendicular to each other, the parallelogram is a rhombus.

    (c) If a diagonal of a parallelogram bisects a vertex angle, then the parallelogram is a rhombus.

    (d) The diagonals of a rhombus divide it into four congruent triangles.

    (e) The diagonals of a rectangle are congruent.

13. Find x and y in each part of Fig. 7-36   (7.10)

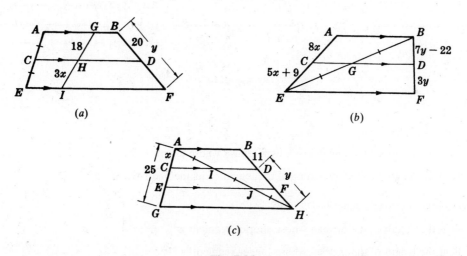

Fig. 7-36

14. Find x and y in each part of Fig. 7-37.   (7.11)

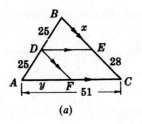

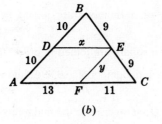

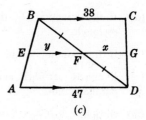

Fig. 7-37

15. If $\overline{MP}$ is the median of trapezoid $ABCD$ in Fig. 7-38,   (7.12)

    (a) Find m if $b = 23$ and $b' = 15$.

    (b) Find b' if $b = 46$ and $m = 41$.

    (c) Find b if $b' = 51$ and $m = 62$.

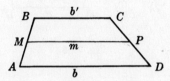

Fig. 7.38

16. Find x and y in each part of Fig. 7-39.   (7.11, 7.12)

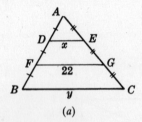 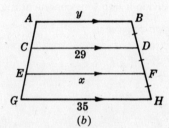 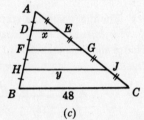

        (a)   (b)   (c)

Fig. 7-39

17. In a right triangle,   (7.13)

    (a) Find the length of the median to a hypotenuse whose length is 45.

    (b) Find the length of the hypotenuse if the length of its median is 35.

18. If the medians of $\triangle ABC$ meet in D,   (7.13)

    (a) Find the length of the median whose shorter segment is 7.

    (b) Find the length of the median whose longer segment is 20.

    (c) Find the length of the shorter segment of the median of length 42.

    (d) Find the length of the longer segment of the median of length 39.

19. Prove each of the following: (7.14)

    (a) If the midpoints of the sides of a rhombus are joined in order, the quadrilateral formed is a rectangle.

    (b) If the midpoints of the sides of a square are joined in order, the quadrilateral formed is a square.

    (c) In $\triangle ABC$, let M, P, and Q be the midpoints of $\overline{AB}$, $\overline{BC}$, and $\overline{AC}$, respectively. Prove that $QMPC$ is a parallelogram.

    (d) In right $\triangle ABC$, $m\angle C = 90°$. If Q, M, and P are the midpoints of $\overline{AC}$, $\overline{AB}$, and $\overline{BC}$, respectively, prove that $QMPC$ is a rectangle.

CHAPTER 8

Circles

8.1 THE CIRCLE; CIRCLE RELATIONSHIPS

The following terms are associated with the circle. Although some have been defined previously, they are repeated here for ready reference.

A *circle* is the set of all points in a plane that are at the same distance from a fixed point called the center. The symbol for circle is ⊙; for circles Ⓢ.

The *circumference* of a circle is the distance around the circle. It contains 360°.

A *radius* of a circle is a line segment joining the center to a point on the circle.

Note: Since all radii of a given circle have the same length, we may at times use the word *radius* to mean the number that is "the length of the radius."

A *central angle* is an angle formed by two radii.

An *arc* is a continuous part of a circle. The symbol for arc is ⌢. A *semicircle* is an arc measuring one-half the circumference of a circle.

A *minor arc* is an arc that is less than a semicircle. A *major arc* is an arc that is greater than a semicircle.

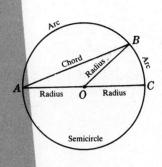

Fig. 8-1

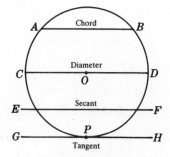

Fig. 8-2

Thus in Fig. 8-1, $\overset{\frown}{BC}$ is a minor arc and $\overset{\frown}{BAC}$ is a major arc. Three letters are needed to indicate a major arc.

To intercept an arc is to cut off the arc.

Thus in Fig. 8-1, ∠BAC and ∠BOC intercept $\overset{\frown}{BC}$.

A *chord* of a circle is a line segment joining two points of the circumference.

Thus in Fig. 8-2, $\overline{AB}$ is a chord.

161

A *diameter* of a circle is a chord through the center. A *secant* of a circle is a line that intersects the circle at two points. A *tangent* of a circle is a line that touches the circle at one and only one point no matter how far produced.

Thus $\overline{CD}$ is a diameter of circle O in Fig. 8-2, $\overleftrightarrow{EF}$ is a secant, and $\overleftrightarrow{GH}$ is a tangent to the circle at P. P is the point of contact or the point of tangency.

An *inscribed polygon* is a polygon all of whose sides are chords of a circle. A *circumscribed circle* is a circle passing through each vertex of a polygon.

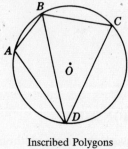

Inscribed Polygons
Circumscribed Circle

Fig. 8-3

Thus $\triangle ABD$, $\triangle BCD$ and quadrilateral $ABCD$ are inscribed polygons of circle O in Fig. 8-3. Circle O is a circumscribed circle of quadrilateral $ABCD$.

A *circumscribed polygon* is a polygon all of whose sides are tangents to a circle. An *inscribed circle* is a circle to which all the sides of a polygon are tangents.

Thus $\triangle ABC$ is a circumscribed polygon of circle O in Fig. 8-4. Circle O is an inscribed circle of $\triangle ABC$.

Concentric circles are circles that have the same center.

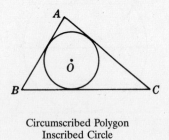

Circumscribed Polygon
Inscribed Circle

Fig. 8-4

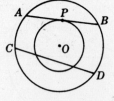

Concentric Circles

Fig. 8-5

Thus the two circles shown in Fig. 8-5 are concentric circles. $\overline{AB}$ is a tangent of the inner circle and a chord of the outer one. $\overline{CD}$ is a secant of the inner circle and a chord of the outer one.

Two circles are *equal* if their radii are equal in length; two circles are *congruent* if their radii are congruent.

Two arcs are congruent if they have equal degree measure and length. We use the notation $m\widehat{AC}$ to denote "measure of arc AC."

8.1A   Circle Principles

PRINCIPLE 1:   *A diameter divides a circle into two equal parts.*

Thus diameter $\overline{AB}$ divides circle O of Fig. 8-6 into two equal semicircles, $\overset{\frown}{ACB}$ and $\overset{\frown}{ADB}$.

PRINCIPLE 2:   *If a chord divides a circle into two equal parts, then it is a diameter.* (This is the converse of Principle 1.)

Thus if $\overset{\frown}{ACB} = \overset{\frown}{ADB}$ in Fig. 8-6, then $\overline{AB}$ is a diameter.

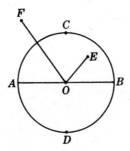

Fig. 8-6

PRINCIPLE 3:   *A point is outside, on, or inside a circle according to whether its distance from the center is greater than, equal to, or smaller than the radius.*

F is outside circle O in Fig. 8-6, since $\overline{FO}$ is greater in length than a radius. E is inside circle O since $\overline{EO}$ is smaller in length than a radius. A is on circle O since $\overline{AO}$ is a radius.

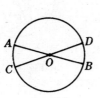

Fig. 8-7

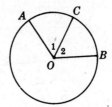

Fig. 8-8

PRINCIPLE 4:   *Radii of the same or congruent circles are congruent.*

Thus in circle O of Fig. 8-7, $\overline{OA} \cong \overline{OC}$.

PRINCIPLE 5:   *Diameters of the same or congruent circles are congruent.*

Thus in circle O of Fig. 8-7, $\overline{AB} \cong \overline{CD}$.

PRINCIPLE 6:   *In the same or congruent circles, congruent central angles have congruent arcs.*

Thus in circle O of Fig. 8-8, if $\angle 1 \cong \angle 2$, then $\overset{\frown}{AC} \cong \overset{\frown}{CB}$.

PRINCIPLE 7:   *In the same or congruent circles, congruent arcs have congruent central angles.*

Thus in circle O of Fig. 8-8, if $\overset{\frown}{AC} \cong \overset{\frown}{CB}$, then $\angle 1 \cong \angle 2$.

(Principles 6 and 7 are converses of each other.)

PRINCIPLE 8:   *In the same or congruent circles, congruent chords have congruent arcs.*

Thus in circle O of Fig. 8-9, if $\overline{AB} \cong \overline{AC}$, then $\overparen{AB} \cong \overparen{AC}$.

PRINCIPLE 9:   *In the same or congruent circles, congruent arcs have congruent chords.*

Thus in circle O of Fig. 8-9, if $\overparen{AB} \cong \overparen{AC}$, then $\overline{AB} \cong \overline{AC}$.

(Principles 8 and 9 are converses of each other.)

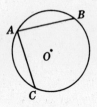

Fig. 8-9

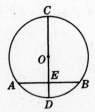

Fig. 8-10

PRINCIPLE 10:   *A diameter perpendicular to a chord bisects the chord and its arcs.*

Thus in circle O of Fig. 8-10, if $\overline{CD} \perp \overline{AB}$, then $\overline{CD}$ bisects $\overline{AB}$, $\overparen{AB}$, and $\overparen{ACB}$.

A proof of this principle is given in Chapter 16.

PRINCIPLE 11:   *A perpendicular bisector of a chord passes through the center of the circle.*

Thus in circle O of Fig. 8-11, if $\overline{PD}$ is the perpendicular bisector of $\overline{AB}$, then $\overline{PD}$ passes through center O.

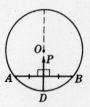

Fig. 8-11

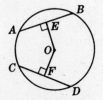

Fig. 8-12

PRINCIPLE 12:   *In the same or congruent circles, congruent chords are equally distant from the center.*

Thus in circle O of Fig. 8-12, if $\overline{AB} \cong \overline{CD}$, if $\overline{OE} \perp \overline{AB}$, and if $\overline{OF} \perp \overline{CD}$, then $\overline{OE} \cong \overline{OF}$.

PRINCIPLE 13:   *In the same or congruent circles, chords that are equally distant from the center are congruent.*

Thus in circle O of Fig. 8-12, if $\overline{OE} \cong \overline{OF}$, $\overline{OE} \perp \overline{AB}$, and $\overline{OF} \perp \overline{CD}$, then $\overline{AB} \cong \overline{CD}$.

(Principles 12 and 13 are converses of each other.)

SOLVED PROBLEMS

8.1   MATCHING TEST OF CIRCLE VOCABULARY
Match each part of Fig. 8-13 on the left with one of the names on the right:

(a) $\overline{OE}$ 1.   Radius

(b) $\overline{FG}$ 2.   Central angle

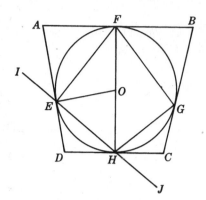

Fig. 8-13

(c) $\overline{FH}$		3.	Semicircle
(d) $\overline{CD}$		4.	Minor arc
(e) $\overline{IJ}$		5.	Major arc
(f) $\widehat{EF}$		6.	Chord
(g) $\widehat{FGH}$		7.	Diameter
(h) $\widehat{FEG}$		8.	Secant
(i) $\angle EOF$		9.	Tangent
(j) Circle O about $EFGH$		10.	Inscribed polygon
(k) Circle O in $ABCD$		11.	Circumscribed polygon
(l) Quadrilateral $EFGH$		12.	Inscribed circle
(m) Quadrilateral $ABCD$		13.	Circumscribed circle

Solutions

(a) 1	(c) 7	(e) 8	(g) 3	(i) 2	(k) 12	(m) 11
(b) 6	(d) 9	(f) 4	(h) 5	(j) 13	(l) 10	

8.2   APPLYING PRINCIPLES 4 AND 5

In Fig. 8-14, (a) what kind of triangle is OCD; (b) what kind of quadrilateral is $ABCD$? (c) In Fig. 8-15 if circle O = circle Q, what kind of quadrilateral is $OAQB$?

Solutions

Radii or diameters of the same or equal circles have equal lengths.

(a) Since $\overline{OC} \cong \overline{OD}$, $\triangle OCD$ is isosceles.

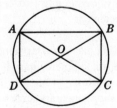

Fig. 8-14

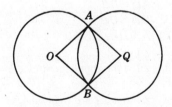

Fig. 8-15

(b) Since diagonals $\overline{AC}$ and $\overline{BD}$ are equal in length and bisect each other, $ABCD$ is a rectangle.

(c) Since the circles are equal, $\overline{OA} \cong \overline{AQ} \cong \overline{QB} \cong \overline{BO}$ and $OAQB$ is a rhombus.

8.3   PROVING A CIRCLE PROBLEM

Given: $\overline{AB} \cong \overline{DE}$
$\overline{BC} \cong \overline{EF}$
To Prove: $\angle B \cong \angle E$
Plan: Prove $\triangle I \cong \triangle II$.

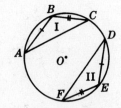

PROOF:

Statements	Reasons
1. $\overline{AB} \cong \overline{DE}$, $\overline{BC} \cong \overline{EF}$	1. Given
2. $\overparen{AB} \cong \overparen{DE}$, $\overparen{BC} \cong \overparen{EF}$	2. In a circle, $\cong$ chords have $\cong$ arcs.
3. $\overparen{ABC} \cong \overparen{DEF}$	3. If equals are added to equals, the sums are equal. Definition of $\cong$ arcs.
4. $\overline{AC} \cong \overline{DF}$	4. In a circle, $\cong$ arcs have $\cong$ chords.
5. $\triangle I \cong \triangle II$	5. s.s.s. $\cong$ s.s.s.
6. $\angle B \cong \angle E$	6. Corresponding parts of congruent $\triangle$ are $\cong$.

8.4   PROVING A CIRCLE PROBLEM STATED IN WORDS

Prove that if a radius bisects a chord, then it is perpendicular to the chord.

Solution

Given: Circle O
$\overline{OC}$ bisects $\overline{AB}$.
To Prove: $\overline{OC} \perp \overline{AB}$
Plan: Prove $\triangle AOD \cong \triangle BOD$
to show $\angle 1 \cong \angle 2$.
Also, $\angle 1$ and $\angle 2$ are
supplementary.

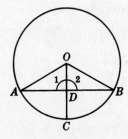

PROOF:

Statements	Reasons
1. Draw $\overline{OA}$ and $\overline{OB}$.	1. A straight line segment may be drawn between any two points.
2. $\overline{OA} \cong \overline{OB}$	2. Radii of a circle are congruent.
3. $\overline{OC}$ bisects $\overline{AB}$.	3. Given
4. $\overline{AD} \cong \overline{DB}$	4. To bisect is to divide into two $\cong$ parts.
5. $\overline{OD} \cong \overline{OD}$	5. Reflexive property
6. $\triangle AOD \cong \triangle BOD$	6. s.s.s. $\cong$ s.s.s.
7. $\angle 1 \cong \angle 2$	7. Corresponding parts of congruent $\triangle$ are $\cong$.
8. $\angle 1$ is the supplement of $\angle 2$.	8. Adjacent $\angle$ are supplementary if exterior sides lie in a straight line.
9. $\angle 1$ and $\angle 2$ are right angles.	9. Congruent supplementary angles are right angles.
10. $\overline{OC} \perp \overline{AB}$.	10. Rt. $\angle$ are formed by perpendiculars.

8.2 TANGENTS

The *length of a tangent* from a point to a circle is the length of the segment of the tangent from the given point to the point of tangency. Thus PA is the length of the tangent from P to circle O in Fig. 8-16.

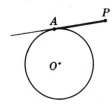

Fig. 8-16

8.2A Tangent Principles

PRINCIPLE 1: *A tangent is perpendicular to the radius drawn to the point of contact.*

Thus if $\overleftrightarrow{AB}$ is a tangent to circle O at P in Fig. 8-17, and $\overline{OP}$ is drawn, then $\overleftrightarrow{AB} \perp \overline{OP}$.

PRINCIPLE 2: *A line is tangent to a circle if it is perpendicular to a radius at its outer end.*

Thus if $\overleftrightarrow{AB} \perp$ radius $\overline{OP}$ at P of Fig. 8-17, then $\overleftrightarrow{AB}$ is tangent to circle O.

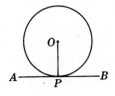

Fig. 8-17

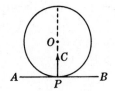

Fig. 8-18

PRINCIPLE 3: *A line passes through the center of a circle if it is perpendicular to a tangent at its point of contact.*

Thus if $\overleftrightarrow{AB}$ is tangent to circle O at P in Fig. 8-18, and $\overline{CP} \perp \overleftrightarrow{AB}$ at P, then $\overleftrightarrow{CP}$ extended will pass through the center O.

PRINCIPLE 4: *Tangents to a circle from an outside point are congruent.*

Thus if $\overline{AP}$ and $\overline{AQ}$ are tangent to circle O at P and Q (Fig. 8-19), then $\overline{AP} \cong \overline{AQ}$.

PRINCIPLE 5: *The segment from the center of a circle to an outside point bisects the angle between the tangents from the point to the circle.*

Thus $\overline{OA}$ bisects $\angle PAQ$ in Fig. 8-19 if $\overline{AP}$ and $\overline{AQ}$ are tangents to circle O.

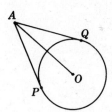

Fig. 8-19

8.2B Two Circles in Varying Relative Positions

The *line of centers of two circles* is the line joining their centers. Thus $\overline{OO'}$ is the line of centers of circles O and O' in Fig. 8-20.

Fig. 8-20   Fig. 8-21

Circles Tangent Externally

Circles O and O' in Fig. 8-21 are tangent externally at P. $\overleftrightarrow{AB}$ is the common internal tangent of both circles. The line of centers $\overline{OO'}$ passes through P, is perpendicular to $\overleftrightarrow{AB}$, and is equal in length to the sum of the radii, $R + r$. Also $\overleftrightarrow{AB}$ bisects each of the common external tangents, $\overline{CD}$ and $\overline{EF}$.

Circles Tangent Internally

Circles O and O' in Fig. 8-22 are tangent internally at P. $\overleftrightarrow{AB}$ is the common external tangent of both circles. The line of centers $\overline{OO'}$ if extended passes through P, is perpendicular to $\overleftrightarrow{AB}$, and is equal in length to the difference of the radii, $R - r$.

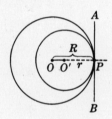

Fig. 8-22

Overlapping Circles

Circles O and O' in Fig. 8-23 overlap. Their common chord is $\overline{AB}$. If the circles are unequal, their (equal) common external tangents $\overleftrightarrow{CD}$ and $\overleftrightarrow{EF}$ meet at P. The line of centers $\overline{OO'}$ is the perpendicular bisector of $\overline{AB}$ and, if extended, passes through P.

Circles Outside Each Other

Circles O and O' in Fig. 8-24 are entirely outside each other. The common internal tangents, $\overline{AB}$ and $\overline{CD}$, meet at P. If the circles are unequal, their common external tangents, $\overline{EF}$ and $\overline{GH}$, if

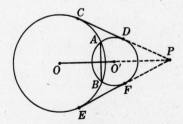

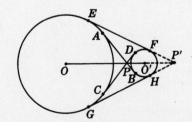

Fig. 8-23   Fig. 8-24

extended, meet at P'. The line of centers $\overline{OO'}$ passes through P and P'. Also, $AB = CD$ and $EF = GH$.

SOLVED PROBLEMS

8.5   TRIANGLES AND QUADRILATERALS HAVING TANGENT SIDES
   Points P, Q, and R in Fig. 8-25 are points of tangency.

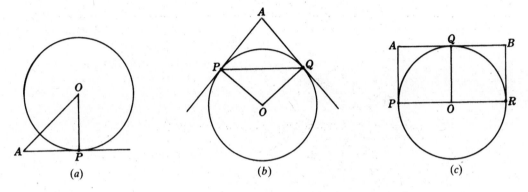

Fig. 8-25

(a) In Fig. 8-25(a), if $AP = OP$, what kind of triangle is OPA?

(b) In Fig. 8-25(b), if $AP = PQ$, what kind of triangle is APQ?

(c) In Fig. 8-25(b), if $AP = OP$, what kind of quadrilateral is $OPAQ$?

(d) In Fig. 8-25(c), if $\overline{OQ} \perp \overline{PR}$, what kind of quadrilateral is $PABR$?

Solutions

(a) $\overline{AP}$ is tangent to the circle at P; then by Principle 1, $\angle OPA$ is a right angle. Also, $AP = OP$. Hence $\triangle OAP$ is an isosceles right triangle.

(b) $\overline{AP}$ and $\overline{AQ}$ are tangents from a point to the circle; hence by Principle 4, $AP = AQ$. Also, $AP = PQ$. Then $\triangle APQ$ is an equilateral triangle.

(c) By Principle 4, $AP = AQ$. Also, $\overline{OP}$ and $\overline{OQ}$ are $\cong$ radii. And $AP = OP$. By Principle 1, $\angle APO$ is a rt. $\angle$. Then $AP = AQ = OP = OQ$; hence $OPAQ$ is a rhombus with a right angle, or a square.

(d) By Principle 1, $\overline{AP} \perp \overline{PR}$ and $\overline{BR} \perp \overline{PR}$. Then $\overline{AP} \| \overline{BR}$, since both are $\perp$ to $\overline{PR}$. By Principle 1, $\overline{AB} \perp \overline{OQ}$; also, $\overline{PR} \perp \overline{OQ}$ (Given). Then $\overline{AB} \| \overline{PR}$, since both are $\perp$ to $\overline{OQ}$. Hence $PABR$ is a parallelogram with a right angle, or a rectangle.

8.6   APPLYING PRINCIPLE 1

(a) In Fig. 8-26(a), $\overline{AP}$ is a tangent. Find $\angle A$ if $m\angle A: m\angle O = 2:3$.

(b) In Fig. 8-26(b), $\overline{AP}$ and $\overline{AQ}$ are tangents. Find $m\angle 1$ if $m\angle O = 140°$.

(c) In Fig. 8-26(c), $\overline{DP}$ and $\overline{CQ}$ are tangents. Find $m\angle 2$ and $m\angle 3$ if $\angle OPD$ is trisected and $\overline{PQ}$ is a diameter.

Solutions

(a) By Principle 1, $m\angle P = 90°$. Then $m\angle A + m\angle O = 90°$. If $m\angle A = 2x$ and $m\angle O = 3x$, then $5x = 90$ and $x = 18$. Hence $m\angle A = 36°$.

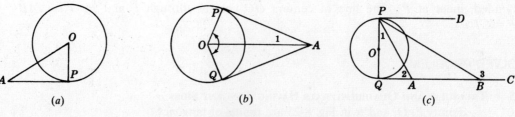

Fig. 8-26

(b) By Principle 1, $m\angle P = m\angle Q = 90°$. Since $m\angle P + m\angle Q + m\angle O + m\angle A = 360°$, $m\angle A + m\angle O = 180°$. Since $m\angle O = 140°$, $m\angle A = 40°$. By Principle 5, $m\angle 1 = \frac{1}{2}m\angle A = 20°$.

(c) By Principle 1, $m\angle DPQ = m\angle PQC = 90°$. Since $m\angle 1 = 30°$, $m\angle 2 = 60°$. Since $\angle 3$ is an exterior angle of $\triangle PBQ$, $m\angle 3 = 90° + 60° = 150°$.

8.7   APPLYING PRINCIPLE 4

(a) $\overline{AP}$, $\overline{BQ}$, and $\overline{AB}$ in Fig. 8-27(a) are tangents. Find y.

(b) $\triangle ABC$ in Fig. 8-27(b) is circumscribed. Find x.

(c) Quadrilateral $ABCD$ in Fig. 8-27(c) is circumscribed. Find x.

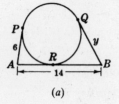   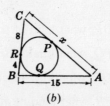   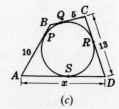

Fig. 8-27

Solutions

(a) By Principle 4, $AR = 6$, and $RB = y$. Then $RB = AB - AR = 14 - 6 = 8$. Hence, $y = RB = 8$.

(b) By Principle 4, $PC = 8$, $QB = 4$, and $AP = AQ$. Then $AQ = AB - QB = 11$. Hence, $x = AP + PC = 11 + 8 = 19$.

(c) By Principle 4, $AS = 10$, $CR = 5$, and $RD = SD$. Then $RD = CD - CR = 8$. Hence, $x = AS + SD = 10 + 8 = 18$.

8.8   FINDING THE LINE OF CENTERS

Two circles have radii of 9 and 4, respectively. Find the length of their line of centers (a) if the circles are tangent externally, (b) if the circles are tangent internally, (c) if the circles are concentric, (d) if the circles are 5 units apart. (See Fig. 8-28.)

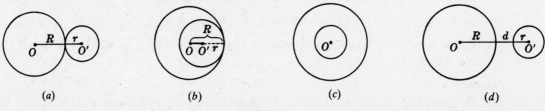

Fig. 8-28

Solutions

Let R = radius of larger circle, r = radius of smaller circle.

(a)   Since $R = 9$ and $r = 4$, $OO' = R + r = 9 + 4 = 13$.

(b)   Since $R = 9$ and $r = 4$, $OO' = R - r = 9 - 4 = 5$.

(c)   Since the circles have the same center, their line of centers has zero length.

(d)   Since $R = 9$, $r = 4$, and $d = 5$, $OO' = R + d + r = 9 + 5 + 4 = 18$.

8.9   PROVING A TANGENT PROBLEM STATED IN WORDS

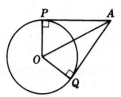

Prove: Tangents to a circle from an outside
point are congruent (Principle 4).
Given: Circle O
$\overline{AP}$ is tangent at P.
$\overline{AQ}$ is tangent at Q.
To Prove: $\overline{AP} \cong \overline{AQ}$
Plan: Draw $\overline{OP}$, $\overline{OQ}$, and $\overline{OA}$ and prove
$\triangle AOP \cong \triangle AOQ$.

PROOF:

Statements	Reasons
1. Draw $\overline{OP}$, $\overline{OQ}$, and $\overline{OA}$.	1. A straight line may be drawn between any two points.
2. $\overline{OP} \cong \overline{OQ}$	2. Radii of a circle are congruent.
3. $\angle P$ and $\angle Q$ are right angles.	3. A tangent is $\perp$ to radius drawn to point of contact.
4. $\overline{OA} \cong \overline{OA}$	4. Reflexive property
5. $\triangle AOP \cong \triangle AOQ$	5. hy.leg $\cong$ hy.leg
6. $\overline{AP} \cong \overline{AQ}$	6. Corresponding parts of congruent $\triangle$ are congruent.

8.3   MEASUREMENT OF ANGLES AND ARCS IN A CIRCLE

A *central angle* has the same number of degrees as the arc it intercepts. Thus, as shown in Fig. 8-29, a central angle which is a right angle intercepts a 90° arc; a 40° central angle intercepts a 40° arc, and a central angle which is a straight angle intercepts a semicircle of 180°.

Since the numerical measures in degrees of both the central angle and its intercepted arc are the same, we may restate the above principle as follows: A central angle is measured by its intercepted arc. The symbol $\stackrel{\circ}{=}$ may be used to mean "is measured by." (Do not say that the central angle equals its intercepted arc. An angle cannot *equal* an arc.)

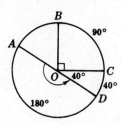

Fig. 8-29

An *inscribed angle* is an angle whose vertex is on the circle and whose sides are chords. *An angle inscribed in an arc* has its vertex on the arc and its sides passing through the ends of the arc. Thus ∠A in Fig. 8-30 is an inscribed angle whose sides are the chords $\overline{AB}$ and $\overline{AC}$. Note that ∠A intercepts $\overparen{BC}$ and is inscribed in $\overparen{BAC}$.

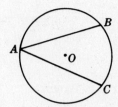

Fig. 8-30

8.3A Angle-Measurement Principles

PRINCIPLE 1: *A central angle is measured by its intercepted arc.*

PRINCIPLE 2: *An inscribed angle is measured by one-half its intercepted arc.*

A proof of this principle is given in Chapter 16.

PRINCIPLE 3: *In the same or congruent circles, congruent inscribed angles have congruent intercepted arcs.*

Thus in Fig. 8-31, if ∠1 ≅ ∠2, then $\overparen{BC} \cong \overparen{DE}$.

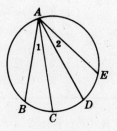

Fig. 8-31

PRINCIPLE 4: *In the same or congruent circles, inscribed angles having congruent intercepted arcs are congruent.* (This is the converse of Principle 3.)

Thus in Fig. 6-31, if $\overparen{BC} \cong \overparen{DE}$, then ∠1 ≅ ∠2.

PRINCIPLE 5: *Angles inscribed in the same or congruent arcs are congruent.*

Thus in Fig. 8-32, if ∠C and ∠D are inscribed in $\overparen{ACB}$, then ∠C ≅ ∠D.

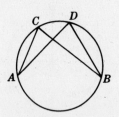

Fig. 8-32

PRINCIPLE 6: *An angle inscribed in a semicircle is a right angle.*

Thus in Fig. 8-33, since $\angle C$ is inscribed in semicircle $\overset{\frown}{ACD}$, $m\angle C = 90°$.

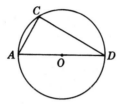

Fig. 8-33

PRINCIPLE 7: *Opposite angles of an inscribed quadrilateral are supplementary.*

Thus in Fig. 8-34, if $ABCD$ is an inscribed quadrilateral, $\angle A$ is the supplement of $\angle C$.

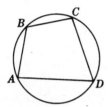   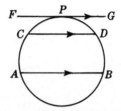

Fig. 8-34   **Fig. 8-35**

PRINCIPLE 8: *Parallel lines intercept congruent arcs on a circle.*

Thus in Fig. 8-35, if $\overline{AB} \parallel \overline{CD}$, then $\overset{\frown}{AC} \cong \overset{\frown}{BD}$. If tangent $\overleftrightarrow{FG}$ is parallel to $\overline{CD}$, then $\overset{\frown}{PC} \cong \overset{\frown}{PD}$.

PRINCIPLE 9: *An angle formed by a tangent and a chord is measured by one-half its intercepted arc.*

PRINCIPLE 10: *An angle formed by two intersecting chords is measured by one-half the sum of the intercepted arcs.*

PRINCIPLE 11: *An angle formed by two secants intersecting outside a circle is measured by one-half the difference of the intercepted arcs.*

PRINCIPLE 12: *An angle formed by a tangent and a secant intersecting outside a circle is measured by one-half the difference of the intercepted arcs.*

PRINCIPLE 13: *An angle formed by two tangents intersecting outside a circle is measured by one-half the difference of the intercepted arcs.*

Proofs of Principles 10 to 13 are given in Chapter 16.

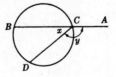

Fig. 8-36

8.3B   Table of Angle-Measurement Principles

Position of Vertex	Kind of Angle	Diagram	Measurement Formula	Method of Measurement
Center of circle	Central angle (apply Principle 1)		$\angle O \stackrel{\circ}{=} \widehat{AB}$ $m\angle O = a°$	By intercepted arc
On the circle	Inscribed angle (apply Principle 2)		$\angle A \stackrel{\circ}{=} \frac{1}{2}\widehat{BC}$ $m\angle A = \frac{1}{2}a°$	By one-half intercepted arc
	Angle formed by a tangent and a chord (apply Principle 9)			
Inside the circle	Angle formed by two intersecting chords (apply Principle 10)		$\angle 1 \stackrel{\circ}{=} \frac{1}{2}(\widehat{AC} + \widehat{BD})$ $m\angle 1 = \frac{1}{2}(a° + b°)$	By one-half sum of intercepted arcs
Outside the circle	Angle formed by two secants (apply Principle 10)		$\angle A \stackrel{\circ}{=} \frac{1}{2}(\widehat{BC} - \widehat{DE})$ $m\angle A = \frac{1}{2}(a° - b°)$	By one-half difference of intercepted arcs
	Angle formed by a secant and a tangent (apply Principle 12)		$\angle A \stackrel{\circ}{=} \frac{1}{2}(\widehat{BC} - \widehat{BD})$ $m\angle A = \frac{1}{2}(a° - b°)$	
	Angle formed by two tangents (apply Principle 13)		$\angle A \stackrel{\circ}{=} \frac{1}{2}(\widehat{BDC} - \widehat{BC})$ $m\angle A = \frac{1}{2}(a° - b°)$ Also, $m\angle A = (180 - b)°$	

Note: To find the angle formed by a secant and a chord meeting on the circle first find the measure of the inscribed angle adjacent to it and then subtract from 180°. Thus if secant $\overline{AB}$ meets chord $\overline{CD}$ at C on the circle in Fig. 8-36, to find $m\angle y$, first find the measure of inscribed $\angle x$. Obtain $m\angle y$ by subtracting $m\angle x$ from 180°.

SOLVED PROBLEMS

8.10   **APPLYING PRINCIPLES 1 AND 2**

(a)   In Fig. 8-37(a), if $m\angle y = 46°$, find $m\angle x$.

(b)   In Fig. 8-39(b), if $m\angle y = 112°$, find $m\angle x$.

(c)   In Fig. 8-37(c), if $m\angle x = 75°$, find $m\widehat{y}$.

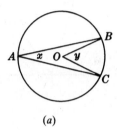

(a)

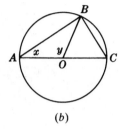

(b)

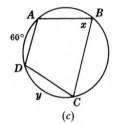

(c)

Fig. 8-37

Solutions

(a) $\angle y \stackrel{\circ}{=} \widehat{BC}$, so $m\widehat{BC} = 46°$. Then $\angle x \stackrel{\circ}{=} \frac{1}{2}\widehat{BC} = \frac{1}{2}(46°) = 23°$, so $m\angle x = 23°$.

(b) $\angle y \stackrel{\circ}{=} \widehat{AB}$ so $m\widehat{AB} = 112°$.
    $m\widehat{BC} = m(\widehat{ABC} - \widehat{AB}) = 180° - 112° = 68°$. Then $\angle x \stackrel{\circ}{=} \frac{1}{2}\widehat{BC} = \frac{1}{2}(68°) = 34°$, so $m\angle x = 34°$.

(c) $\angle x \stackrel{\circ}{=} \frac{1}{2}\widehat{ADC}$, so $m\widehat{ADC} = 150°$. Then $m\widehat{y} = m(\widehat{ADC} - \widehat{AD}) = 150° - 60° = 90°$.

8.11   **APPLYING PRINCIPLES 3 TO 8**
    Find x and y in each part of Fig. 8-38.

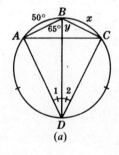

(a)

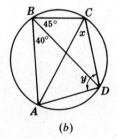

(b)

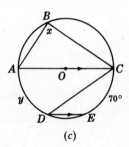

(c)

Fig. 8-38

Solutions

(a)   Since $m\angle 1 = m\angle 2$, $mx = m\widehat{AB} = 50°$. Since $\widehat{AD} \cong \widehat{CD}$, $m\angle y = m\angle ABD = 65°$.

(b) $\angle ABD$ and $\angle x$ are inscribed in $\widehat{ABD}$; hence $m\angle x = m\angle ABD = 40°$.
    $ABCD$ is an inscribed quadrilateral; hence $m\angle y = 180° - m\angle B = 95°$.

(c)   Since $\angle x$ is inscribed in a semicircle, $m\angle x = 90°$. Since $\overline{AC} \| \overline{DE}$, $m\widehat{y} = m\widehat{CE} = 70°$.

8.12 APPLYING PRINCIPLE 9
In each part of Fig. 8-39, CD is a tangent at P.

(a) If $m\widehat{y} = 220°$ in part (a), find $m\angle x$.

(b) If $m\widehat{y} = 140°$ in part (b), find $m\angle x$.

(c) If $m\angle y = 75°$ in part (c), find $m\angle x$.

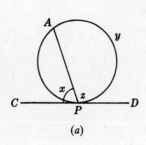
(a)

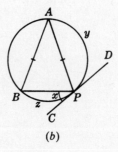
(b)

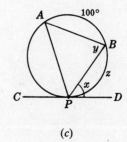
(c)

Fig. 8-39

Solutions

(a) $\angle z \doteq \frac{1}{2}\widehat{y} = \frac{1}{2}(220°) = 110°$. So $m\angle x = 180° - 110° = 70°$.

(b) Since $AB = AP$, $m\widehat{AB} = m\widehat{y} = 140°$. Then $m\widehat{z} = 360° - 140° - 140° = 80°$.
Since $\angle x \doteq \frac{1}{2}\widehat{z} = 40°$, $m\angle x = 40°$.

(c) $\angle y \doteq \frac{1}{2}\widehat{AP}$, so $m\widehat{AP} = 150°$. Then $m\widehat{z} = 360° - 100° - 150° = 110°$.
Since $\angle x \doteq \frac{1}{2}\widehat{z} = 55°$, $m\angle x = 55°$.

8.13 APPLYING PRINCIPLE 10

(a) If $m\angle x = 95°$ in Fig. 8-40(a), find $m\widehat{y}$.

(b) If $m\widehat{y} = 80°$ in Fig. 8-40(b), find $m\angle x$.

(c) If $m\widehat{x} = 78°$ in Fig. 8-40(c), find $m\angle y$.

Solutions

(a) $\angle x \doteq \frac{1}{2}(\widehat{AC} + \widehat{y})$; thus $95° = \frac{1}{2}(70° + m\widehat{y})$, so $m\widehat{y} = 120°$.

(b) $\angle z \doteq \frac{1}{2}(\widehat{y} + \widehat{AB}) = \frac{1}{2}(80° + 120°) = 100°$. Then $m\angle x = 180° - m\angle z = 80°$.

(c) $\overline{BC} \| \overline{AD}$, because $m\widehat{CD} = m\widehat{x} = 78°$. Also, $\angle z \doteq \frac{1}{2}(\widehat{x} + \widehat{CD}) = 78°$. Then $m\angle y = 180° - m\angle z = 102°$.

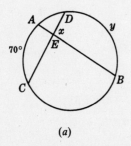
(a)

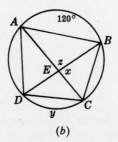
(b)

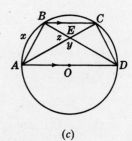
(c)

Fig. 8-40

8.14   APPLYING PRINCIPLES 11 TO 13

(a)   If $m\angle x = 40°$ in Fig. 8-41(a), find $m\widehat{y}$.

(b)   If $m\angle x = 67°$ in Fig. 8-41(b), find $m\widehat{y}$.

(c)   If $m\angle x = 61°$ in Fig. 8-41(c), find $m\widehat{y}$.

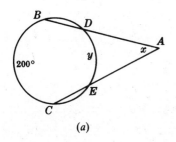
(a)

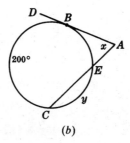
(b)

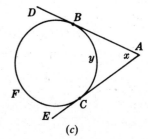
(c)

Fig. 8-41

Solutions

(a) $\angle x \doteq \frac{1}{2}(\widehat{BC} - \widehat{y})$, so $40° = \frac{1}{2}(200° - m\widehat{y})$ or $m\widehat{y} = 120°$.

(b) $\angle x \doteq \frac{1}{2}(\widehat{BC} - \widehat{BE})$, so $67° = \frac{1}{2}(200° - m\widehat{BE})$ or $m\widehat{BE} = 66°$.
Then $m\widehat{y} = 360° - 200° - 66° = 94°$.

(c) $\angle x \doteq \frac{1}{2}(\widehat{BFC} - \widehat{y})$, and $m\widehat{BFC} = 360° - m\widehat{y}$. Then $61° = \frac{1}{2}[(360° - m\widehat{y}) - m\widehat{y}] = 180° - m\widehat{y}$. Thus $m\widehat{y} = 119°$.

8.15   USING EQUATIONS IN TWO UNKNOWNS TO FIND ARCS
In each part of Fig. 8-42, find x and y using equations in two unknowns.

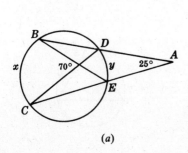
(a)

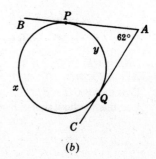
(b)

Fig. 8-42

Solutions

(a)   By Principle 10, $70° = \frac{1}{2}(m\widehat{x} + m\widehat{y})$
By Principle 11, $25° = \frac{1}{2}(m\widehat{x} - m\widehat{y})$
If we add these two equations, we get $m\widehat{x} = 95°$. If we subtract one from the other, we get $m\widehat{y} = 45°$.

(b)   Since $m\widehat{x} + m\widehat{y} = 360°$, $\frac{1}{2}(m\widehat{x} + m\widehat{y}) = 180°$
By Principle 13, $\frac{1}{2}(m\widehat{x} - m\widehat{y}) = 62°$
If we add these two equations, we find that $m\widehat{x} = 242°$. If we subtract one from the other, we get $m\widehat{y} = 118°$.

8.16 MEASURING ANGLES AND ARCS IN GENERAL
Find x and y in each part of Fig. 8-43.

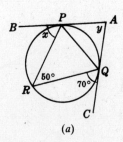

(a)

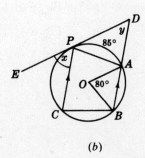

(b)

Fig. 8-43

Solutions

(a) By Principle 2, $50° = \frac{1}{2}m\widehat{PQ}$ or $m\widehat{PQ} = 100°$. Also, by Principle 9, $70° = \frac{1}{2}m\widehat{QR}$ or $m\widehat{QR} = 140°$.
Then $m\widehat{PR} = 360° - m\widehat{PQ} - m\widehat{QR} = 120°$.
By Principle 9, $x = \frac{1}{2}m\widehat{PR} = 60°$.
By Principle 13, $y = \frac{1}{2}(m\widehat{PRQ} - m\widehat{PQ}) = \frac{1}{2}(260° - 100°) = 80°$.

(b) By Principle 1, $m\widehat{AB} = 80°$. Also, by Principle 8, $m\widehat{BC} = m\widehat{PA} = 85°$. Then $m\widehat{PC} = 360° - m\widehat{PA} - m\widehat{AB} - m\widehat{BC} = 100°$.
By Principle 9, $x = \frac{1}{2}m\widehat{PC} = 55°$.
By Principle 12, $y = \frac{1}{2}(m\widehat{PCB} - m\widehat{PA}) = \frac{1}{2}(195° - 85°) = 55°$.

8.17 PROVING AN ANGLE MEASUREMENT PROBLEM

Given: $\widehat{BD} = \widehat{CE}$
To Prove: $AB = AC$
Plan: First prove $\widehat{CD} = \widehat{BE}$.
Use this to show that $\angle B \cong \angle C$.

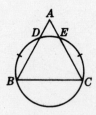

PROOF:

Statements	Reasons
1. $\widehat{BD} = \widehat{CE}$	1. Given
2. $\widehat{DE} = \widehat{DE}$	2. Reflexive property
3. $\widehat{BE} = \widehat{CD}$	3. If =s are added to =s, the sums are =.
4. $\angle B \cong \angle C$	4. In a circle, inscribed angles having equal intercepted arcs are $\cong$.
5. $AB = AC$	5. In a triangle, sides opposite $\cong$ angles are equal in length.

8.18 **PROVING AN ANGLE MEASUREMENT PROBLEM STATED IN WORDS**

Prove that parallel chords drawn at the ends of a diameter are equal in length.

Solutions

Given: Circle O
$\overline{AB}$ is a diameter.
$\overline{AC} \parallel \overline{BD}$

To Prove: $AC = BD$

Plan: Prove $\overparen{AC} \cong \overparen{BD}$

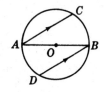

PROOF:

Statements	Reasons
1. $\overline{AB}$ is a diameter.	1. Given
2. $\overparen{ACB} \cong \overparen{ADB}$	2. A diameter cuts a circle into two equal semicircles.
3. $\overline{AC} \parallel \overline{BD}$	3. Given
4. $\overparen{AD} \cong \overparen{BC}$	4. Parallel lines intercept $\cong$ arcs on a circle.
5. $\overparen{AC} \cong \overparen{BD}$	5. If equals are subtracted from equals, the differences are equal. Definition of $\cong$ arcs.
6. $AC = BD$	6. In a circle, equal arcs have chords which are equal in length.

Supplementary Problems

1. Provide the proofs requested in Fig. 8-44. (8.3)

(a) **Given:** $AB = DE$
$AC = DF$
To Prove: $\angle B \cong \angle E$

(b) **Given:** Circle O, $AB = BC$
Diameter $\overline{BD}$
To Prove: $\overline{BD}$ bisects $\angle AOC$.

(c) **Given:** Circle O
$\overparen{AB} = \overparen{CD}$
To Prove: $\angle AOC \cong \angle BOD$

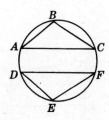

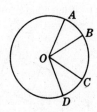

Fig. 8-44

2. Provide the proofs requested in Fig. 8-45. (8.3)

(a) **Given:** $AB = AC$
To Prove: $\overparen{ABC} \cong \overparen{ACB}$

(b) **Given:** $\overparen{ABC} \cong \overparen{ACB}$
To Prove: $AB = AC$

(c) **Given:** Circle O, $AB = AD$
Diameter $\overline{AC}$
To Prove: $BC = CD$

(d) **Given:** Circle O
$AB = AD$, $BC = CD$
To Prove: $\overline{AC}$ is a diameter.

(e) **Given:** $AD = BC$
To Prove: $AC = BD$

(f) **Given:** $AC = BD$
To Prove: $AD = BC$

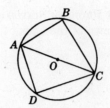

Fig 8-45

3. Prove each of the following: (8.4)

   (a) If a radius bisects a chord, then it bisects its arcs.

   (b) If a diameter bisects the major arc of a chord, then it is perpendicular to the chord.

   (c) If a diameter is perpendicular to a chord, it bisects the chord and its arcs.

4. Prove each of the following: (8.4)

   (a) A radius through the point of intersection of two congruent chords bisects an angle formed by them.

   (b) If chords drawn from the ends of a diameter make congruent angles with the diameter, the chords are congruent.

   (c) In a circle, congruent chords are equally distant from the center of the circle.

   (d) In a circle, chords that are equally distant from the center are congruent.

5. Determine each of the following, assuming t, t', and t'' in Fig. 8-46 are tangents.   (8.5)

   (a) If $m\angle A = 90°$ in Fig. 8-46(a), what kind of quadrilateral is $PAQO$?

   (b) If $BR = RC$ in Fig. 8-46(b), what kind of triangle is ABC?

   (c) What kind of quadrilateral is $PABQ$ in Fig. 8-46(c) if $\overline{PQ}$ is a diameter?

   (d) What kind of triangle is AOB in Fig. 8-46(c)?

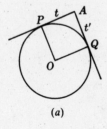 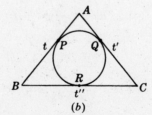 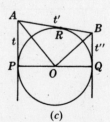

        (a)   (b)   (c)

Fig. 8-46

6. In circle O, radii $\overline{OA}$ and $\overline{OB}$ are drawn to the points of tangency of $\overline{PA}$ and $\overline{PB}$. Find $m\angle AOB$ if $m\angle APB$ equals (a) 40°; (b) 120°; (c) 90°; (d) $x°$; (e) $(180 - x)°$; (f) $(90 - x)°$.   (8.6)

7. Find each of the following (t and t' in Fig. 8-47 are tangents).   (8.6)

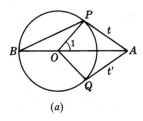

(a)

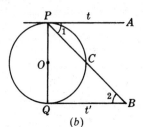

(b)

Fig. 8-47

In Fig. 8-47(a):

(a) If $m\angle POQ = 80°$, find $m\angle PAQ$.

(b) If $m\angle PBO = 25°$, find $m\angle 1$ and $m\angle PAQ$.

(c) If $m\angle PAQ = 72°$, find $m\angle 1$ and $m\angle PBO$.

In Fig. 8-47(b):

(d) If $\overline{PB}$ bisects $\angle APQ$, find $m\angle 2$.

(e) If $m\angle 1 = 35°$, find $m\angle 2$.

(f) If $PQ = QB$, find $m\angle 1$.

8. In Fig. 8-48(a), $\triangle ABC$ is circumscribed. (a) If $y = 9$, find x. (b) If $x = 25$, find y.   (8.7)
In Fig. 8-48(b), quadrilateral $ABCD$ is circumscribed. (c) Find $AB + CD$. (d) Find perimeter of $ABCD$.
In Fig. 8-48(c), quadrilateral $ABCD$ is circumscribed. (e) If $r = 10$, find x. (f) If $x = 25$, find r.

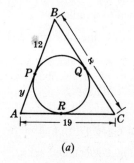

(a)

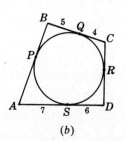

(b)

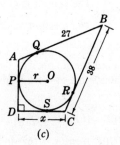

(c)

Fig. 8-48

9. If two circles have radii of 20 and 13, respectively, find their line of centers: (8.8)

(a) If the circles are concentric   (c) If the circles are tangent externally

(b) If the circles are 7 units apart   (d) If the circles are tangent internally

10. If the line of centers of two circles measures 30, what is the relation between the two circles: (8.8)

(a) If their radii are 25 and 5?   (c) If their radii are 20 and 5?

(b) If their radii are 35 and 5?   (d) If their radii are 25 and 10?

11. What is the relation between two circles if the length of their line of centers is (*a*) 0; (*b*) equal to the difference of their radii; (*c*) equal to the sum of their radii; (*d*) greater than the sum of their radii, (*e*) less than the difference of their radii and greater than 0; (*f*) greater than the difference and less than the sum of their radii.   (8.8)

12. Prove each of the following: (8.9)

   (*a*) The line from the center of a circle to an outside point bisects the angle between the tangents from the point to the circle.

   (*b*) If two circles are tangent externally, their common internal tangent bisects a common external tangent.

   (*c*) If two circles are outside each other, their common internal tangents are congruent.

   (*d*) In a circumscribed quadrilateral, the sum of the lengths of the two opposite sides equals the sum of the lengths of the other two.

13. Find the number of degrees in a central angle which intercepts an arc of (*a*) 40°; (*b*) 90°; (*c*) 170°; (*d*) 180°; (*e*) $2x°$; (*f*) $(180 - x)°$; (*g*) $(2x - 2y)°$.   (8.10)

14. Find the number of degrees in an inscribed angle which intercepts an arc of (*a*) 40°; (*b*) 90°; (*c*) 170°; (*d*) 180°; (*e*) 260°; (*f*) 348°; (*g*) $2x°$; (*h*) $(180 - x)°$; (*i*) $(2x - 2y)°$.   (8.10)

15. Find the number of degrees in the arc intercepted by: (8.10)

   (*a*) A central angle of 85°

   (*b*) An inscribed angle of 85°

   (*c*) A central angle of $c°$

   (*d*) An inscribed angle of $i°$

   (*e*) The central angle of a triangle formed by two radii and a chord equal to a radius

   (*f*) The smallest angle of an inscribed triangle whose angles intercept arcs in the ratio of 1:2:3

16. Find the number of degrees in each of the arcs intercepted by the angles of an inscribed triangle if the measures of these angles are in the ratio of (*a*) 1:2:3; (*b*) 2:3:4; (*c*) 5:6:7; (*d*) 1:4:5.   (8.10)

17. (*a*) If $m\widehat{y} = 140°$ in Fig. 8-49(*a*), find $m\angle x$.   (*d*) If $m\angle x = 108°$ in Fig. 8-49(*b*), find $m\angle y$.   (8.10)

   (*b*) If $m\angle x = 165°$ in Fig. 8-49(*a*), find $m\widehat{y}$.   (*e*) If $m\widehat{y} = 105°$ in Fig. 8-49(*c*), find $m\angle x$.

   (*c*) If $m\angle y = 115°$ in Fig. 8-49(*b*), find $m\angle x$.   (*f*) If $m\angle x = 96°$ in Fig. 8-49(*c*), find $m\widehat{y}$.

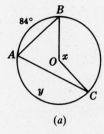

(*a*)

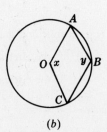

(*b*)

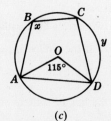

(*c*)

Fig. 8-49

18. If quadrilateral *ABCD* is inscribed in a circle in Fig. 8-50, find: (8.11)

    (*a*) $m\angle A$ if $m\angle C = 45°$ (*e*) $m\angle A$ if $m\widehat{BAD} = 160°$

    (*b*) $m\angle B$ if $m\angle D = 90°$ (*f*) $m\angle B$ if $m\widehat{ABC} = 200°$

    (*c*) $m\angle C$ if $m\angle A = x°$ (*g*) $m\angle C$ if $m\widehat{BC} = 140°$ and $m\widehat{CD} = 110°$

    (*d*) $m\angle D$ if $m\angle B = (90 - x)°$ (*h*) $m\angle D$ if $m\angle D : m\angle B = 2:3$

        Fig. 8-50   **Fig. 8-51**

19. If *BC* and *AD* are the parallel sides of inscribed trapezoid *ABCD* in Fig. 8-51, find: (8.11)

    (*a*) $m\widehat{AB}$ if $m\widehat{CD} = 85°$ (*e*) $m\angle A$ if $m\angle D = 72°$

    (*b*) $m\widehat{CD}$ if $m\widehat{AB} = y°$ (*f*) $m\angle A$ if $m\angle C = 130°$

    (*c*) $m\widehat{AB}$ if $m\widehat{BC} = 60°$ and $m\widehat{AD} = 80°$ (*g*) $m\angle B$ if $m\angle C = 145°$

    (*d*) $m\widehat{CD}$ if $m\widehat{AD} + m\widehat{BC} = 170°$ (*h*) $m\angle B$ if $m\widehat{AD} = 90°$ and $m\widehat{AB} = 84°$

20. A diameter is parallel to a chord. Find the number of degrees in an arc between the diameter and chord if the chord intercepts (*a*) a minor arc of 80°; (*b*) a major arc of 300°.   (8.11)

21. Find *x* and *y* in each part of Fig. 8-52.   (8.11)

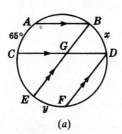   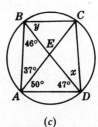

      (*a*)   (*b*)   (*c*)

        Fig. 8-52

22. Find the number of degrees in the angle formed by a tangent and a chord drawn to the point of tangency if the intercepted arc has measure (*a*) 38°; (*b*) 90°; (*c*) 138°; (*d*) 180°; (*e*) 250°; (*f*) 334°; (*g*) *x*°; (*h*) $(360 - x)°$; (*i*) $(2x + 2y)°$.   (8.12)

23. Find the number of degrees in the arc intercepted by an angle formed by a tangent and a chord drawn to the point of tangency if the angle measures (*a*) 55°; (*b*) $67\frac{1}{2}°$; (*c*) 90°; (*d*) 135°; (*e*) $(90 - x)°$; (*f*) $(180 - x)°$; (*g*) $(x - y)°$; (*h*) $3\frac{1}{2}x°$.   (8.12)

24. Find the number of degrees in the acute angle formed by a tangent through one vertex and an adjacent side of an inscribed (*a*) square; (*b*) equilateral triangle; (*c*) regular hexagon; (*d*) regular decagon.   (8.12)

25. Find x and y in each part of Fig. 8-53 (t and t' are tangents).   (8.12)

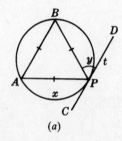

(a)

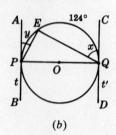

(b)

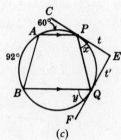

(c)

Fig. 8-53

26. If $\overline{AC}$ and $\overline{BD}$ are chords intersecting in a circle as shown in Fig. 8-54, find: (8.13)

(a) $m\angle x$ if $m\widehat{AB} = 90°$ and $m\widehat{CD} = 60°$ (e) $m\widehat{AB} + m\widehat{CD}$ if $m\angle x = 70°$

(b) $m\angle x$ if $m\widehat{AB}$ and $m\widehat{CD}$ each equals $75°$ (f) $m\widehat{BC} + m\widehat{AD}$ if $m\angle x = 65°$

(c) $m\angle x$ if $m\widehat{AB} + m\widehat{CD} = 230°$ (g) $m\widehat{BC}$ if $m\angle x = 60°$ and $m\widehat{AD} = 160°$

(d) $m\angle x$ if $m\widehat{BC} + m\widehat{AD} = 160°$ (h) $m\widehat{BC}$ if $m\angle y = 72°$ and $m\widehat{AD} = 2m\widehat{BC}$

Fig. 8-54

Fig. 8-55

27. If $\overline{AC}$ and $\overline{BD}$ are diagonals of an inscribed quadrilateral $ABCD$ as shown in Fig. 8-55, find: (8.13)

(a) $m\angle 1$ if $m\widehat{a} = 95°$ and $m\widehat{c} = 75°$ (e) $m\angle 2$ if $m\widehat{b} + m\widehat{d} = m\widehat{a} + m\widehat{c}$

(b) $m\angle 1$ if $m\widehat{b} = 88°$ and $m\widehat{d} = 66°$ (f) $m\angle 2$ if $\widehat{BC} \| \widehat{AD}$ and $m\widehat{a} = 70°$

(c) $m\angle 1$ if $m\widehat{b}$ and $m\widehat{d}$ each equals $100°$ (g) $m\angle 2$ if $\overline{AD}$ is a diameter and $m\widehat{b} = 80°$

(d) $m\angle 1$ if $m\widehat{a}:m\widehat{b}:m\widehat{c}:m\widehat{d} = 1:2:3:4$ (h) $m\angle 1$ if $ABCD$ is a rectangle and $m\widehat{a} = 70°$

28. Find x and y in each part of Fig. 8-56.   (8.13)

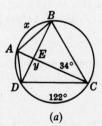

(a)

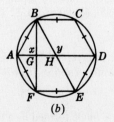

(b)

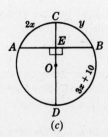

(c)

Fig. 8-56

29. If $\overline{AB}$ and $\overline{AC}$ are intersecting secants as shown in Fig. 8-57, find: (8.14)

    (a) $m\angle A$ if $m\widehat{c} = 100°$ and $m\widehat{a} = 40°$ (e) $m\widehat{a}$ if $m\widehat{c} = 160°$ and $m\angle A = 20°$

    (b) $m\angle A$ if $m\widehat{c} - m\widehat{a} = 74°$ (f) $m\widehat{c}$ if $m\widehat{a} = 60°$ and $m\angle A = 35°$

    (c) $m\angle A$ if $m\widehat{c} = m\widehat{a} + 40°$ (g) $m\widehat{c} - m\widehat{a}$ if $m\angle A = 47°$

    (d) $m\angle A$ if $m\widehat{a}:m\widehat{b}:m\widehat{c}:m\widehat{d} = 1:4:3:2$ (h) $m\widehat{a}$ if $m\widehat{c} = 3\widehat{a}$ and $m\angle A = 25°$

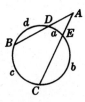

               Fig. 8-57   **Fig. 8-58**

30. If tangent $\overline{AP}$ and secant $\overline{AB}$ intersect as shown in Fig. 8-58, find: (8.14)

    (a) $m\angle A$ if $m\widehat{c} = 150°$ and $m\widehat{a} = 60°$ (f) $m\widehat{a}$ if $m\widehat{c} = 220°$ and $m\angle A = 40°$

    (b) $m\angle A$ if $m\widehat{c} = 200°$ and $m\widehat{b} = 110°$ (g) $m\widehat{c}$ if $m\widehat{a} = 55°$ and $m\angle A = 30°$

    (c) $m\angle A$ if $m\widehat{b} = 120°$ and $m\widehat{a} = 70°$ (h) $m\widehat{a}$ if $m\widehat{c} = 3m\widehat{a}$ and $m\angle A = 45°$

    (d) $m\angle A$ if $m\widehat{c} - m\widehat{a} = 73°$ (i) $m\widehat{a}$ if $m\widehat{b} = 100°$ and $m\angle A = 50°$

    (e) $m\angle A$ if $m\widehat{a}:m\widehat{b}:m\widehat{c} = 1:4:7$

31. If $\overrightarrow{AP}$ and $\overrightarrow{AQ}$ are intersecting tangents as shown in Fig. 8-59, find: (8.14)

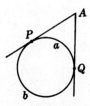

                           Fig. 8-59

    (a) $m\angle A$ if $m\widehat{b} = 200°$ (i) $m\angle A$ if $m\widehat{b}:m\widehat{a} = 7:3$

    (b) $m\angle A$ if $m\widehat{a} = 95°$ (j) $m\angle A$ if $m\widehat{b} = 5m\widehat{a} - 60°$

    (c) $m\angle A$ if $m\widehat{a} = x°$ (k) $m\widehat{a}$ if $m\angle A = 35°$

    (d) $m\angle A$ if $m\widehat{a} = (90 - x)°$ (l) $m\widehat{a}$ if $m\angle A = y°$

    (e) $m\angle A$ if $m\widehat{b} = 3m\widehat{a}$ (m) $m\widehat{b}$ if $m\angle A = 60°$

    (f) $m\widehat{A}$ if $m\widehat{b} = m\widehat{a} + 50°$ (n) $m\widehat{b}$ if $m\angle A = x°$

    (g) $m\angle A$ if $m\widehat{b} - m\widehat{a} = 84°$ (o) $m\widehat{b}$ if $\overrightarrow{AP} \perp \overrightarrow{AQ}$

    (h) $m\angle A$ if $m\widehat{b}:m\widehat{a} = 5:1$

32. Find x and y in each part of Fig. 8-60 (t and t' are tangents).   (8.14)

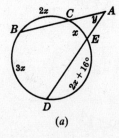

(a)

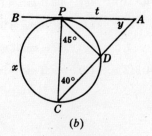

(b)

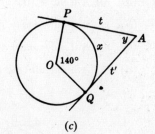

(c)

Fig. 8-60

33. If $\overline{AB}$ and $\overline{AC}$ are intersecting secants as shown in Fig. 8-61, find: (8.15)

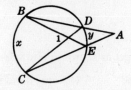

Fig. 8-61

(a) $m\widehat{x}$ if $m\angle 1 = 80°$ and $m\angle A = 40°$ (d) $m\widehat{y}$ if $m\angle 1 = 95°$ and $m\angle A = 45°$

(b) $m\widehat{x}$ if $m\angle 1 + m\angle A = 150°$ (e) $m\widehat{y}$ if $m\angle 1 = m\angle A = 22\frac{1}{2}°$

(c) $m\widehat{x}$ if $\angle 1$ and $\angle A$ are supplementary   (f) $m\widehat{y}$ if $m\widehat{x} + m\widehat{y} = 190°$ and $m\angle A = 50°$

34. Find x and y in each part of Fig. 8-62 (t and t' are tangents).   (8.15)

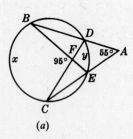

(a)

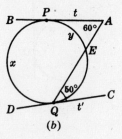

(b)

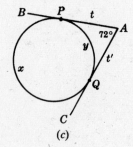

(c)

Fig. 8-62

35. If ABC is an inscribed triangle as shown in Fig. 8-63, find: (8.16)

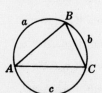

Fig. 8-63

(a) $m\angle A$ if $m\widehat{a} = 110°$ and $m\widehat{c} = 200°$

(f) $m\angle B$ if $m\widehat{ABC} = 208°$

(b) $m\angle A$ if $\overline{AB} \perp \overline{BC}$ and $m\widehat{a} = 102°$

(g) $m\angle B$ if $m\widehat{a} + m\widehat{b} = 3m\widehat{c}$

(c) $m\angle A$ if $\overline{AC}$ is a diameter and $m\widehat{a} = 80°$

(h) $m\angle B$ if $m\widehat{a} = 75°$ and $m\widehat{c} = 2m\widehat{b}$

(d) $m\angle A$ if $m\widehat{a}:m\widehat{b}:m\widehat{c} = 3:1:2$

(i) $m\angle C$ if $\overline{AB} \perp \overline{BC}$ and $m\widehat{a} = 5m\widehat{b}$

(e) $m\angle A$ in $\overline{AC}$ is a diameter and $m\widehat{a}:m\widehat{b} = 5:4$

(j) $m\widehat{c}$ if $m\angle A:m\angle B:m\angle C = 5:4:3$

36. If $ABCP$ is an inscribed quadrilateral, $\overleftrightarrow{PD}$ a tangent, and $\overrightarrow{AF}$ a secant in Fig. 8-64, find: (8.16)

(a) $m\angle 1$ if $m\widehat{a} = 94°$ and $m\widehat{c} = 54°$

(g) $m\widehat{a}$ if $\overline{BC} \parallel \overline{AP}$ and $m\angle 6 = 42°$

(b) $m\angle 2$ if $\overline{AP}$ is a diameter

(h) $m\widehat{a}$ if $\overline{AC}$ is a diameter and $m\angle 5 = 35°$

(c) $m\angle 3$ if $m\widehat{CPA} = 250°$

(i) $m\widehat{b}$ if $\overline{AC} \perp \overline{BP}$ and $m\angle 2 = 57°$

(d) $m\angle 3$ if $m\angle ABC = 120°$

(j) $m\widehat{c}$ if $\overline{AC}$ and $\overline{BP}$ are diameters and $m\angle 5 = 41°$

(e) $m\angle 4$ if $m\widehat{BCP} = 130°$ and $m\widehat{b} = 50°$

(k) $m\widehat{d}$ if $m\angle 1 = 95°$ and $m\widehat{b} = 95°$

(f) $m\angle 4$ if $\overline{BC} \parallel \overline{AP}$ and $m\widehat{a} = 74°$

(l) $m\angle CPB$ if $m\angle 3 = 79°$

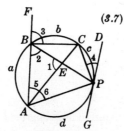

(8.7)

Fig. 8-64

37. Find x and y in each part of Fig. 8-65 (t and t' are tangents).   (8.16)

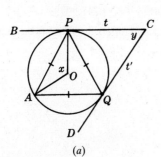

(a)

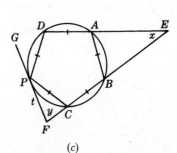

(b) $PBCA$ is an inscribed square.

(c)

Fig. 8-65

38. Find x and y in each part of Fig. 8-66.   (8.16)

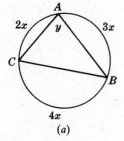

(a)

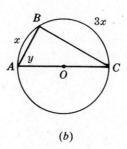

(b)

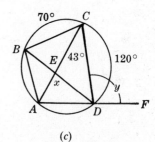

(c)

Fig. 8-66

39. Provide the proofs requested in Fig. 8-67.   (8.17)

(a) **Given:** $\overline{AC}$ bisects $\angle A$
     To Prove: $\overline{BC} \cong \overline{CD}$

(b) **Given:** $\overline{BC} \cong \overline{CD}$
     To Prove: $\overline{AC}$ bisects $\angle A$

(c) **Given:** $\overline{AB} \| \overline{CD}$
     $\overleftrightarrow{AB}$ is a tangent.
     To Prove: $PC = PD$

(d) **Given:** $PC = PD$
     $\overleftrightarrow{AB}$ is a tangent.
     To Prove: $\overline{AB} \| \overline{CD}$

(e) **Given:** $m\widehat{AC} = m\widehat{BD}$
     To Prove: $CE = ED$

(f) **Given:** $CE = EB$
     To Prove: $\widehat{AC} \cong \widehat{BD}$

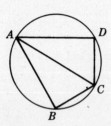   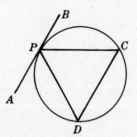   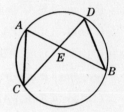

Fig. 8-67

40. Prove each of the following: (8.18)

(a) The base angles of an inscribed trapezoid are congruent.

(b) A parallelogram inscribed in a circle is a rectangle.

(c) In a circle, parallel chords intercept equal arcs.

(d) Diagonals drawn from a vertex of a regular inscribed pentagon trisect the vertex angle.

(e) If a tangent through a vertex of an inscribed triangle is parallel to its opposite side, the triangle is isosceles.

CHAPTER 9

Similarity

9.1 RATIOS

Ratios are used to compare quantities by division: the ratio of two quantities is the first divided by the second. A ratio is an abstract number, that is, a number without a unit of measure. Thus the ratio of 10 ft to 5 ft is 10 ft ÷ 5 ft, which equals 2.

A ratio can be expressed in the following ways: (1) using a colon, as in $3:4$; (2) using "to" as in 3 to 4; (3) as a common fraction, as in $\frac{3}{4}$; (4) as a decimal, 0.75; and (5) as a percent, 75%.

The quantities involved in a ratio must have the same unit. A ratio should be simplified by reducing to lowest terms and eliminating fractions. Thus to find the ratio of 1 ft to 4 in, we first change the foot to 12 inches, and then take the ratio of 12 inches to 4 inches; the result is a ratio of 3 to 1, or 3. Also, the ratio of $2\frac{1}{2}:\frac{1}{2}$ would be restated as $5:1$ or 5.

The ratio of three or more quantities may be expressed as a *continued ratio*. Thus the ratio of $2 to $3 to $5 is the continued ratio $2:3:5$. This enlarged ratio is a combination of three separate ratios; these are $2:3$, $3:5$, and $2:5$.

Throughout this chapter, the reader should use a calculator whenever s/he chooses. See Chapter 2 for an introduction to the calculator and its use in arithmetic and algebra.

SOLVED PROBLEMS

9.1 RATIO OF TWO QUANTITIES WITH THE SAME UNIT

Express each of the following ratios in lowest terms: (*a*) 15° to 3°, (*b*) $1.25 to $5; (*c*) $2\frac{1}{2}$ years to 2 years.

Solutions

(*a*) $\dfrac{15}{3} = 5$ (*b*) $\dfrac{1.25}{5} = \dfrac{1}{4}$ (*c*) $\dfrac{2\frac{1}{2}}{2} = \dfrac{5}{4}$

9.2 RATIO OF TWO QUANTITIES WITH DIFFERENT UNITS

Express each of the following ratios in lowest terms: (*a*) 2 years to 3 months; (*b*) 80 cents to $3.20.

Solutions

(*a*)   2 years to 3 months = 24 months to 3 months $= \dfrac{24}{3} = 8$

(*b*)   80 cents to $3.20 = 80 cents to 320 cents $= \dfrac{80}{320} = \dfrac{1}{4}$

9.3   CONTINUED RATIO OF THREE QUANTITIES

Express each of the following ratios in lowest terms: (a) 1 gal to 2 qt to 2 pt; (b) 1 ton to 1 lb to 8 oz.

Solutions

(a)   1 gal to 2 qt to 2 pt = 4 qt to 2 qt to 1 qt = 4:2:1

(b)   1 ton to 1 lb to 8 oz = 2000 lb to 1 lb to $\frac{1}{2}$ lb = $2000:1:\frac{1}{2}$ = 4000:2:1

9.4   NUMERICAL AND ALGEBRAIC RATIOS

Express each of the following ratios in lowest terms: (a) 50 to 60; (b) 6.3 to 0.9; (c) 12 to $\frac{3}{8}$; (d) $2x$ to $5x$; (e) $5s^2$ to s^3; (f) x to $5x$ to $7x$.

Solutions

(a) $\dfrac{50}{60} = \dfrac{5}{6}$ (d) $\dfrac{2x}{5x} = \dfrac{2}{5}$

(b) $\dfrac{6.3}{0.9} = 7$ (e) $\dfrac{5s^2}{s^3} = \dfrac{5}{s}$

(c) $12 \div \dfrac{3}{8} = 32$ (f) $x:5x:7x = 1:5:7$

9.5   USING RATIOS IN ANGLE PROBLEMS

If two angles are in the ratio of 3:2, find the angles if (a) they are adjacent and form an angle measuring 40°; (b) they are acute angles of a right triangle; (c) they are two angles of a triangle whose third angle measures 70°.

Solutions

Let the measures of the angles be $3x$ and $2x$. Then:

(a) $3x + 2x = 40$, so that $5x = 40$ or $x = 8$; hence the angles measure 24° and 16°.

(b) $3x + 2x = 90$, so $5x = 90$ or $x = 18$; hence the angles measure 54° and 36°.

(c) $3x + 2x + 70 = 180$, so $5x = 110$ or $x = 22$; hence the angles measure 66° and 44°.

9.6   THREE ANGLES HAVING A FIXED RATIO

Three angles are in the ratio of 4:3:2. Find the angles if (a) the first and the third are supplementary; (b) the angles are the three angles of a triangle.

Solutions

Let the measures of the angles be $4x$, $3x$, and $2x$. Then:

(a) $4x + 2x = 180$, so that $6x = 180$ for $x = 30$; hence the angles measure 120°, 90°, and 60°.

(b) $4x + 3x + 2x = 180$, so $9x = 180$ or $x = 20$; hence the angles measure 80°, 60°, and 40°.

9.2   PROPORTIONS

A *proportion* is an equality of two ratios. Thus 2:5 = 4:10 (or $\frac{2}{5} = \frac{4}{10}$) is a proportion.

The fourth term of a proportion is the *fourth proportional* to the other three taken in order. Thus in 2:3 = 4:x, x is the fourth proportional to 2, 3, and 4.

The *means* of a proportion are its middle terms, that is, its second and third terms. The *extremes* of a proportion are its outside terms, that is, its first and fourth terms. Thus in $a:b = c:d$, the means are b and c, and the extremes are a and d.

If the two means of a proportion are the same, either mean is the *mean proportional* between the first and fourth terms. Thus in $9:3 = 3:1$, 3 is the mean proportional between 9 and 1.

9.2A Proportion Principles

PRINCIPLE 1:   *In any proportion, the product of the means equals the product of the extremes.*

Thus if $a:b = c:d$, then $ad = bc$.

PRINCIPLE 2:   *If the product of two numbers equals the product of two other numbers, either pair may be made the means of a proportion and the other pair may be made the extremes.*

Thus if $3x = 5y$, then $x:y = 5:3$ or $y:x = 3:5$ or $3:y = 5:x$ or $5:x = 3:y$.

9.2B Methods of Changing a Proportion Into an Equivalent Proportion

PRINCIPLE 3:   (Inversion method) *A proportion may be changed into an equivalent proportion by inverting each ratio.*

Thus if $\dfrac{1}{x} = \dfrac{4}{5}$, then $\dfrac{x}{1} = \dfrac{5}{4}$.

PRINCIPLE 4:   (Alternation method) *A proportion may be changed into an equivalent proportion by interchanging the means or by interchanging the extremes.*

Thus if $\dfrac{x}{3} = \dfrac{y}{2}$, then $\dfrac{x}{y} = \dfrac{3}{2}$ or $\dfrac{2}{3} = \dfrac{y}{x}$.

PRINCIPLE 5:   (Addition method) *A proportion may be changed into an equivalent proportion by adding terms in each ratio to obtain new first and third terms.*

Thus if $\dfrac{a}{b} = \dfrac{c}{d}$, then $\dfrac{a+b}{b} = \dfrac{c+d}{d}$. If $\dfrac{x-2}{2} = \dfrac{9}{1}$, then $\dfrac{x}{2} = \dfrac{10}{1}$.

PRINCIPLE 6:   (Subtraction method) *A proportion may be changed into an equivalent proportion by subtracting terms in each ratio to obtain new first and third terms.*

Thus if $\dfrac{a}{b} = \dfrac{c}{d}$, then $\dfrac{a-b}{b} = \dfrac{c-d}{d}$. If $\dfrac{x+3}{3} = \dfrac{9}{1}$, then $\dfrac{x}{3} = \dfrac{8}{1}$.

9.2C Other Proportion Principles

PRINCIPLE 7:   *If any three terms of one proportion equal the corresponding three terms of another proportion, the remaining terms are equal.*

Thus if $\dfrac{x}{y} = \dfrac{3}{5}$ and $\dfrac{x}{4} = \dfrac{3}{5}$, then $y = 4$.

PRINCIPLE 8:   *In a series of equal ratios, the sum of any of the numerators is to the sum of the corresponding denominators as any numerator is to its denominator.*

Thus if $\dfrac{a}{b} = \dfrac{c}{d} = \dfrac{e}{f}$, then $\dfrac{a+c+e}{b+d+f} = \dfrac{a}{b}$. If $\dfrac{x-y}{4} = \dfrac{y-3}{5} = \dfrac{3}{1}$, then $\dfrac{x-y+y-3+3}{4+5+1} = \dfrac{3}{1}$ or $\dfrac{x}{10} = \dfrac{3}{1}$.

SOLVED PROBLEMS

9.7   **FINDING UNKNOWNS IN PROPORTIONS**
Solve the following proportions for x:

(a) $x:4 = 6:8$ (c) $x:5 = 2x:(x+3)$ (e) $\dfrac{x}{2x-3} = \dfrac{3}{5}$

(b) $3:x = x:27$ (d) $\dfrac{3}{x} = \dfrac{2}{5}$ (f) $\dfrac{x-2}{4} = \dfrac{7}{x+2}$

Solutions

(a)   Since $4(6) = 8x$, $8x = 24$ or $x = 3$.

(b)   Since $x^2 = 3(27)$, $x^2 = 81$ or $x = \pm 9$.

(c)   Since $5(2x) = x(x+3)$, we have $10x = x^2 + 3x$. Then $x^2 - 7x = 0$, so $x = 0$ or 7.

(d)   Since $2x = 3(5)$, $2x = 15$ or $x = 7\frac{1}{2}$.

(e)   Since $3(2x-3) = 5x$, we have $6x - 9 = 5x$, so $x = 9$.

(f)   Since $4(7) = (x-2)(x+2)$, we have $28 = x^2 - 4$. Then $x^2 = 32$, so $x = \pm 4\sqrt{2}$.

9.8   **FINDING FOURTH PROPORTIONALS TO THREE GIVEN NUMBERS**
Find the fourth proportional to (a) 2, 4, 6; (b) 4, 2, 6; (c) $\frac{1}{2}$, 3, 4; (d) b, d, c.

Solutions

(a)   We have $2:4 = 6:x$, so $2x = 24$ or $x = 12$.

(b)   We have $4:2 = 6:x$, so $4x = 12$ or $x = 3$.

(c)   We have $\frac{1}{2}:3 = 4:x$, so $\frac{1}{2}x = 12$ or $x = 24$.

(d)   We have $b:d = c:x$, so $bx = cd$ or $x = cd/b$.

9.9   **FINDING THE MEAN PROPORTIONAL TO TWO GIVEN NUMBERS**
Find the positive mean proportional x between (a) 5 and 20; (b) $\frac{1}{2}$ and $\frac{8}{9}$.

Solutions

(a)   We have $5:x = x:20$, so $x^2 = 100$ or $x = 10$.

(b)   We have $\frac{1}{2}:x = x:\frac{8}{9}$, so $x^2 = \frac{4}{9}$ or $x = \frac{2}{3}$.

9.10   **CHANGING EQUAL PRODUCTS INTO PROPORTIONS**

(a)   Form a proportion whose fourth term is x and such that $2bx = 3s^2$.

(b)   Find the ratio x to y if $ay = bx$.

Solutions

(a) $2b:3s = s:x$ or $2b:3 = s^2:x$ or $2b:s^2 = 3:x$ (b) $x:y = a:b$

9.11   CHANGING PROPORTIONS INTO NEW PROPORTIONS

Use each of the following to form a new proportion whose first term is x:

(a) $\dfrac{15}{x} = \dfrac{3}{4}$ (b) $\dfrac{x-6}{6} = \dfrac{5}{3}$ (c) $\dfrac{x+8}{8} = \dfrac{4}{3}$ (d) $\dfrac{5}{2} = \dfrac{15}{x}$

Solutions

(a) By Principle 3, $\dfrac{x}{15} = \dfrac{4}{3}$.   (c) By Principle 6, $\dfrac{x}{8} = \dfrac{1}{3}$.

(b) By Principle 5, $\dfrac{x}{6} = \dfrac{8}{3}$.   (d) By Principle 4, $\dfrac{x}{2} = \dfrac{15}{5}$.

9.12   COMBINING NUMERATORS AND DENOMINATORS OF PROPORTIONS

Use Principle 8 to find x in each of the following proportions:

(a) $\dfrac{x-2}{9} = \dfrac{2}{3}$ (b) $\dfrac{x+y}{8} = \dfrac{x-y}{4} = \dfrac{2}{3}$ (c) $\dfrac{3x-y}{15} = \dfrac{y-3}{10} = \dfrac{3}{5}$.

Solutions

(a) Adding numerators and denominators yields $\dfrac{x-2+2}{9+3} = \dfrac{2}{3}$ or $\dfrac{x}{12} = \dfrac{2}{3}$, so $x = 8$.

(b) Here we have $\dfrac{(x+y)+(x-y)}{8+4} = \dfrac{2}{3}$, which gives $\dfrac{2x}{12} = \dfrac{2}{3}$, so $x = 4$.

(c) We use all three ratios to get $\dfrac{(3x-y)+(y-3)+3}{15+10+5} = \dfrac{3}{5}$ or $\dfrac{3x}{30} = \dfrac{3}{5}$, so $x = 6$.

9.3   PROPORTIONAL SEGMENTS

If two segments are divided proportionately, (1) the corresponding new segments are in proportion, and (2) the two original segments and either pair of corresponding new segments are in proportion.

Thus if $\overline{AB}$ and $\overline{AC}$ in Fig. 9-1 are divided proportionately by $\overline{DE}$, we may write a proportion such as $\dfrac{a}{b} = \dfrac{c}{d}$ using the four segments; or we may write a proportion such as $\dfrac{a}{AB} = \dfrac{c}{AC}$ using the two original segments and two of their new segments.

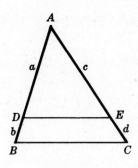

Fig. 9-1

9.3A Obtaining the Eight Arrangements of Any Proportion

A proportion such as $\dfrac{a}{b} = \dfrac{c}{d}$ can be arranged in eight ways. To obtain the eight variations, we let each term of the proportion represent one of the new segments of Fig. 9-1. Two of the possible proportions are then obtained from each direction, as follows:

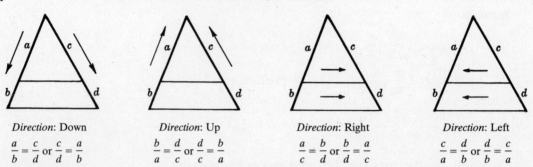

Direction: Down	*Direction:* Up	*Direction:* Right	*Direction:* Left
$\dfrac{a}{b} = \dfrac{c}{d}$ or $\dfrac{c}{d} = \dfrac{a}{b}$	$\dfrac{b}{a} = \dfrac{d}{c}$ or $\dfrac{d}{c} = \dfrac{b}{a}$	$\dfrac{a}{c} = \dfrac{b}{d}$ or $\dfrac{b}{d} = \dfrac{a}{c}$	$\dfrac{c}{a} = \dfrac{d}{b}$ or $\dfrac{d}{b} = \dfrac{c}{a}$

9.3B Principles of Proportional Segments

PRINCIPLE 1: *If a line is parallel to one side of a triangle, then it divides the other two sides proportionately.*

Thus in $\triangle ABC$ of Fig. 9-2, if $\overline{DE} \,\|\, \overline{BC}$, then $\dfrac{a}{b} = \dfrac{c}{d}$.

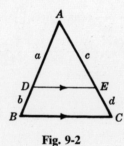

Fig. 9-2

PRINCIPLE 2: *If a line divides two sides of a triangle proportionately, it is parallel to the third side.* (Principles 1 and 2 are converses.)

Thus in $\triangle ABC$ (Fig. 9-2), if $\dfrac{a}{b} = \dfrac{c}{d}$, then $\overline{DE} \,\|\, \overline{BC}$.

PRINCIPLE 3: *Three or more parallel lines divide any two transversals proportionately.*

Thus if $\overleftrightarrow{AB} \,\|\, \overleftrightarrow{EF} \,\|\, \overleftrightarrow{CD}$ in Fig. 9-3, then $\dfrac{a}{b} = \dfrac{c}{d}$.

PRINCIPLE 4: *A bisector of an angle of a triangle divides the opposite side into segments which are proportional to the adjacent sides.*

Thus in $\triangle ABC$ of Fig. 9-4, if $\overline{CD}$ bisects $\angle C$, then $\dfrac{a}{b} = \dfrac{c}{d}$.

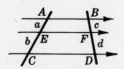

Fig. 9-3

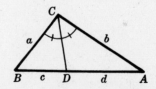

Fig. 9-4

SOLVED PROBLEMS

9.13 APPLYING PRINCIPLE 1

Find x in each part of Fig. 9-5.

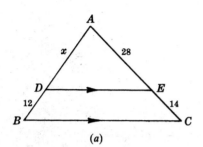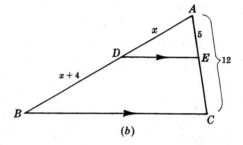

(a)   (b)

Fig. 9-5

Solutions

(a) $\overline{DE} \| \overline{BC}$; hence $\dfrac{x}{12} = \dfrac{28}{14}$, so that $x = 24$.

(b) We have $EC = 7$ and $\overline{DE} \| \overline{BC}$; hence $\dfrac{x}{x+4} = \dfrac{5}{7}$. Then $7x = 5x + 20$ and $x = 10$.

9.14 APPLYING PRINCIPLE 3

Find x in each part of Fig. 9-6.

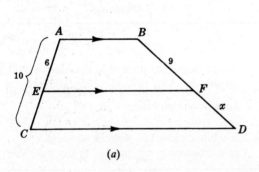

(a)   (b)

Fig. 9-6

Solutions

(a) We have $EC = 4$ and $\overline{AB} \| \overline{EF} \| \overline{CD}$; hence $\dfrac{x}{9} = \dfrac{4}{6}$ and $x = 6$.

(b) $\overleftrightarrow{AB} \| \overleftrightarrow{CD} \| \overleftrightarrow{EF}$; hence $\dfrac{5x-5}{2x+1} = \dfrac{7}{4}$, from which $20x - 20 = 14x + 7$. Then $6x = 27$ and $x = 4\frac{1}{2}$.

9.15 APPLYING PRINCIPLE 4

Find x in each part of Fig. 9-7.

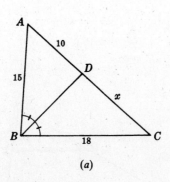

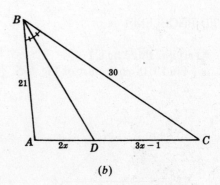

(a)   (b)

Fig. 9-7

Solutions

(a) $\overline{BD}$ bisects $\angle B$; hence $\dfrac{x}{10} = \dfrac{18}{15}$ and $x = 12$.

(b) $\overline{BD}$ bisects $\angle B$; hence $\dfrac{3x-1}{2x} = \dfrac{30}{21} = \dfrac{10}{7}$. Thus $21x - 7 = 20x$ and $x = 7$.

9.16   PROVING A PROPORTIONAL-SEGMENTS PROBLEM

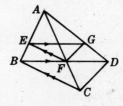

Given: $\overline{EG} \parallel \overline{BD}$, $\overline{EF} \parallel \overline{BC}$
To Prove: $\overline{FG} \parallel \overline{CD}$
Plan: Prove that $\overline{FG}$ divides $\overline{AC}$
and $\overline{AD}$ proportionately.

PROOF:

Statements	Reasons
1. $\overline{EG} \parallel \overline{BD}$, $\overline{EF} \parallel \overline{BC}$	1. Given
2. $\dfrac{AE}{EB} = \dfrac{AG}{GD}$, $\dfrac{AE}{EB} = \dfrac{AF}{FC}$	2. A line (segment) parallel to one side of a triangle divides the other two sides proportionately.
3. $\dfrac{AF}{FC} = \dfrac{AG}{GD}$	3. Substitution postulate.
4. $\overline{FG} \parallel \overline{CD}$	4. If a line divides two sides of a triangle proportionately, it is parallel to the third side.

9.4   SIMILAR TRIANGLES

Similar polygons are polygons whose corresponding angles are congruent and whose corresponding sides are in proportion. Similar polygons have the same shape although not necessarily the same size.

The symbol for "similar" is ~. The notation $\triangle ABC \sim \triangle A'B'C'$ is read "triangle ABC is similar to triangle A-prime B-prime C-prime." As in the case of congruent triangles, *corresponding sides of similar triangles are opposite congruent angles.* (Note that corresponding sides and angles are usually designated by the same letter and primes.)

In Fig. 9-8 $\triangle ABC \sim \triangle A'B'C'$ because

$$m\angle A = m\angle A' = 37° \qquad m\angle B = m\angle B' = 53° \qquad m\angle C = m\angle C' = 90°$$

and
$$\frac{a}{a'} = \frac{b}{b'} = \frac{c}{c'} \qquad \text{or} \qquad \frac{6}{3} = \frac{8}{4} = \frac{10}{5}$$

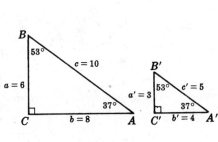
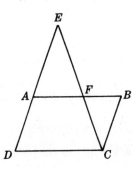

Fig. 9-8   **Fig. 9-9**

9.4A Selecting Similar Triangles to Prove a Proportion

In Solved Problem 9.25, it is given that $ABCD$ in a figure like Fig. 9-9 is a parallelogram, and we must prove that $\frac{AE}{BC} = \frac{AF}{FB}$. To prove this proportion, it is necessary to find similar triangles whose sides are in the proportion. This can be done simply by selecting the triangle whose letters A, E, and F are in the numerators and the triangle whose letters B, C, and F are in the denominators. Hence, we would prove $\triangle AEF \sim BCF$.

Suppose that the proportion to be proved is $\frac{AE}{AF} = \frac{BC}{FB}$. In such a case, interchanging the means leads to $\frac{AE}{BC} = \frac{AF}{FB}$. The needed triangles can then be selected based on the numerators and the denominators.

Suppose that the proportion to be proved is $\frac{AE}{AD} = \frac{AF}{FB}$. Then our method of selecting triangles could not be used until the term AD were replaced by BC. This is possible, since $\overline{AD}$ and $\overline{BC}$ are opposite sides of the parallelogram $ABCD$ and therefore are congruent.

9.4B Principles of Similar Triangles

PRINCIPLE 1: *Corresponding angles of similar triangles are congruent.* (By the definition)

PRINCIPLE 2: *Corresponding sides of similar triangles are in proportion.* (By the definition)

PRINCIPLE 3: *Two triangles are similar if two angles of one triangle are congruent respectively to two angles of the other.*

Thus in Fig. 9-10, if $\angle A \cong \angle A'$ and $\angle B \cong \angle B'$, then $\triangle ABC \sim \triangle A'B'C'$.

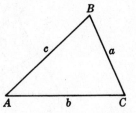

Fig. 9-10

PRINCIPLE 4:   *Two triangles are similar if an angle of one triangle is congruent to an angle of the other and the sides including these angles are in proportion.*

Thus in Fig. 9-10, if $\angle C \cong \angle C'$ and $\dfrac{a}{a'} = \dfrac{b}{b'}$, then $\triangle ABC \sim \triangle A'B'C'$.

PRINCIPLE 5:   *Two triangles are similar if their corresponding sides are in proportion.*

Thus in Fig. 9-10, if $\dfrac{a}{a'} = \dfrac{b}{b'} = \dfrac{c}{c'}$, then $\triangle ABC \sim A'B'C'$.

PRINCIPLE 6:   *Two right triangles are similar if an acute angle of one is congruent to an acute angle of the other.* (Corollary of Principle 3)

PRINCIPLE 7:   *A line parallel to a side of a triangle cuts off a triangle similar to the given triangle.*

Thus in Fig. 9-11, if $\overline{DE} \| \overline{BC}$, then $\triangle ADE \sim \triangle ABC$.

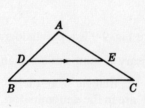

Fig. 9-11

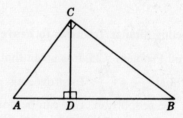

Fig. 9-12

PRINCIPLE 8:   *Triangles similar to the same triangle are similar to each other.*

PRINCIPLE 9:   *The altitude to the hypotenuse of a right triangle divides it into two triangles which are similar to the given triangle and to each other.*

Thus in Fig. 9-12, $\triangle CDA \sim \triangle CDB \sim \triangle ABC$.

PRINCIPLE 10:   *Triangles are similar if their sides are respectively parallel to each other.*

Thus in Fig. 9-13, $\triangle ABC \sim \triangle A'B'C'$.

PRINCIPLE 11:   *Triangles are similar if their sides are respectively perpendicular to each other.*

Thus in Fig. 9-14, $\triangle ABC \sim \triangle A'B'C'$.

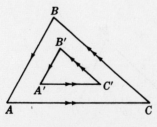

Fig. 9-13

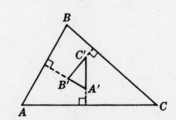

Fig. 9-14

SOLVED PROBLEMS

9.17 APPLYING PRINCIPLE 2

In similar triangles ABC and $A'B'C'$ (Fig. 9-15), find x and y if $\angle A \cong \angle A'$ and $\angle B \cong \angle B'$.

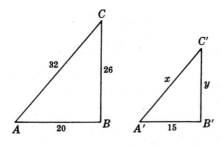

Fig. 9-15

Solution

Since $\angle A \cong \angle A'$ and $\angle B \cong \angle B'$, x and y correspond to 32 and 26, respectively. Hence $\dfrac{x}{32} = \dfrac{15}{20}$, from which $x = 24$; also $\dfrac{y}{26} = \dfrac{15}{20}$ so $y = 19\frac{1}{2}$.

9.18 APPLYING PRINCIPLE 3

In each part of Fig. 9-16, two pairs of congruent angles can be used to prove the indicated triangles similar. Determine the congruent angles and state the reason they are congruent.

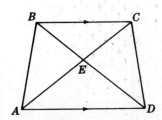 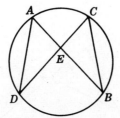 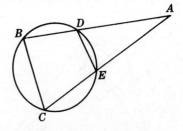

(a) $\triangle BEC \sim \triangle AED$ (b) $\triangle AED \sim \triangle CEB$ (c) $\triangle ADE \sim \triangle ABC$
      $ABCD$ is a trapezoid.

Fig. 9-16

Solutions

(a) $\angle CBD \cong \angle BDA$ and $\angle BCA \cong \angle CAD$, since alternate interior angles of parallel lines are congruent $(\overline{BC} \| \overline{AD})$. Also, $\angle BEC$ and $\angle AED$ are congruent vertical angles.

(b) $\angle A \cong \angle C$ and $\angle B \cong \angle D$, since angles inscribed in the same arc are congruent. Also, $\angle AED$ and $\angle CEB$ are congruent vertical angles.

(c) $\angle ABC \cong \angle AED$, since each is a supplement of $\angle DEC$. $\angle ACB \cong \angle ADE$, since each is a supplement of $\angle BDE$. Also, $\angle A \cong \angle A$.

9.19   APPLYING PRINCIPLE 6

In each part of Fig. 9-17, determine the angles that can be used to prove the indicated triangles similar.

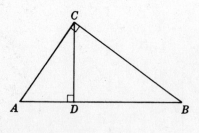

(a)   △ACD ~ △ACB

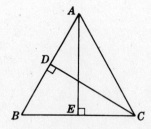

(b)   △AEC ~ △CDB
     AB = AC

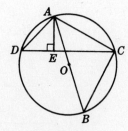

(c)   △ADE ~ △ABC
     $\overline{AB}$ is a diameter.

Fig. 9-17

Solutions

(a)   ∠ACB and ∠ADC are right angles. ∠A ≅ ∠A.

(b)   ∠AEC and ∠BDC are right angles. ∠B ≅ ∠ACE, since angles in a triangle opposite congruent sides are congruent.

(c)   ∠ACB is a right angle, since it is inscribed in a semicircle. Hence, ∠AED ≅ ∠ACB. ∠D ≅ ∠B, since angles inscribed in the same arc are congruent.

9.20   APPLYING PRINCIPLE 4

In each part of Fig. 9-18, determine the pair of congruent angles and the proportion needed to prove the indicated triangles similar.

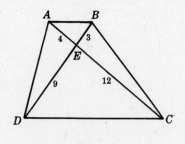

(a)   △AEB ~ △DEC

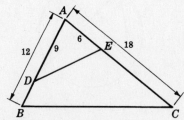

(b)   △AED ~ △ABC

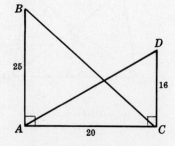

(c)   △ABC ~ △ADC

Fig. 9-18

Solutions

(a) $\angle AEB \cong \angle DEC; \dfrac{3}{9} = \dfrac{4}{12}$ (c) $\angle BAC \cong \angle ACD; \dfrac{20}{16} = \dfrac{25}{20}$

(b) $\angle A \cong \angle A; \dfrac{6}{12} = \dfrac{9}{18}$

9.21   APPLYING PRINCIPLE 5

In each part of Fig. 9-19, determine the proportion needed to prove the indicated triangles similar.

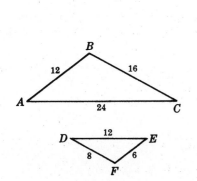

(a) $\triangle ABC \sim \triangle DEF$

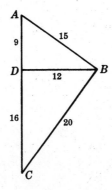

(b) $\triangle ABD \sim \triangle BDC$

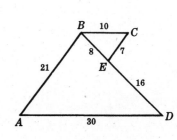

(c) $\triangle ABD \sim \triangle BEC$

Fig. 9-19

Solutions

(a) $\dfrac{6}{12} = \dfrac{8}{16} = \dfrac{12}{24}$ (c) $\dfrac{7}{21} = \dfrac{8}{24} = \dfrac{10}{30}$

(b) $\dfrac{9}{12} = \dfrac{12}{16} = \dfrac{15}{20}$

9.22   PROPORTIONS OBTAINED FROM SIMILAR TRIANGLES

Find x in each part of Fig. 9-20.

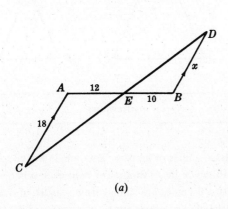

(a)

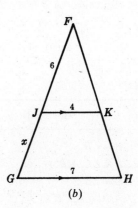

(b)

Fig. 9-20

Solutions

(a)   Since $\overline{BD} \| \overline{AC}$, $\angle A \cong \angle B$ and $\angle C \cong \angle D$; hence $\triangle AEC \sim \triangle DEB$. Then $\dfrac{x}{18} = \dfrac{10}{12}$ and $x = 15$.

(b)   Since $\overline{JK} \| \overline{GH}$, $\triangle FJK \sim \triangle FGH$ by Principle 7. Hence $\dfrac{6}{x+6} = \dfrac{4}{7}$ and $x = 4\frac{1}{2}$.

9.23 **FINDING HEIGHTS USING GROUND SHADOWS**

A tree casts a 15-ft shadow at a time when a nearby upright pole 6 ft high casts a shadow of 2 ft. Find the height of the tree if both tree and pole make right angles with the ground.

Solution

At the same time in localities near each other, the rays of the sun strike the ground at equal angles; hence $\angle B \cong \angle B'$ in Fig. 9-21. Since the tree and the pole make right angles with the ground, $\angle C \cong \angle C'$. Hence $\triangle ABC \sim \triangle A'B'C'$, so $\dfrac{h}{6} = \dfrac{15}{2}$ and $h = 45$ ft.

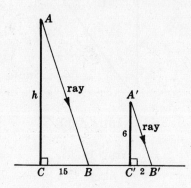

Fig. 9-21

9.24 **PROVING A SIMILAR-TRIANGLE PROBLEM STATED IN WORDS**

Prove that two isosceles triangles are similar if a base angle of one is congruent to a base angle of the other.

Solution

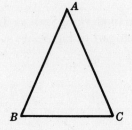
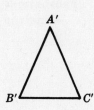

Given: Isosceles $\triangle ABC$ ($AB = AC$)
            Isosceles $\triangle A'B'C'$ ($A'B' = A'C'$)
            $\angle B \cong \angle B'$
To Prove: $\triangle ABC \sim \triangle A'B'C'$
Plan: Prove $\angle C \cong \angle C'$ and use Principle 3.

PROOF:

Statements	Reasons
1. $\angle B \cong \angle B'$	1. Given
2. $\angle B \cong \angle C$, $\angle B' \cong \angle C'$	2. Base angles of an isosceles triangle are congruent.
3. $\angle C \cong \angle C'$	3. Things $\cong$ to $\cong$ things are $\cong$ to each other.
4. $\triangle ABC \sim \triangle A'B'C'$	4. Two triangles are similar if two angles of one triangle are congruent to two angles of the other.

9.25 **PROVING A PROPORTION PROBLEM INVOLVING SIMILAR TRIANGLES**

Given: Parallelogram $ABCD$

To Prove: $\dfrac{AE}{BC} = \dfrac{AF}{BF}$

Plan: Prove $\triangle AEF \sim \triangle BFC$

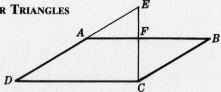

PROOF:

Statements	Reasons
1. $ABCD$ is a parallelogram.	1. Given
2. $\overline{ED} \parallel \overline{BC}$	2. Opposite sides of a parallelogram are parallel.
3. $\angle DEC \cong \angle ECB$	3. Alternate interior angles of parallel lines are congruent.
4. $\angle EFA \cong \angle BFC$	4. Vertical angles are congruent.
5. $\triangle AEF \sim \triangle BFC$	5. Two triangles are similar if two angles of one triangle are congruent to two angles of the other.
6. $\dfrac{AE}{BC} = \dfrac{AF}{BF}$	6. Corresponding sides of similar triangles are in proportion.

9.5 EXTENDING A BASIC PROPORTION PRINCIPLE

PRINCIPLE 1: *Corresponding sides of similar triangles are in proportion.*

PRINCIPLE 2: *Corresponding segments of similar triangles are in proportion.*

PRINCIPLE 3: *Corresponding segments of similar polygons are in proportion.*

When *segments* replaces *sides*, Principle 1 becomes the more general Principle 2. When *polygons* replaces *triangles*, Principle 2 becomes the even more general Principle 3.

By *segments* we mean straight or curved segments such as altitudes, medians, angle bisectors, radii of inscribed or circumscribed circles, and circumferences of inscribed or circumscribed circles.

The *ratio of similitude* of two similar polygons is the ratio of any pair of corresponding lines.

Corollaries of Principles 2 and 3, such as the following, can be devised for any combination of corresponding lines:

1. Corresponding *altitudes* of similar triangles have the same ratio as any two corresponding *medians*. Thus if $\triangle ABC \sim \triangle A'B'C'$ in Fig. 9-22, then $\dfrac{h}{h'} = \dfrac{m}{m'}$.

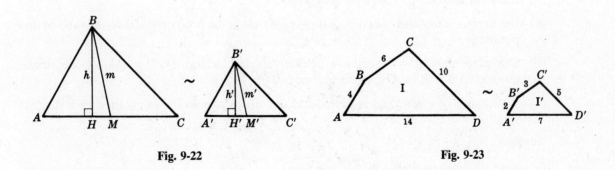

Fig. 9-22   Fig. 9-23

2. *Perimeters* of similar polygons have the same ratio as any two corresponding *sides*. Thus in Fig. 9-23, if quadrilateral I $\sim$ quadrilateral I′, then $\dfrac{34}{17} = \dfrac{4}{2} = \dfrac{6}{3} = \dfrac{10}{5} = \dfrac{14}{7}$.

SOLVED PROBLEMS

9.26 LINE RATIOS FROM SIMILAR TRIANGLES

(a) In two similar triangles, corresponding sides are in the ratio 3:2. Find the ratio of corresponding medians [see Fig. 9-24(a)].

(b) The sides of a triangle are 4, 6, and 7 [Fig. 9-24(b)]. If the perimeter of a similar triangle is 51, find its longest side.

(c) In $\triangle ABC$ of Fig. 9-24(c), $BC = 25$ and the measure of the altitude to $\overline{BC}$ is 10. A line segment terminating in the sides of the triangle is parallel to $\overline{BC}$ and 3 units from A. Find its length.

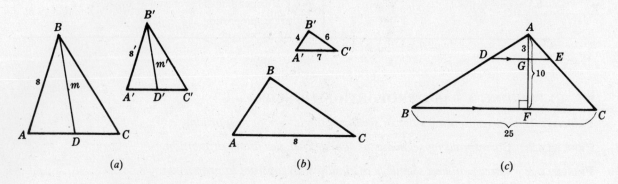

(a)   (b)   (c)

Fig. 9-24

Solutions

(a) If $\triangle ABC \sim \triangle A'B'C'$ and $\dfrac{s}{s'} = \dfrac{3}{2}$, then $\dfrac{m}{m'} = \dfrac{3}{2}$.

(b) The perimeter of $\triangle A'B'C'$ is $4 + 6 + 7 = 17$. Since $\triangle ABC \sim \triangle A'B'C'$, $\dfrac{s}{7} = \dfrac{51}{17}$ and $s = 21$.

(c) Since $\triangle ADE \sim ABC$, $\dfrac{DE}{25} = \dfrac{3}{10}$ and $DE = 7\frac{1}{2}$.

9.27 LINE RATIOS FROM SIMILAR POLYGONS

Complete each of the following statements:

(a) If corresponding sides of two similar polygons are in the ratio of 4:3, then the ratio of their perimeter is __?__.

(b) The perimeters of two similar quadrilaterals are 30 and 24. If a side of the smaller quadrilateral is 8, the corresponding side of the larger is __?__.

(c) If each side of a pentagon is tripled and the angles remain the same, then each diagonal is __?__.

Solutions

(a) Since the polygons are similar, $\dfrac{p}{p'} = \dfrac{s}{s'} = \dfrac{4}{3}$.

(b) Since the quadrilaterals are similar, $\dfrac{s}{s'} = \dfrac{p}{p'}$. Then $\dfrac{s}{8} = \dfrac{30}{24}$ and $s = 10$.

(c) Tripled, since polygons are similar if their corresponding angles are congruent and their corresponding sides are in proportion.

9.6　PROVING EQUAL PRODUCTS OF LENGTHS OF SEGMENTS

In a problem, to prove that the product of the lengths of two segments equals the product of the lengths of another pair of segments, it is necessary to set up the proportion which will lead to the two equal products.

SOLVED PROBLEM

9.28　**PROVING AN EQUAL-PRODUCTS PROBLEM**

Prove that if two secants intersect outside a circle, the product of the lengths of one of the secants and its external segment equals the product of the lengths of the other secant and its external segment.

Solution

Given:　Secants $\overline{AB}$ and $\overline{AC}$.
To Prove:　$AB \times AD = AC \times AE$
Plan:　Prove $\triangle ABE \sim \triangle ACD$ to obtain

$$\frac{AB}{AC} = \frac{AE}{AD}.$$

PROOF:

Statements	Reasons
1.　Draw $\overline{BE}$ and $\overline{CD}$.	1.　A segment may be drawn between any two points.
2.　$\angle A \cong \angle A$	2.　Reflexive property.
3.　$\angle B \cong \angle C$	3.　Angles inscribed in the same arc are congruent.
4.　$\triangle AEB \sim \triangle ADC$	4.　Two triangles are similar if two angles of one triangle are congruent respectively to two angles of the other.
5.　$\dfrac{AB}{AC} = \dfrac{AE}{AD}$	5.　Corresponding sides of similar triangles are in proportion.
6.　$AB \times AD = AC \times AE$	6.　In a proportion, the product of the means equals the product of the extremes.

9.7　SEGMENTS INTERSECTING INSIDE AND OUTSIDE A CIRCLE

PRINCIPLE 1:　*If two chords intersect within a circle, the product of the lengths of the segments of one chord equals the product of the lengths of the segments of the other.*

Thus in Fig. 9-25, $AE \times EB = CE \times ED$.

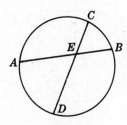

Fig. 9-25

PRINCIPLE 2:   *If a tangent and a secant intersect outside a circle, the tangent is the mean proportional between the secant and its external segment.*

Thus in Fig. 9-26, if $\overline{PA}$ is a tangent, then $\dfrac{AB}{AP} = \dfrac{AP}{AC}$.

PRINCIPLE 3:   *If two secants intersect outside a circle, the product of the lengths of one of the secants and its external segment equals the product of the lengths of the other secant and its external segment.*

Thus in Fig. 9-27 $AB \times AD = AC \times AE$.

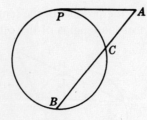

Fig. 9-26

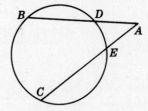

Fig. 9-27

SOLVED PROBLEMS

9.29   APPLYING PRINCIPLE 1

Find x in each part of Fig. 9-28 if chords $\overline{AB}$ and $\overline{CD}$ intersect in E.

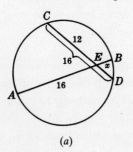

(a)

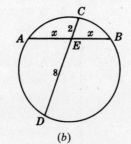

(b)

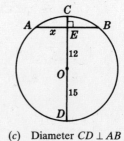

(c)   Diameter $CD \perp AB$

Fig. 9-28

Solutions

(a) $ED = 4$. Then $16x = 4(12)$, so that $16x = 48$ or $x = 3$.

(b) $AE = EB = x$. Then $x^2 = 8(2)$, so $x^2 = 16$ and $x = 4$.

(c) $CE = 3$ and $AE = EB = x$. Then $x^2 = 27(3)$ or $x^2 = 81$, and $x = 9$.

9.30   APPLYING PRINCIPLE 2

Find x in each part of Fig. 9-29 if tangent $\overline{AP}$ and $\overline{AB}$ intersect at A.

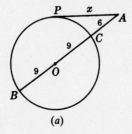

(a)

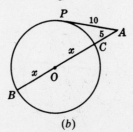

(b)

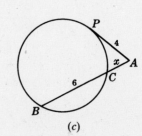

(c)

Fig. 9-29

Solutions

(a) $AB = 9 + 9 + 6 = 24$. Then $x^2 = 24(6)$ or $x^2 = 144$, and $x = 12$.

(b) $AB = 2x + 5$. Then $5(2x + 5) = 100$ and $x = 7\frac{1}{2}$.

(c) $AB = x + 6$. Then $x(x + 6) = 16$ or $x^2 + 6x - 16 = 0$. Factoring gives $(x + 8)(x - 2) = 0$ and $x = 2$.

9.31   APPLYING PRINCIPLE 3

Find x in each part of Fig. 9-30 if secants $\overline{AB}$ and $\overline{AC}$ intersect in A.

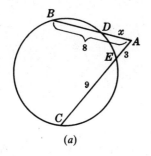   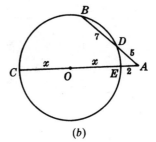

(a)   (b)

Fig. 9-30

Solutions

(a) $AC = 12$. Then $8x = 12(3)$ and $x = 4\frac{1}{2}$.

(b) $AC = 2x + 2$ and $AB = 12$. Then $2(2x + 2) = 12(5)$ and $x = 14$.

9.8   MEAN PROPORTIONALS IN A RIGHT TRIANGLE

PRINCIPLE 1:   *The length of the altitude to the hypotenuse of a right triangle is the mean proportional between the lengths of the segments of the hypotenuse.*

Thus in right $\triangle ABC$ (Fig. 9-31), $\dfrac{BD}{CD} = \dfrac{CD}{DA}$.

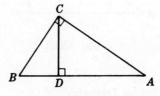

Fig. 9-31

PRINCIPLE 2:   *In a right triangle, the length of either leg is the mean proportional between the length of the hypotenuse and the length of the projection of that leg on the hypotenuse.*

Thus in right $\triangle ABC$, $\dfrac{AB}{BC} = \dfrac{BC}{BD}$ and $\dfrac{AB}{AC} = \dfrac{AC}{AD}$.

A proof of this principle is given in Chapter 16.

SOLVED PROBLEMS

9.32   FINDING MEAN PROPORTIONALS IN A RIGHT TRIANGLE

In each triangle in Fig. 9-32, find x and y.

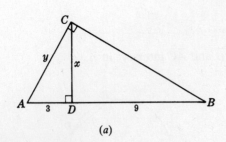   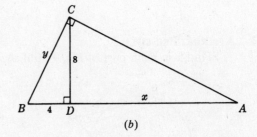

(a)   (b)

Fig. 9-32

Solutions

(a)   By Principle 1, $\dfrac{3}{x} = \dfrac{x}{9}$ or $x^2 = 27$, and $x = 3\sqrt{3}$. By Principle 2, $\dfrac{12}{y} = \dfrac{y}{3}$, so $y^2 = 36$ and $y = 6$.

(b)   By Principle 1, $\dfrac{x}{8} = \dfrac{8}{4}$ and $x = 16$. By Principle 2, $\dfrac{20}{y} = \dfrac{y}{4}$, so $y^2 = 80$ and $y = 4\sqrt{5}$.

9.9   PYTHAGOREAN THEOREM

In a right triangle, the square of the length of the hypotenuse equals the sum of the squares of the lengths of the legs. Thus in Fig. 9-33, $c^2 = a^2 + b^2$.

A proof of the Pythagorean Theorem is given in Chapter 16.

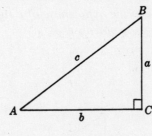

Fig. 9-33

9.9A   Tests for Right, Acute, and Obtuse Triangles

If $c^2 = a^2 + b^2$ applies to the three sides of a triangle, then the triangle is a right triangle; but if $c^2 \neq a^2 + b^2$, then the triangle is not a right triangle.

In $\triangle ABC$, if $c^2 < a^2 + b^2$ where c is the longest side of the triangle, then the triangle is an acute triangle.

Thus in Fig. 9-34, $9^2 < 6^2 + 8^2$ (that is, $81 < 100$); hence $\triangle ABC$ is an acute triangle.

In $\triangle ABC$, if $c^2 > a^2 + b^2$ where c is the longest side of the triangle, then the triangle is an obtuse triangle.

Thus in Fig. 9-35, $11^2 > 6^2 + 8^2$ (that is, $121 > 100$); hence $\triangle ABC$ is an obtuse triangle.

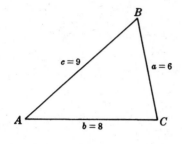

Fig. 9-34

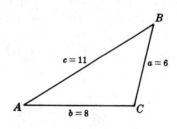

Fig. 9-35

SOLVED PROBLEMS

9.33   FINDING THE SIDES OF A RIGHT TRIANGLE

In Fig. 9-36, (a) find the length of hypotenuse c if $a = 12$ and $b = 9$; (b) find a if $b = 6$ and $c = 8$; (c) find b if $a = 4\sqrt{3}$ and $c = 8$.

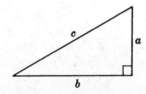

Fig. 9-36

Solutions

(a) $c^2 = a^2 + b^2 = 12^2 + 9^2 = 225$ and $c = 15$.

(b) $a^2 = c^2 - b^2 = 8^2 - 6^2 = 28$ and $a = 2\sqrt{7}$.

(c) $b^2 = c^2 - a^2 = 8^2 - (4\sqrt{3})^2 = 64 - 48 = 16$ and $b = 4$.

9.34   RATIOS IN A RIGHT TRIANGLE

In a right triangle, the hypotenuse has length 20 and the ratio of the two arms is $3:4$. Find each arm.

Solution

Let the lengths of the two arms be denoted by $3x$ and $4x$. Then $20^2 = (3x)^2 + (4x)^2$.

Multiplying out, we get $400 = 9x^2 + 16x^2$ or $400 = 25x^2$, and $x = 4$; hence the arms have lengths 12 and 16.

9.35   APPLYING THE PYTHAGOREAN THEOREM TO AN ISOSCELES TRIANGLE

Find the length of the altitude to the base of an isosceles triangle if the base is 8 and the equal sides are 12.

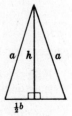

Solution

The altitude h of an isosceles triangle bisects the base (Fig. 9-37).
Then $h^2 = a^2 - (\tfrac{1}{2}b)^2 = 12^2 - 4^2 = 128$ and $h = 8\sqrt{2}$.

Fig. 9-37

9.36   APPLYING THE PYTHAGOREAN THEOREM TO A RHOMBUS

In a rhombus, find (*a*) the length of a side *s* if the diagonals are 30 and 40; (*b*) the length of a diagonal *d* if a side is 26 and the other diagonal is 20.

Solution

The diagonals of a rhombus are perpendicular bisectors of each other; hence $s^2 = (\frac{1}{2}d)^2 + (\frac{1}{2}d')^2$ in Fig. 9-38.

(*a*)   If $d = 30$ and $d' = 40$, then $s^2 = 15^2 + 20^2 = 625$ or $s = 25$.

(*b*)   If $s = 26$ and $d' = 20$, then $26^2 = (\frac{1}{2}d)^2 + 10^2$ or $576 = (\frac{1}{2}d)^2$. Thus $\frac{1}{2}d = 24$ or $d = 48$.

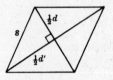

Fig. 9-38

9.37   APPLYING THE PYTHAGOREAN THEOREM TO A TRAPEZOID

Find *x* in each part of Fig. 9-39 if *ABCD* is a trapezoid.

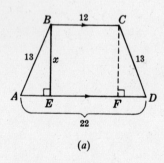

(*a*)   (*b*)

Fig. 9-39

Solutions

The dashed perpendiculars in the diagrams are additional segments needed only for the solutions. Note how rectangles are formed by these added segments.

(*a*) $EF = BC = 12$ and $AE = \frac{1}{2}(22 - 12) = 5$. Then $x^2 = 13^2 - 5^2 = 144$ or $x = 12$.

(*b*) $b^2 = 25^2 - 7^2 = 576$ or $b = 24$; also, $BE = b = 24$ and $CE = 17 - 7 = 10$. Then $x^2 = 24^2 + 10^2$ or $x = 26$.

9.38   APPLYING THE PYTHAGOREAN THEOREM TO A CIRCLE

(*a*)   Find the distance *d* from the center of a circle of radius 17 to a chord whose length is 30 [Fig. 9-40(*a*)].

(*b*)   Find the length of a common external tangent to two externally tangent circles with radii 4 and 9 [Fig. 9-40(*b*)].

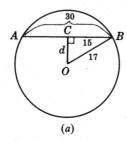

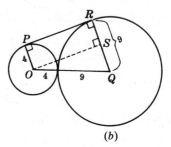

(a) (b)

Fig. 9-40

Solutions

(a) $BC = \frac{1}{2}(30) = 15$. Then $d^2 = 17^2 - 15^2 = 64$ and $d = 8$.

(b) $\overline{OS} \cong \overline{PR}$, $RS = 4$, $OQ = 13$, and $SQ = 9 - 4 = 5$. Then in right $\triangle OSQ$, $(OS)^2 = 13^2 - 5^2 = 144$ so $OS = 12$; hence $PR = 12$.

9.10   SPECIAL RIGHT TRIANGLES

9.10A   The 30°-60°-90° Triangle

A 30°-60°-90° triangle is one-half an equilateral triangle. Thus in right $\triangle ABC$ (Fig. 9-41), $a = \frac{1}{2}c$. Consider that $c = 2$; then $a = 1$, and the Pythagorean Theorem gives

$$b^2 = c^2 - a^2 = 2^2 - 1^2 = 3 \qquad \text{or} \qquad b = \sqrt{3}$$

The ratio of the sides is then $a:b:c = 1:\sqrt{3}:2$.

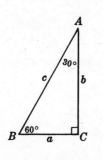

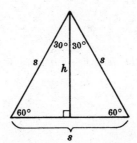

Fig. 9-41

Principles of the 30°-60°-90° Triangle

PRINCIPLE 1:   *The length of the leg opposite the 30° angle equals one-half the length of the hypotenuse.*

   In Fig. 9-41, $a = \frac{1}{2}c$.

PRINCIPLE 2:   *The length of the leg opposite the 60° angle equals one-half the length of the hypotenuse times the square root of 3.*

   In Fig. 9-41, $b = \frac{1}{2}c\sqrt{3}$.

PRINCIPLE 3:   *The length of the leg opposite the 60° angle equals the length of the leg opposite the 30° angle times the square root of 3.*

   In Fig. 9-41, $b = a\sqrt{3}$.

Equilateral-Triangle Principle

PRINCIPLE 4:   *The length of the altitude of an equilateral triangle equals one-half the length of a side times the square root of 3. (Principle 4 is a corollary of Principle 2.)*

In Fig. 9-41, $h = \frac{1}{2}s\sqrt{3}$.

9.10B   The 45°-45°-90° Triangle

A 45°-45°-90° triangle is one-half a square. In right triangle ABC (Fig. 9-42), $c^2 = a^2 + a^2$ or $c = a\sqrt{2}$. Hence the ratio of the sides is $a:a:c = 1:1:\sqrt{2}$.

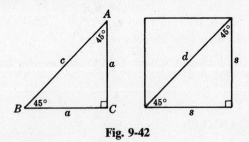

Fig. 9-42

Principles of the 45°-45°-90° Triangle

PRINCIPLE 5:   *The length of a leg opposite a 45° angle equals one-half the length of the hypotenuse times the square root of 2.*

In Fig. 9-42, $a = \frac{1}{2}c\sqrt{2}$.

PRINCIPLE 6:   *The length of the hypotenuse equals the length of a side times the square root of 2.*

In Fig. 9-42, $c = a\sqrt{2}$.

Square Principle

PRINCIPLE 7:   *In a square, the length of a diagonal equals the length of a side times the square root of 2.*

In Fig. 9-42, $d = s\sqrt{2}$.

SOLVED PROBLEMS

9.39   APPLYING PRINCIPLES 1 TO 4

(a)   If the length of the hypotenuse of a 30°-60°-90° triangle is 12, find the lengths of its legs [Fig. 9-43(a)].

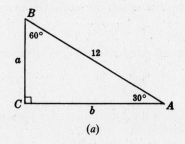

(a)

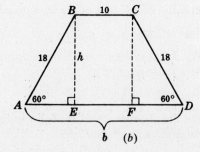

(b)

Fig. 9-43

(b) Each leg of an isosceles trapezoid has length 18. If the base angles are 60° and the upper base is 10, find the lengths of the altitude and the lower base [Fig. 9-43(b)].

Solutions

(a) By Principle 1, $a = \frac{1}{2}(12) = 6$. By Principle 2, $b = \frac{1}{2}(12)\sqrt{3} = 6\sqrt{3}$.

(b) By Principle 2, $h = \frac{1}{2}(18)\sqrt{3} = 9\sqrt{3}$. By Principle 1, $AE = FD = \frac{1}{2}(18) = 9$; hence $b = 9 + 10 + 9 = 28$.

9.40   APPLYING PRINCIPLES 5 AND 6

(a) Find the length of the leg of an isosceles right triangle whose hypotenuse has length 28 [Fig. 9-44(a)].

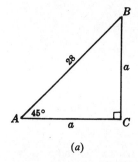

(a)

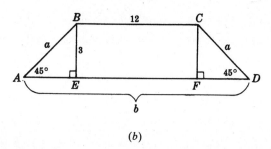

(b)

Fig. 9-44

(b) An isosceles trapezoid has base angles measuring 45°. If the upper base has length 12 and the altitude has length 3, find the lengths of the lower base and each leg [Fig. 9-44(b)].

Solutions

(a) By Principle 5, $a = \frac{1}{2}(28)\sqrt{2} = 14\sqrt{2}$.

(b) By Principle 6, $a = 3\sqrt{2}$. $AE = BE = 3$ and $EF = 12$; hence $b = 3 + 12 + 3 = 18$.

Supplementary Problems

1. Express each of the following ratios in lowest terms: (9.1)

   (a) 20 cents to 5 cents   (f) 50% to 25%   (k) $\frac{1}{2}$ lb to $\frac{1}{4}$ lb

   (b) 5 dimes to 15 dimes   (g) 15° to 75°   (l) $2\frac{1}{2}$ days to $3\frac{1}{2}$ days

   (c) 30 lb to 25 lb   (h) 33% to 77%   (m) 5 ft to $\frac{1}{4}$ ft

   (d) 20° to 14°   (i) $2.20 to $3.30   (n) $\frac{1}{2}$ yd to $1\frac{1}{2}$ yd

   (e) 27 min to 21 min   (j) $.84 to $.96   (o) $16\frac{1}{2}$ m to $5\frac{1}{2}$ m

2. Express each of the following ratios in lowest terms: (9.2)

   (a) 1 year to 2 months   (e) 2 yd to 2 ft   (i) 100 lb to 1 ton

   (b) 2 weeks to 5 days   (f) $2\frac{1}{3}$ yd to 2 ft   (j) $2 to 25 cents

   (c) 3 days to 3 weeks   (g) $1\frac{1}{2}$ ft to 9 in   (k) 2 quarters to 3 dimes

   (d) $\frac{1}{2}$ h to 20 min   (h) 2 g to 8 mg   (l) 1 yd² to 2 ft²

3. Express each of the following ratios in lowest terms: (9.3)

   (*a*)   20 cents to 30 cents to $1   (*f*)   2 h to $\frac{1}{2}$ h to 15 min

   (*b*) $3 to $1.50 to 25 cents   (*g*)   1 ton to 200 lb to 40 lb

   (*c*)   1 quarter to 1 dime to 1 nickel   (*h*)   3 lb to 1 lb to 8 oz

   (*d*)   1 day to 4 days to 1 week   (*i*)   1 gal to 1 qt to 1 pt

   (*e*) $\frac{1}{2}$ day to 9 h to 3 h

4. Express each of the following ratios in lowest terms: (9.4)

   (*a*)   60 to 70   (*e*)   630 to 105   (*i*)   0.002 to 0.007   (*m*) $7\frac{1}{2}$ to $2\frac{1}{2}$

   (*b*)   84 to 7   (*f*)   1760 to 990   (*j*)   0.055 to 0.005   (*n*) $1\frac{1}{2}$ to 10

   (*c*)   65 to 15   (*g*)   0.7 to 2.1   (*k*)   6.4 to 8   (*o*) $\frac{5}{6}$ to $1\frac{2}{3}$

   (*d*)   125 to 500   (*h*)   0.36 to 0.24   (*l*)   144 to 2.4   (*p*) $\frac{7}{4}$ to $\frac{1}{8}$

5. Express each of the following ratios in lowest terms: (9.4)

   (*a*) x to $3x$ (*d*) $2\pi r$ to πD (*g*) S^3 to $6S^2$ (*j*) $15y$ to $10y$ to $5y$

   (*b*) $15c$ to 5 (*e*) πab to πa^2 (*h*) $9r^2$ to $6rt$ (*k*) x^3 to x^2 to x

   (*c*) $11d$ to 22 (*f*) $4S$ to S^2 (*i*) x to $4x$ to $10x$ (*l*) $12w$ to $10w$ to $8w$ to $2w$

6. Use x as the common factor to represent the following numbers and their sum: (9.4)

   (*a*)   Two numbers whose ratio is 5:4   (*c*)   Three numbers whose ratio is 2:5:11

   (*b*)   Two numbers whose ratio is 9 to 1   (*d*)   Five numbers whose ratio is 1:2:2:3:7

7. If two angles in the ratio of 5:4 are represented by $5x$ and $4x$, express each of the following statements as an equation; then find x and the angles: (9.5)

   (*a*)   The angles are adjacent and together form an angle measuring 45°.

   (*b*)   The angles are complementary.

   (*c*)   The angles are supplementary.

   (*d*)   The angles are two angles of a triangle whose third angle is their difference.

8. If three angles in the ratio of 7:6:5 are represented by $7x$, $6x$, and $5x$, express each of the following statements as an equation; then find x and the angles: (9.6)

   (*a*)   The first and second are adjacent and together form an angle measuring 91°.

   (*b*)   The first and third are supplementary.

   (*c*)   The first and one-half the second are complementary.

   (*d*)   The angles are the three angles of a triangle.

9. Solve the following proportions for x: (9.7)

   (*a*) $x:6 = 8:3$ (*d*) $x:2 = 10:x$ (*g*) $a:b = c:x$

   (*b*) $5:4 = 20:x$ (*e*) $(x+4):3 = 3:(x-4)$ (*h*) $x:2y = 18y:x$

   (*c*) $9:x = x:4$ (*f*) $(2x+8):(x+2) = (2x+5):(x+1)$

10. Solve the following proportions for x: (9.7)

(a) $\dfrac{5}{7}=\dfrac{15}{x}$ (c) $\dfrac{3}{x}=\dfrac{x}{12}$ (e) $\dfrac{x+2}{5}=\dfrac{6}{3}$ (g) $\dfrac{2x}{x+7}=\dfrac{3}{5}$

(b) $\dfrac{7}{x}=\dfrac{3}{2}$ (d) $\dfrac{x}{5}=\dfrac{15}{x}$ (f) $\dfrac{x-1}{3}=\dfrac{5}{x+1}$ (h) $\dfrac{a}{x}=\dfrac{x}{b}$

11. Find the fourth proportional each of the following sets of numbers: (9.8)

(a) 1, 3, 5   (c) 2, 3, 4   (e) 3, 2, 5   (g) 2, 8, 8

(b) 8, 6, 4   (d) 3, 4, 2   (f) $\frac{1}{3}$, 2, 5   (h) b, $2a$, $3b$

12. Find the positive mean proportional between each of the following pairs of numbers: (9.9)

(a) 4 and 9   (c) $\frac{1}{3}$ and 27   (e) 2 and 5   (g) p and q

(b) 12 and 3   (d) $2b$ and $8b$ (f) 3 and 9   (h) a^2 and b

13. From each of these equations, form a proportion whose fourth term is x: (a) $cx = bd$; (b) $pq = ax$; (c) $hx = a^2$; (d) $3x = 7$; (e) $x = ab/c$.   (9.10)

14. In each of the following equations, find the ratio of x to y: (a) $2x = y$; (b) $3y = 4x$; (c) $x = \frac{1}{2}y$; (d) $ax = hy$; (e) $x = by$.   (9.10)

15. Which of the following is not a proportion?   (9.11)

(a) $\dfrac{4}{3}\overset{?}{=}\dfrac{24}{18}$ (b) $\dfrac{3}{5}\overset{?}{=}\dfrac{7}{12}$ (c) $\dfrac{25}{45}\overset{?}{=}\dfrac{10}{18}$ (d) $\dfrac{.2}{.3}\overset{?}{=}\dfrac{6}{9}$ (e) $\dfrac{x}{8}\overset{?}{=}\dfrac{3}{4}$ when $x = 6$.

16. From each of the following, form a new proportion whose first term is x. Then find x.   (9.11)

(a) $\dfrac{3}{2}=\dfrac{9}{x}$ (b) $\dfrac{1}{x}=\dfrac{5}{4}$ (c) $\dfrac{a}{x}=\dfrac{2}{b}$ (d) $\dfrac{x+5}{5}=\dfrac{11}{10}$ (e) $\dfrac{x-20}{20}=\dfrac{1}{4}$

17. Find x in each of these pairs of proportions: (9.11)

(a) $a:b = c:x$ and $a:b = c:d$ (c) $2:3x = 4:5y$ and $2:15 = 4:5y$

(b) $5:7 = x:42$ and $5:7 = 35:42$ (d) $7:5x - 2 = 14:3y$ and $7:18 = 14:3y$

18. Find x in each of the following proportions: (9.12)

(a) $\dfrac{x-7}{8}=\dfrac{7}{4}$ (b) $\dfrac{x+y}{6}=\dfrac{x-y}{3}=\dfrac{1}{3}$ (c) $\dfrac{2x-y}{8}=\dfrac{y-1}{10}=\dfrac{1}{2}$

19. Find x in each part of Fig. 9-45.   (9.13)

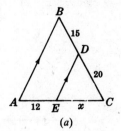   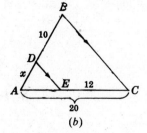   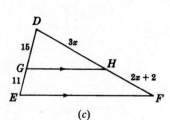

(a)   (b)   (c)

Fig. 9-45

20. In which parts of Fig. 9-46 is a line parallel to one side of the triangle? (9.13)

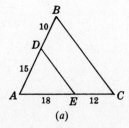 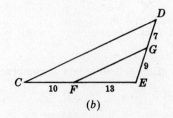 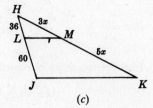

(a)   (b)   (c)

Fig. 9-46

21. Find x in each part of Fig. 9-47. (9.14)

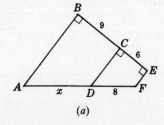 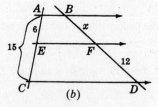 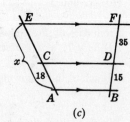

(a)   (b)   (c)

Fig. 9-47

22. Find x in each part of Fig. 9-48. (9.15)

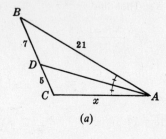 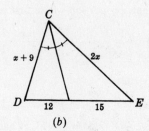 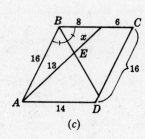

(a)   (b)   (c)

Fig. 9-48

23. Prove that three or more parallel lines divide any two transversals proportionately. (9.16)

24. In similar triangles ABC and $A'B'C'$ of Fig. 9-49, $\angle B$ and $\angle B'$ are corresponding angles. Find $m\angle B$ if (a) $m\angle A' = 120°$ and $m\angle C' = 25°$; (b) $m\angle A' + m\angle C' = 127°$. (9.17)

25. In similar triangles ABC and $A'B'C'$ of Fig. 9-50, $\angle A \cong \angle A'$ and $\angle B \cong \angle B'$. (a) Find a if $c = 24$; (b) find b if $a = 20$; (c) find c if $b = 63$. (9.17)

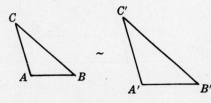 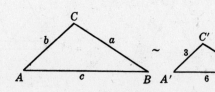

Fig. 9-49   **Fig. 9-50**

26. In each part of Fig. 9-51, show that the indicated triangles are similar.   (9.18)

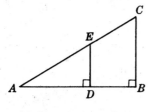

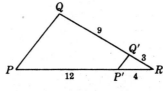

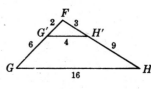

(a)   △ADE ~ △ABC   (b)   △RQP ~ △RQ'P'   (c)   △FG'H' ~ △FGH

Fig. 9-51

27. In each part of Fig. 9-52, two pairs of congruent angles can be used to prove the indicated triangles similar. Find the congruent angles.   (9.18)

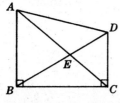

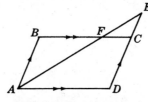

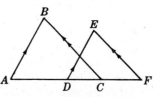

(a)   △AEB ~ △DEC   (b)   △BFA ~ AED   (c)   △ABC ~ △DEF

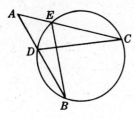

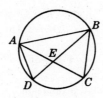

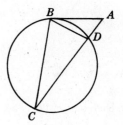

(d)   △AEB ~ △ADC   (e)   △ADE ~ △ACB   (f)   △ABC ~ △ABD
                                 $\overarc{BC} \cong \overarc{CD}$

Fig. 9-52

28. In each part of Fig. 9-53, determine the angles that can be used to prove the indicated triangles similar.   (9.19)

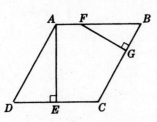

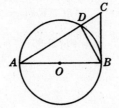

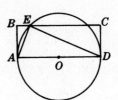

(a)   △AED ~ △FGB   (b)   △ABD ~ △ABC   (c)   △AEB ~ △AED
ABCD is a parallelogram.   AB is a diameter.   AD is a diameter.
                                 BC is a tangent.   ABCD is a rectangle.

Fig. 9-53

29. In each part of Fig. 9-54, determine the pair of congruent angles and the proportion needed to prove the indicated triangles similar.   (9.20)

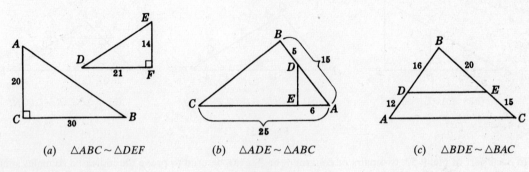

    (a) $\triangle ABC \sim \triangle DEF$ (b) $\triangle ADE \sim \triangle ABC$ (c) $\triangle BDE \sim \triangle BAC$

Fig. 9-54

30. In each part of Fig. 9-55, state the proportion needed to prove the indicated triangles similar.   (9.21)

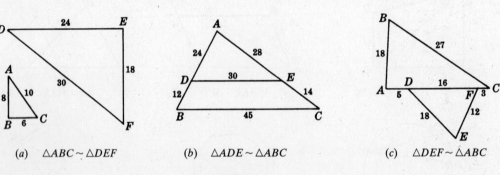

    (a) $\triangle ABC \sim \triangle DEF$ (b) $\triangle ADE \sim \triangle ABC$ (c) $\triangle DEF \sim \triangle ABC$

Fig. 9-55

31. In each part of Fig. 9-56, prove the indicated proportion.   (9.25)

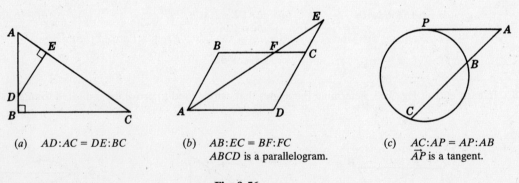

    (a) $AD:AC = DE:BC$ (b) $AB:EC = BF:FC$ (c) $AC:AP = AP:AB$
                                $ABCD$ is a parallelogram. $\overleftrightarrow{AP}$ is a tangent.

Fig. 9-56

32. In $\triangle ABC$ (Fig. 9-57), $\overline{DE} \parallel \overline{BC}$.   (9.22)

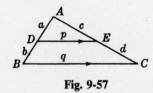

Fig. 9-57

(a) Let $a = 4$, $AB = 8$, $p = 10$. Find q.   (d) Let $b = 9$, $p = 20$, $q = 35$. Find a.

(b) Let $c = 5$, $AC = 15$, $q = 24$. Find p.   (e) Let $a = 10$, $p = 24$, $q = 84$. Find AB.

(c) Let $a = 7$, $p = 11$, $q = 22$. Find b.   (f) Let $c = 3$, $p = 4$, $q = 7$. Find d.

33. Find x in each part of Fig. 9-58.   (9.22)

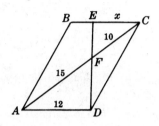
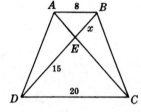
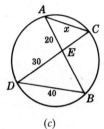

(a) $ABCD$ is a parallelogram.   (b) $ABCD$ is a trapezoid.   (c)

Fig. 9-58

34. A 7-ft upright pole near a vertical tree casts a 6-ft shadow. At that same time, find (a) the height of the tree if its shadow is 36 ft long; (b) the length of the shadow of the tree if its height is 77 ft.   (9.23)

35. Prove each of the following: (9.23)

(a) In $\triangle ABC$, if $\overline{AD}$ and $\overline{CE}$ are altitudes, then $AD:CE = AB:BC$.

(b) In circle O, diameter $\overline{AB}$ and tangent $\overline{BC}$ are sides of $\triangle ABC$. If $\overline{AC}$ intersects the circle in D, then $AD:AB = AB:AC$.

(c) The diagonals of a trapezoid divide each other into proportional segments.

(d) In right $\triangle ABC$, $\overline{CD}$ is the altitude to the hypotenuse $\overline{AB}$, then $AC:CD = AB:BC$.

36. Prove each of the following: (9.24)

(a) A line parallel to one side of a triangle cuts off a triangle similar to the given triangle.

(b) Isosceles right triangles are similar to each other.

(c) Equilateral triangles are similar to each other.

(d) The bases of a trapezoid form similar triangles with the segments of the diagonals.

37. Complete each of the following statements: (9.26)

(a) In similar triangles, if corresponding sides are in the ratio $8:5$, then corresponding altitudes are in the ratio __?__ .

(b) In similar triangles, if corresponding angle bisectors are in the ratio $3:5$, then their perimeters are in the ratio __?__ .

(c) If the sides of a triangle are halved, then the perimeter is __?__ , the angle bisectors are __?__ , the medians are __?__ , and the radii of the circumscribed circle are __?__ .

38. (a) Corresponding sides of two similar triangles have lengths 18 and 12. If an altitude of the smaller has length 10, find the length of the corresponding altitude of the larger.   (9.26)

(b) Corresponding medians of two similar triangles have lengths 25 and 15. Find the perimeter of the larger if the perimeter of the smaller is 36.

(*c*) The sides of a triangle have lengths 5, 7, and 8. If the perimeter of a similar triangle is 100, find its sides.

(*d*) The bases of a trapezoid have lengths 5 and 20, and the altitude has length 12. Find the length of the altitude of the triangle formed by the shorter base and the nonparallel sides extended to meet.

(*e*) The bases of a trapezoid have lengths 11 and 22. Its altitude has length 9. Find the distance from the point of intersection of the diagonals to each of the bases.

39. Complete each of the following statements: (9.27)

(*a*) If corresponding sides of two similar polygons are in the ratio 3:7, then the ratio of their corresponding altitudes is __?__ .

(*b*) If the perimeters of two similar hexagons are in the ratio of 56 to 16, then the ratio of their corresponding diagonals is __?__ .

(*c*) If each side of an octagon is quadrupled and the angles remain the same, then its perimeter is __?__ .

(*d*) The base of a rectangle is twice that of a similar rectangle. If the radius of the circumscribed circle of the first rectangle is 14, then the radius of the circumscribed circle of the second is __?__ .

40. Prove each of the following: (9.28)

(*a*) Corresponding angle bisectors of two triangles have the same ratio as a pair of corresponding sides.

(*b*) Corresponding medians of similar triangles have the same ratio as a pair of corresponding sides.

41. Provide the proofs requested in Fig. 9-59.   (9.28)

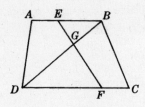

(*a*) **Given:** Trapezoid *ABCD*
       To Prove:
       $GB \times DF = GD \times EB$

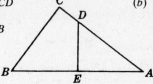

(*b*) **Given:** $\overline{BC} \perp \overline{AC}$
       $\overline{DE} \perp \overline{AB}$
       To Prove:
       $DE \times AC = BC \times AE$

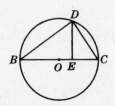

(*c*) **Given:** Diameter $\overline{BC}$
       $\overline{DE} \perp \overline{BC}$
       To Prove:
       $(BD^2) = BE \times BC$

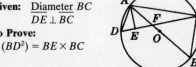

(*d*) **Given:** Circle *O*
       Diameter $\overline{AB}$
       $\overline{AE} \perp \overline{CD}$
       To Prove:
       $AD \times BC = AB \times DE$

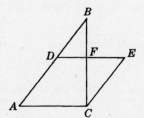

(*e*) **Given:** $\overline{AB} \parallel \overline{CE}$
       $\overline{AC} \perp \overline{BC}$
       $\overline{DE} \perp \overline{BC}$
       To Prove:
       $AB \times CF = BC \times EC$

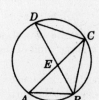

(*f*) **Given:** $\overarc{BC} \cong \overarc{CD}$
       To Prove:
       $(BC^2) = AC \times EC$

Fig. 9-59

42. Prove each of the following: (9.28)

  (a) If two chords intersect in a circle, the product of the lengths of the segments of one chord equals the product of the lengths of the segments of the other.

  (b) In a right triangle, the product of the lengths of the hypotenuse and the altitude upon it equals the product of the lengths of the legs.

  (c) If in inscribed $\triangle ABC$ the bisector of $\angle A$ intersects $\overline{BC}$ in D and the circle in E, then $BD \times AC = AD \times EC$.

43. In Fig. 9-60:   In Fig. 9-61, diameter $\overline{CD} \perp$ chord $\overline{AB}$:   (9.29)

  (a) Let $AE = 10$, $EB = 6$, $CE = 12$. Find ED.   (e) Let $OD = 10$, $OE = 8$. Find AB.

  (b) Let $AB = 15$, $EB = 8$, $ED = 4$. Find CE.   (f) Let $AB = 24$, $OE = 5$. Find OD.

  (c) Let $AE = 6$, $ED = 4$, $CD = 13$. Find EB.   (g) Let $OD = 25$, $EC = 18$. Find AB.

  (d) Let $ED = 5$, $EB = 2(AE)$, $CD = 15$. Find AE.   (h) Let $AB = 8$, $OD = 5$. Find EC.

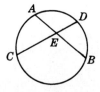

Fig. 9-60

Fig. 9-61

44. A point is 12 in from the center of a circle whose radius is 15 in. Find the lengths of the longest and shortest chords that can be drawn through this point. (*Hint*: The longest chord is a diameter, and the shortest chord is perpendicular to this diameter.)   (9.29)

45. In Fig. 9-62, $\overline{AB}$ is a tangent: In Fig. 9-63, $\overline{CD}$ is a diameter; $\overline{AB}$ is a tangent: (9.29)

  (a) Let $AC = 16$, $AD = 4$. Find AB.   (f) Let $AD = 6$, $OD = 9$. Find AB.

  (b) Let $CD = 5$, $AD = 4$. Find AB.   (g) Let $AD = 2$, $AB = 8$. Find CD.

  (c) Let $AB = 6$, $AD = 3$. Find AC.   (h) Let $AD = 5$, $AB = 19$. Find OD.

  (d) Let $AC = 20$, $AB = 10$. Find AD.   (i) Let $AB = 12$, $AC = 18$. Find OD.

  (e) Let $AB = 12$, $AD = 9$. Find CD.   (j) Let $OD = 5$, $AB = 12$. Find AD.

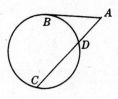

Fig. 9-62

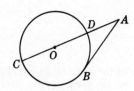

Fig. 9-63

46. In Fig. 9-64.

(a) Let $AB = 14$, $AD = 4$, $AE = 7$. Find AC.

(b) Let $AC = 8$, $AE = 6$, $AD = 3$. Find BD.

(c) Let $BD = 5$, $AD = 7$, $AE = 4$. Find AC.

(d) Let $AD = DB$, $EC = 14$, $AE = 4$. Find AD.

In Fig. 9-65, $\overline{CE}$ is a diameter: (9.31)

(e) Let $OC = 3$, $AE = 6$, $AD = 8$. Find AB.

(f) Let $BD = 7$, $AD = 5$, $AE = 2$. Find OC.

(g) Let $OC = 11$, $AB = 15$, $AD = 5$. Find AE.

(h) Let $OC = 5$, $AE = 6$, $BD = 4$. Find AD.

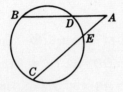

Fig. 9-64

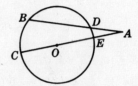

Fig. 9-65

47. $\overline{CD}$ is the altitude to hypotenuse $\overline{AB}$ in Fig. 9-66. (9.32)

(a) If $p = 2$ and $q = 6$, find a and h.

(b) If $p = 4$ and $a = 6$, find c and h.

(c) If $p = 16$ and $h = 8$, find q and b.

(d) If $b = 12$ and $q = 6$, find p and h.

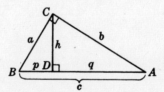

Fig. 9-66

48. In a right triangle whose arms have lengths a and b, find the length of the hypotenuse c when: (9.33)

(a) $a = 15$, $b = 20$

(b) $a = 15$, $b = 36$

(c) $a = 5$, $b = 4$

(d) $a = 5$, $b = 5\sqrt{3}$

(e) $a = 7$, $b = 7$

49. In the right triangle in Fig. 9-67, find the length of each missing arm when: (9.33)

(a) $a = 12$, $c = 20$

(b) $b = 6$, $c = 8$

(c) $b = 15$, $c = 17$

(d) $a = 2$, $c = 4$

(e) $a = 5\sqrt{2}$, $c = 10$

(f) $a = \sqrt{5}$, $c = 2\sqrt{2}$

Fig. 9-67

50. Find the lengths of the arms of a right triangle whose hypotenuse has length c if these arms have a ratio of (a) 3:4 and $c = 15$; (b) 5:12 and $c = 26$; (c) 8:15 and $c = 170$; (d) 1:2 and $c = 10$. (9.34)

51. In a rectangle, find the length of the diagonal if its sides have lengths (a) 9 and 40; (b) 5 and 10. (9.33)

52. In a rectangle, find the length of one side if the diagonal has length 15 and the other side has length (*a*) 9; (*b*) 5; (*c*) 10.   (9.33)

53. Of triangles having sides with lengths as follows, which are right triangles?

  (*a*)   33, 55, 44   (*c*)   4, $7\frac{1}{2}$, $8\frac{1}{2}$ (*e*)   5 in, 1 ft, 1 ft 1 in   (*g*)   11 mi, 60 mi, 61 mi

  (*b*)   120, 130, 50   (*d*)   25, 7, 24   (*f*)   1 yd, 1 yd 1 ft, 1 yd 2 ft   (*h*)   5 cm, 5 cm, 7 cm

54. Is a triangle a right triangle if its sides have the ratio of (*a*) 3:4:5; (*b*) 2:3:4?

55. Find the length of the altitude of an isosceles triangle if each of its two congruent sides has length 10 and its base has length (*a*) 12; (*b*) 16; (*c*) 18; (*d*) 10.   (9.35)

56. In a rhombus, find the length of a side if the diagonals have lengths (*a*) 18 and 24; (*b*) 4 and 8; (*c*) 6 and $6\sqrt{3}$.   (9.36)

57. In a rhombus, find the length of a diagonal if a side and the other diagonal have lengths, respectively, (*a*) 10 and 12; (*b*) 17 and 16; (*c*) 4 and 4; (*d*) 10 and $10\sqrt{3}$.   (9.36)

58. In isosceles trapezoid *ABCD* in Fig. 9-68,   (9.37)

  (*a*)   Find *a* if *b* = 32, *b'* = 20, and *h* = 8.   (*c*)   Find *b* if *a* = 15, *b'* = 10, and *h* = 12.

  (*b*)   Find *h* if *b* = 24, *b'* = 14, and *a* = 13.   (*d*)   Find *b'* if *a* = 6, *b* = 21, and *h* = $3\sqrt{3}$.

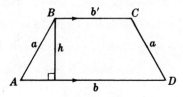

Fig. 9-68

59. In a trapezoid *ABCD* in Fig. 9-69,   (9.37)

  (*a*)   Find *d* if *a* = 11, *b* = 3, and *c* = 15.   (*c*)   Find *d* if *a* = 5, *p* = 13, and *c* = 14.

  (*b*)   Find *a* if *d* = 20, *b* = 12, and *c* = 36.   (*d*)   Find *p* if *a* = 20, *c* = 28, and *d* = 17.

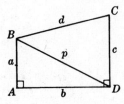

Fig. 9-69

60. The radius of a circle is 15. Find (*a*) the distance from its center to a chord whose length is 18; (*b*) the length of a chord whose distance from its center is 9.   (9.38)

61. In a circle, a chord whose length is 16 is at a distance of 6 from the center. Find the length of a chord whose distance from the center is 8.   (9.38)

62. Two externally tangent circles have radii of 25 and 9. Find the length of a common external tangent.

(9.38)

63. In a 30°-60°-90° triangle, find the lengths of (*a*) the legs if the hypotenuse has length 20; (*b*) the other leg and hypotenuse if the leg opposite 30° has length 7; (*c*) the other leg and hypotenuse if the leg opposite 60° has length $5\sqrt{3}$.

(9.39)

64. In an equilateral triangle, find the length of the altitude if the side has length (*a*) 22; (*b*) 2*a*. Find the side if the altitude has length (*c*) $24\sqrt{3}$; (*d*) 24.

(9.39)

65. In a rhombus which has an angle measuring 60°, find the lengths of (*a*) the diagonals if a side has length 25; (*b*) the side and larger diagonal if the smaller diagonal has length 35.

(9.39)

66. In an isosceles trapezoid which has base angles measuring 60°, find the lengths of:

(9.39)

(*a*)   The lower base and altitude if the upper base has length 12 and the legs have length 16

(*b*)   The upper base and altitude if the lower base has length 45 and the legs have length 28

67. In an isosceles right triangle, find the length of each leg if the hypotenuse has length (*a*) 34; (*b*) 2*a*. Find the length of the hypotenuse if each leg has length (*c*) 34; (*d*) $15\sqrt{2}$.

(9.40)

68. In a square, find the length of (*a*) the side if the diagonal has length 40; (*b*) the diagonal if the side has length 40.

(9.40)

69. In an isosceles trapezoid which has base angles of measure 45°, find the lengths of:

(9.40)

(*a*)   The lower base and each leg if the altitude has length 13 and the upper base has length 19

(*b*)   The upper base and each leg if the altitude has length 27 and the lower base has length 65

(*c*)   Each leg and the lower base if the upper base has length 25 and the altitude has length 15

70. A parallelogram has an angle measuring 45°. Find the distances between its pairs of opposite sides if its sides have lengths 10 and 12.

(9.40)

CHAPTER 10

Areas

10.1 AREA OF A RECTANGLE AND OF A SQUARE

A *square unit* is the surface enclosed by a square whose side is 1 unit (Fig. 10-1).

The *area of a closed plane figure*, such as a polygon, is the number of square units contained in its surface. Since a rectangle 5 units long and 4 units wide can be divided into 20 unit squares, its area is 20 square units (Fig. 10-2).

The area of a rectangle equals the product of the length of its base and the length of its altitude (Fig. 10-3). Thus if $b = 8$ in and $h = 3$ in, then $A = 24$ in^2.

The area of a square equals the square of the length of a side (Fig. 10-4). Thus if $s = 6$, then $A = s^2 = 36$.

It follows that the area of a square also equals one-half the square of the length of a diagonal. Since $A = s^2$ and $s = d/\sqrt{2}$, $A = \frac{1}{2}d^2$.

Note that we sometimes use the letter A for both a vertex of a figure and its area. You should have no trouble determining which is meant.

The reader should feel free to use a calculator for the work in this chapter. See Chapter 2 for an introduction to the use of calculators.

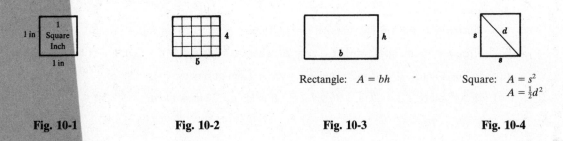

Rectangle: $A = bh$

Square: $A = s^2$
$A = \frac{1}{2}d^2$

| Fig. 10-1 | Fig. 10-2 | Fig. 10-3 | Fig. 10-4 |

SOLVED PROBLEMS

10.1 AREA OF A RECTANGLE

(*a*) Find the area of a rectangle if the base has length 15 and the perimeter is 50.

(*b*) Find the area of a rectangle if the altitude has length 10 and the diagonal has length 26.

(*c*) Find the lengths of the base and altitude of a rectangle if its area is 70 and its perimeter is 34.

Solutions

See Fig. 10-5.

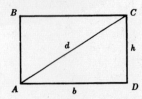

Fig. 10-5

(a) Here $p = 50$ and $b = 15$. Since $p = 2b + 2h$, we have $50 = 2(15) + 2h$ so $h = 10$.
     Hence $A = bh = 15(10) = 150$.

(b) Here $d = 26$ and $h = 10$. In right $\triangle ACD$, $d^2 = b^2 + h^2$, so $26^2 = b^2 + 10^2$ or $b = 24$.
     Hence $A = bh = 24(10) = 240$.

(c) Here $A = 70$ and $p = 34$. Since $p = 2b + 2h$, we have $34 = 2(b + h)$ or $h = 17 - b$.
     Since $A = bh$, we have $70 = b(17 - b)$, so $b^2 - 17b + 70 = 0$ and $b = 7$ or 10. Then since $h = 17 - b$,
     we obtain $h = 10$ or 7.
     Ans. 10 and 7, or 7 and 10.

10.2 AREA OF A SQUARE

(a) Find the area of a square whose perimeter is 30.

(b) Find the area of a square if the radius of the circumscribed circle is 10.

(c) Find the side and the perimeter of a square whose area is 20.

(d) Find the number of square inches in a square foot.

Solutions

(a) Since $p = 4s = 30$ in Fig. 10-6(a), $s = 7\frac{1}{2}$. Then $A = s^2 = (7\frac{1}{2})^2 = 56\frac{1}{4}$.

(b) Since $r = 10$ in Fig. 10-6(b), $d = 2r = 20$. Then $A = \frac{1}{2}d^2 = \frac{1}{2}(20)^2 = 200$.

(c) In Fig. 10-6(a), $A = s^2 = 20$; hence $s = 2\sqrt{5}$. Then perimeter $= 4s = 8\sqrt{5}$.

(d) $A = s^2$. Since 1 ft = 12 in, 1 ft^2 = 1 ft $\times$ 1 ft = 12 in $\times$ 12 in = 144 in^2.

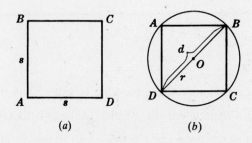

(a)   (b)

Fig. 10-6

10.2   AREA OF A PARALLELOGRAM

The area of a parallelogram equals the product of the length of a side and the length of the altitude to that side. (A proof of this theorem is given in Chapter 16.) Thus in $\square ABCD$ (Fig. 10-7), if $b = 10$ and $h = 2.7$, then $A = 10(2.7) = 27$.

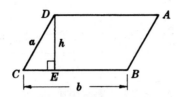

Parallelogram: $A = bh$

Fig. 10-7

SOLVED PROBLEMS

10.3   AREA OF A PARALLELOGRAM

(a)   Find the area of a parallelogram if the area is represented by $x^2 - 4$, the length of a side by $x + 4$, and the length of the altitude to that side by $x - 3$.

(b)   In a parallelogram, find the length of the altitude if the area is 54 and the ratio of the altitude to the base is $2 : 3$.

Solutions

See Fig. 10-7.

(a) $A = x^2 - 4$, $b = x + 4$, $h = x - 3$. Since $A = bh$, $x^2 - 4 = (x + 4)(x - 3)$ or $x^2 - 4 = x^2 + x - 12$ and $x = 8$.
Hence $A = x^2 - 4 = 64 - 4 = 60$.

(b)   Let $h = 2x$, $b = 3x$. Then $A = bh$ or $54 = (3x)(2x) = 6x^2$, so $9 = x^2$ and $x = 3$.
Hence $h = 2x = 2(3) = 6$.

10.3   AREA OF A TRIANGLE

The area of a triangle equals one-half the product of the length of a side and the length of the altitude to that side. (A proof of this theorem is given in Chapter 16.)

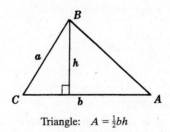

Triangle: $A = \frac{1}{2}bh$

Fig. 10-8

SOLVED PROBLEMS

10.4   AREA OF A TRIANGLE
Find the area of the triangle in Fig. 10-9.

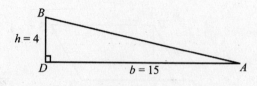

Fig. 10-9

Solution

Here, $b = 15$ and $h = 4$. Thus, $A = \frac{1}{2}bh = \frac{1}{2}(15)(4) = 30$.

10.5   FORMULAS FOR THE AREA OF AN EQUILATERAL TRIANGLE
Derive the formula for the area of an equilateral triangle (*a*) whose side has length s; (*b*) whose altitude has length h.

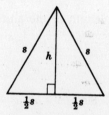

Equilateral Triangle:
$A = \frac{1}{4}s^2\sqrt{3}$
$A = \frac{1}{3}h^2\sqrt{3}$

Fig. 10-10

Solutions

See Fig. 10-10.

(*a*)   Here $A = \frac{1}{2}bh$, where $b = s$ and $h^2 = s^2 - (\frac{1}{2}s)^2 = \frac{3}{4}s^2$ or $h = \frac{1}{2}s\sqrt{3}$.
Then $A = \frac{1}{2}bh = \frac{1}{2}s(\frac{1}{2}s\sqrt{3}) = \frac{1}{4}s^2\sqrt{3}$.

(*b*)   Here $A = \frac{1}{2}bh$, where $b = s$ and $h = \frac{1}{2}s\sqrt{3}$ or $s = \dfrac{2h}{\sqrt{3}}$.

Then $A = \frac{1}{2}bh = \frac{1}{2}sh = \dfrac{1}{2}\left(\dfrac{2h}{\sqrt{3}}\right)^h = \frac{1}{3}h^2\sqrt{3}$.

10.6   AREA OF AN EQUILATERAL TRIANGLE
In Fig. 10.11, find the area of (*a*) an equilateral triangle whose perimeter is 24; (*b*) a rhombus in which the shorter diagonal has length 12 and an angle measures 60°; (*c*) a regular hexagon with a side of length 6.

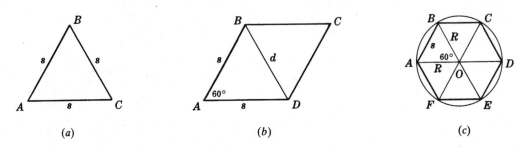

(a)   (b)   (c)

Fig. 10-11

Solutions

(a) Since $p = 3s = 24$, $s = 8$. Then $A = \frac{1}{4}s^2\sqrt{3} = \frac{1}{4}(64)\sqrt{3} = 16\sqrt{3}$.

(b) Since $m\angle A = 60°$, $\triangle ADB$ is equilateral and $s = d = 12$. The area of the rhombus is twice the area of $\triangle ABD$. Hence $A = 2(\frac{1}{4}s^2\sqrt{3}) = 2(\frac{1}{4})(144)\sqrt{3} = 72\sqrt{3}$.

(c) A side s of the inscribed hexagon subtends a central angle of measure $\frac{1}{6}(360°) = 60°$. Then, since $OA = OB = $ radius R of the circumscribed circle, $m\angle OAB = m\angle OBA = 60°$. Thus $\triangle AOB$ is equilateral.
Area of hexagon $= 6(\text{area of } \triangle AOB) = 6(\frac{1}{4}s^2\sqrt{3}) = 6(\frac{1}{4})(36\sqrt{3}) = 54\sqrt{3}$.

10.4 AREA OF A TRAPEZOID

The area of a trapezoid equals one-half the product of the length of its altitude and the sum of the lengths of its bases. (A proof of this theorem is given in Chapter 16.) Thus if $h = 20$, $b = 27$, and $b' = 23$ in Fig. 10-12, then $A = \frac{1}{2}(20)(27 + 23) = 500$.

The area of a trapezoid equals the product of the lengths of its altitude and median. Since $A = \frac{1}{2}h(b + b')$ and $m = \frac{1}{2}(b + b')$, $A = hm$.

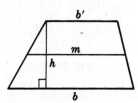

Trapezoid: $A = \frac{1}{2}h(b + b')$
$A = hm$

Fig. 10-12

SOLVED PROBLEMS

10.7 AREA OF A TRAPEZOID

(a) Find the area of a trapezoid if the bases have lengths 7.3 and 2.7, and the altitude has length 3.8.

(b) Find the area of an isosceles trapezoid if the bases have lengths 22 and 10, and the legs have length 10.

(c) Find the bases of an isosceles trapezoid if the area is $52\sqrt{3}$, the altitude has length $4\sqrt{3}$, and each leg has length 8.

Solutions

See Fig. 10-13.

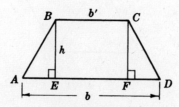

Fig. 10-13

(a)   Here $b = 7.3$, $b' = 2.7$, $h = 3.8$. Then $A = \frac{1}{2}h(b + b') = \frac{1}{2}(3.8)(7.3 + 2.7) = 19$.

(b)   Here $b = 22$, $b' = 10$, $AB = 10$. Also $EF = b' = 10$ and $AE = \frac{1}{2}(22 - 10) = 6$.
      In $\triangle BEA$, $h^2 = 10^2 - 6^2 = 64$ so $h = 8$. Then $A = \frac{1}{2}h(b + b') = \frac{1}{2}(8)(22 + 10) = 128$.

(c) $AE = \sqrt{(AB)^2 - h^2} = \sqrt{64 - 48} = 4$. Also $FD = AE = 4$, and $b' = b - (AE + FD) = b - 8$. Then
      $A = \frac{1}{2}h(b + b') = \frac{1}{2}h(2b - 8)$ or $52\sqrt{3} = \frac{1}{2}(4\sqrt{3})(2b - 8)$, from which $26 = 2b - 8$ or $b = 17$. Then
      $b' = b - 8 = 17 - 8 = 9$.

10.5   AREA OF A RHOMBUS

The area of a rhombus equals one-half the product of the lengths of its diagonals.

Since each diagonal is the perpendicular bisector of the other, the area of triangle I in Fig. 10-14 is $\frac{1}{2}(\frac{1}{2}d)(\frac{1}{2}d') = \frac{1}{8}dd'$. Thus the rhombus, which consists of four triangles congruent to $\triangle I$, has an area of $4(\frac{1}{8}dd')$ or $\frac{1}{2}dd'$.

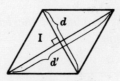

Rhombus: $A = \frac{1}{2}dd'$

Fig. 10-14

SOLVED PROBLEMS

10.8   AREA OF A RHOMBUS

(a)   Find the area of a rhombus if one diagonal has length 30 and a side has length 17.

(b)   Find the length of a diagonal of a rhombus if the other diagonal has length 8 and the area of the rhombus is 52.

Solutions

See Fig. 10-15.

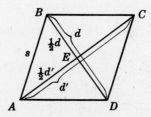

Fig. 10-15

(a) In right $\triangle AEB$, $s^2 = (\frac{1}{2}d)^2 + (\frac{1}{2}d')^2$ or $17^2 = (\frac{1}{2}d)^2 + 15^2$. Then $\frac{1}{2}d = 8$ and $d = 16$. Now $A = \frac{1}{2}dd' = \frac{1}{2}(16)(30) = 240$.

(b) We have $d' = 8$ and $A = 52$. Then $A = \frac{1}{2}dd'$ or $52 = \frac{1}{2}(d)(8)$ and $d = 13$.

10.6 POLYGONS OF THE SAME SIZE OR SHAPE

Figure 10-16 shows what we mean when we say that two polygons are of equal area, or are similar, or are congruent.

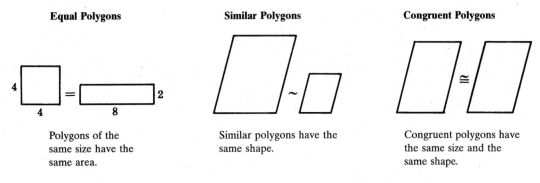

Equal Polygons	Similar Polygons	Congruent Polygons
Polygons of the same size have the same area.	Similar polygons have the same shape.	Congruent polygons have the same size and the same shape.

Fig. 10-16

PRINCIPLE 1: *Parallelograms have equal areas if they have congruent bases and congruent altitudes.*

Thus the two parallelograms shown in Fig. 10-17 are equal.

Fig. 10-17   **Fig. 10-18**

PRINCIPLE 2: *Triangles have equal areas if they have congruent bases and congruent altitudes.*

Thus in Fig. 10-18, the area of $\triangle CAB$ equals the area of $\triangle CAD$.

PRINCIPLE 3: *A median divides a triangle into two triangles with equal areas.*

Thus in Fig. 10-19, where $\overline{BM}$ is a median, the area of $\triangle AMB$ equals the area of $\triangle BMC$ since they have congruent bases ($\overline{AM} \cong \overline{MC}$) and common altitude $\overline{BD}$.

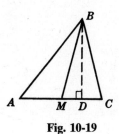

Fig. 10-19

PRINCIPLE 4: *Triangles are equal in area if they have a common base and their vertices lie on a line parallel to the base.*

Thus in Fig. 10-20, the area of △*ABC* is equal to the area of △*ADC*.

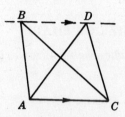

Fig. 10-20

SOLVED PROBLEMS

10.9   PROVING AN EQUAL-AREAS PROBLEM

Given: Trapezoid *ABCD* ($\overline{BC} \| \overline{AD}$)
Diagonals $\overline{AC}$ and $\overline{BD}$

To Prove: Area(△*AEB*) = area(△*DEC*)

Plan: Use Principle 4 to obtain
area(△*ABD*) = area(△*ACD*).
Then use the Subtraction Postulate.

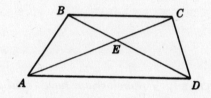

PROOF:

Statements	Reasons
1. $\overline{BC} \| \overline{AD}$	1. Given
2. Area(△*ABD*) = area(△*ACD*)	2. Triangles have equal area if they have a common base and their vertices lie on a line parallel to the base.
3. Area(△*AED*) = area(△*AED*)	3. Identity Postulate
4. Area(△*AED*) = area(△*DEC*)	4. Subtraction Postulate

10.10   PROVING AN EQUAL-AREAS PROBLEM STATED IN WORDS

Prove that if *M* is the midpoint of diagonal $\overline{AC}$ in quadrilateral *ABCD*, and $\overline{BM}$ and $\overline{DM}$ are drawn, then the area of quadrilateral *ABMD* equals the area of quadrilateral *CBMD*.

Solution

Given: Quadrilateral *ABCD*
M is midpoint of diagonal $\overline{AC}$.

To Prove: Area of quadrilateral *ABMD* equals area of quadrilateral *CBMD*.

Plan: Use Principle 3 to obtain two pairs of triangles which are equal in area. Then use the Addition Postulate.

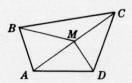

PROOF:

Statements	Reasons
1. M is the midpoint of $\overline{AC}$.	1. Given
2. $\overline{BM}$ is a median of $\triangle ACB$. $\overline{DM}$ is a median of $\triangle ACD$.	2. A line from a vertex of a triangle to the midpoint of the opposite side is a median.
3. Area $(\triangle AMB)$ = area$(\triangle BMC)$, Area$(\triangle AMD)$ = area$(\triangle DMC)$.	3. A median divides a triangle into two triangles of equal area.
4. Area of quadrilateral $ABMD$ equals area of quadrilateral $CBMD$.	4. If equals are added to equals, the results are equal.

10.7   COMPARING AREAS OF SIMILAR POLYGONS

The areas of similar polygons are to each other as the squares of any two corresponding segments.

Thus if $\triangle ABC \sim \triangle A'B'C'$ and the area of $\triangle ABC$ is 25 times the area of $\triangle A'B'C'$, then the ratio of the lengths any two corresponding sides, medians, altitudes, radii of inscribed or circumscribed circles, and such is 5:1.

SOLVED PROBLEMS

10.11 RATIO OF AREAS AND SEGMENTS OF SIMILAR TRIANGLES

Find the ratio of the areas of two similar triangles (a) if the ratio of the lengths of two corresponding sides is $3:5$; (b) if their perimeters are 12 and 7. Find the ratio of the lengths of a pair of (c) corresponding sides if the ratio of the areas is $4:9$; (d) corresponding medians if the areas are 250 and 10.

Solutions

(a) $\dfrac{A}{A'} = \left(\dfrac{s}{s'}\right)^2 = \left(\dfrac{3}{5}\right)^2 = \dfrac{9}{25}$ (c) $\left(\dfrac{s}{s'}\right)^2 = \dfrac{A}{A'} = \dfrac{4}{9}$ or $\dfrac{s}{s'} = \dfrac{2}{3}$

(b) $\dfrac{A}{A'} = \left(\dfrac{p}{p'}\right)^2 = \left(\dfrac{12}{7}\right)^2 = \dfrac{144}{49}$ (d) $\left(\dfrac{m}{m'}\right)^2 = \dfrac{A}{A'} = \dfrac{250}{10}$ or $\dfrac{m}{m'} = 5$

10.12 PROPORTIONS DERIVED FROM SIMILAR POLYGONS

(a) The areas of two similar polygons are 80 and 5. If a side of the smaller polygon has length 2, find the length of the corresponding side of the larger polygon.

(b) The corresponding diagonals of two similar polygons have lengths 4 and 5. If the area of the larger polygon is 75, find the area of the smaller polygon.

Solutions

(a) $\left(\dfrac{s}{s'}\right)^2 = \dfrac{A}{A'}$, so $\left(\dfrac{s}{2}\right)^2 = \dfrac{80}{5} = 16$. Then $\dfrac{s}{2} = 4$ and $s = 8$.

(b) $\dfrac{A}{A'} = \left(\dfrac{d}{d'}\right)^2$, so $\dfrac{A}{75} = \left(\dfrac{4}{5}\right)^2$. Then $A = 75\left(\dfrac{16}{25}\right) = 48$.

10.8   AREAS IN ANALYTIC GEOMETRY

10.8A   Area of a Triangle

If one side of a triangle is parallel to either coordinate axis, the length of that side and the length of the altitude to that side can be found readily. Then the formula $A = \frac{1}{2}bh$ can be used.

If no side of a triangle is parallel to either axis, then either

1. The triangle can be enclosed in a rectangle whose sides are parallel to the axes (Fig. 10-21), or

2. Trapezoids whose bases are parallel to the y-axis can be formed by dropping perpendiculars to the x-axis (Fig. 10-22).

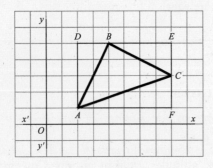

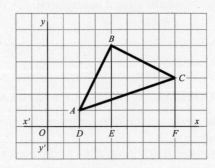

Fig. 10-21   **Fig. 10-22**

The area of the triangle can then be found from the areas of the figures so formed:

1. In Fig. 10-21, area($\triangle ABC$) = area(rectangle $ADEF$) − [area($\triangle ABD$) + area($\triangle BCE$) + area($\triangle ACF$)].

2. In Fig. 10-22, area($\triangle ABC$) = area(trapezoid $ABED$) + area(trapezoid $BEFC$) − area(trapezoid $DFCA$).

10.8B   Area of a Quadrilateral

The trapezoid method described above can be extended to finding the area of a quadrilateral if its vertices are given.

SOLVED PROBLEMS

10.13   Area of a Triangle Having No Side Parallel to an Axis

Find the area of $\triangle ABC$ whose vertices are $A(2,4)$, $B(5,8)$, and $C(8,2)$ (*a*) using the rectangle method; (*b*) using the trapezoid method.

Solutions

See Fig. 10-23.

(*a*)   Area of rectangle $DEFC = bh = 6(6) = 36$. Then:
Area of $\triangle DAC = \frac{1}{2}bh = \frac{1}{2}(2)(6) = 6$.
Area of $\triangle ABE = \frac{1}{2}bh = \frac{1}{2}(3)(4) = 6$.
Area of $\triangle BCF = \frac{1}{2}bh = \frac{1}{2}(3)(6) = 9$.
So area($\triangle ABC$) = area($DEFC$) − area($\triangle DAC + \triangle ABE + \triangle BCF$) = $36 - (6 + 6 + 9) = 15$.

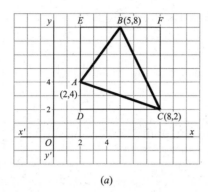

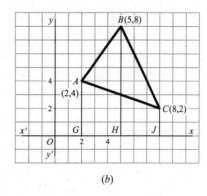

$$(a) \qquad\qquad\qquad\qquad\qquad (b)$$

Fig. 10-23

(b)   Area of trapezoid $ABHG = \frac{1}{2}h(b + b') = \frac{1}{2}(3)(4 + 8) = 18$.
Area of trapezoid $BCJH = \frac{1}{2}(3)(2 + 8) = 15$.
Area of trapezoid $ACJG = \frac{1}{2}(6)(2 + 4) = 18$.
Then area$(\triangle ABC) =$ area$(ABHG) +$ area$(BCJH) -$ area$(ACJG) = 18 + 15 - 18 = 15$.

Supplementary Problems

1. Find the area of a rectangle   (10.1)

(a)   If the base has length 11 in and the altitude has length 9 in

(b)   If the base has length 2 ft and the altitude has length 1 ft 6 in

(c)   If the base has length 25 and the perimeter is 90

(d)   If the base has length 15 and the diagonal has length 17

(e)   If the diagonal has length 12 and the angle between the diagonal and the base measures 60°

(f)   If the diagonal has length 20 and the angle between the diagonal and the base measures 30°

(g)   If the diagonal has length 25 and the lengths of the sides are in the ratio of 3:4

(h)   If the perimeter is 50 and the lengths of the sides are in the ratio of 2:3

2. Find the area of a rectangle inscribed in a circle   (10.1)

(a)   If the radius of the circle is 5 and the base has length 6

(b)   If the radius of the circle is 15 and the altitude has length 24

(c)   If the radius and the altitude both have length 5

(d)   If the diameter has length 26 and the base and altitude are in the ratio of 5:12

3. Find the base and altitude of a rectangle   (10.1)

(a)   If its area is 28 and the base has a length of 3 more than the altitude

(b)   If its area is 72 and the base is twice the altitude

(c)   If its area is 54 and the ratio of the base to the altitude is 3:2

(d)   If its area is 12 and the perimeter is 16

(e)   If its area is 70 and the base and altitude are represented by $2x$ and $x + 2$

(f)   If its area is 160 and the base and altitude are represented by $3x - 4$ and x

4. Find the area of (a) a square yard in square inches; (b) a square meter in square decimeters ($1\,\text{m} = 10\,\text{dm}$).   (10.2)

5. Find the area of a square if (a) a side has length 15; (b) a side has length $3\frac{1}{2}$; (c) a side has length 1.8; (d) a side has length $8a$; (e) the perimeter is 44; (f) the perimeter is 10; (g) the perimeter is $12b$; (h) the diagonal has length 8; (i) the diagonal has length 9; (j) the diagonal has length $8\sqrt{2}$.   (10.2)

6. Find the area of a square if (a) the radius of the circumscribed circle is 8; (b) the diameter of the circumscribed circle is 12; (c) the diameter of the circumscribed circle is $10\sqrt{2}$; (d) the radius of the inscribed circle is $3\frac{1}{2}$; (e) the diameter of the inscribed circle is 20.   (10.2)

7. If a floor is 20 m long and 80 m wide, how many tiles are needed to cover it if (a) each tile is $1\,\text{m}^2$; (b) each tile is a square 2 m on a side; (c) each tile is a square 4 m on a side.   (10.2)

8. If the area of a square is 81, find the length of (a) its side; (b) its perimeter; (c) its diagonal; (d) the radius of the inscribed circle; (e) the radius of the circumscribed circle.   (10.2)

9. (a)   Find the length of the side of a square whose area is $6\frac{1}{4}$.   (10.2)

(b)   Find the perimeter of a square whose area is 169.

(c)   Find the length of the diagonal of a square whose area is 50.

(d)   Find the length of the diagonal of a square whose area is 25.

(e)   Find the radius of the inscribed circle of a square whose area is 144.

(f)   Find the radius of the circumscribed circle of a square whose area is 32.

10. Find the area of a parallelogram if the base and altitude have lengths, respectively, of (a) 3 ft and $5\frac{1}{3}$ ft; (b) 4 ft and 1 ft 6 in; (c) 20 and 3.5; (d) 1.8 m and 0.9 m.   (10.3)

11. Find the area of a parallelogram if the base and altitude have lengths, respectively, of (a) $3x$ and x; (b) $x + 3$ and x; (c) $x - 5$ and $x + 5$; (d) $4x + 1$ and $3x + 2$.   (10.3)

12. Find the area of a parallelogram if

(a)   The area is represented by x^2, the base by $x + 3$, and the altitude by $x - 2$.

(b)   The area is represented by $x^2 - 10$, the base by x, and the altitude by $x - 2$.

(c)   The area is represented by $2x^2 - 34$, the base by $x + 3$, and the altitude by $x - 3$.

13. In a parallelogram, find   (10.3)

(a)   The base if the area is 40 and the altitude has length 15

(b)   The length of the altitude if the area is 22 and the base has length 1.1

(c)   The length of the base if the area is 27 and the base is three times the altitude

(d)   The length of the altitude if the area is 21 and the base has length four more than the altitude

(e)   The base if the area is 90 and the ratio of the base to the altitude is $5:2$

(f)   The length of the altitude to a side of length 20 if the altitude to a side of length 15 is 16

(g)   The length of the base if the area is 48, the base is represented by $x + 3$, and the altitude by $x + 1$

(h)   The length of the base if the area is represented by $x^2 + 17$, the base by $2x - 3$, and the altitude by $x + 1$

14. Find the area of a triangle if the lengths of the base and altitude are, respectively, (a) 6 in and $3\frac{2}{3}$ in; (b) 1 yd and 2 ft; (c) 8 and $x - 7$; (d) $5x$ and $4x$; (e) $4x$ and $x + 9$; (f) $x + 4$ and $x - 4$; (g) $2x - 6$ and $x + 3$.   (10.4)

15. Find the area of   (10.4)

(a)   A triangle if two sides have lengths 13 and 15 and the altitude to the third side has length 12

(b)   A triangle whose sides have lengths 10, 10, and 16

(c)   A triangle whose sides have lengths 5, 12, and 13

(d)   An isosceles triangle whose base has length 30 and whose legs each have length 17

(e)   An isosceles triangle whose base has length 20 and whose vertex angle measures 68°

(f)   An isosceles triangle whose base has length 30 and whose base angle measures 62°

(g)   A triangle inscribed in a circle of radius 4 if one side is a diameter and another side makes an angle measuring 30° with the diameter

(h)   A triangle cut off by a line parallel to the base of a triangle if the base and altitude of the larger triangle have lengths 10 and 5, respectively, and the line parallel to the base is 6.

16. Find the altitude of a triangle if   (10.4)

(a)   Its base has length 10 and the triangle is equal in area to a parallelogram whose base and altitude have lengths 15 and 8.

(b)   Its base has length 8 and the triangle is equal in area to a square whose diagonal has length 4.

(c)   Its base has length 12 and the triangle is equal in area to another triangle whose sides have lengths 6, 8, and 10.

17. In a triangle, find the length of   (10.4)

(a)   A side if the area is 40 and the altitude to that side has length 10

(b)   An altitude if the area is 25 and the side to which the altitude is drawn has length 5

(c)   A side if the area is 24 and the side has length 2 more than its altitude

(d)   A side if the area is 108 and the side and its altitude are in the ratio 3:2

(e)   The altitude to a side of length 20, if the sides of the triangle have lengths 12, 16, and 20

(f)   The altitude to a side of length 12 if another side and its altitude have lengths 10 and 15

(g)   A side represented by $4x$ if the altitude to that side is represented by $x + 7$ and the area is 60

(h)   A side if the area is represented by $x^2 = 55$, the side by $2x - 2$, and its altitude by $x - 5$

18. Find the area of an equilateral triangle if (a) a side has length 10; (b) the perimeter is 36; (c) an altitude has length 6; (d) an altitude has length $5\sqrt{3}$; (e) a side has length $2b$; (f) the perimeter is $12x$; (g) an altitude has length $3r$.   (10.6)

19. Find the area of a rhombus having an angle of 60° if (a) a side has length 2; (b) the shorter diagonal has length 7; (c) the longer diagonal has length 12; (d) the longer diagonal has length $6\sqrt{3}$.   (10.6)

20. Find the area of a regular hexagon if (a) a side is 4; (b) the radius of the circumscribed circle is 6; (c) the diameter of the circumscribed circle is 20.   (10.6)

21. Find the side of an equilateral triangle whose area equals   (10.6)

(a) The sum of the areas of two equilateral triangles whose sides have lengths 9 and 12

(b) The difference of the areas of two equilateral triangles whose sides have lengths 17 and 15

(c) The area of a trapezoid whose bases have lengths 6 and 2 and whose altitude has length $9\sqrt{3}$

(d) Twice the area of a right triangle having a hypotenuse of length 5 and an acute angle of measure 30°

22. Find the area of trapezoid $ABCD$ in Fig. 10-24 if: (10.7)

(a) $b = 25$, $b' = 15$, and $h = 7$ (d) $AB = 12$, $m\angle A = 45°$, $b = 13$, and $b' = 7$

(b) $m = 10$ and $h = 6.9$ (e) $AB = 10$, $m\angle A = 70°$, and $b + b' = 20$

(c) $AB = 30$, $m\angle A = 30°$, $b = 24$, and $b' = 6$

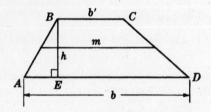   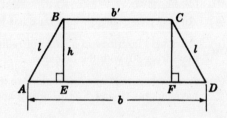

Fig. 10-24   Fig. 10-25

23. Find the area of isosceles trapezoid $ABCD$ in Fig. 10-25 if   (10.7)

(a) $b' = 17$, $l = 10$, and $h = 6$ (d) $b = 20$, $l = 8$, and $m\angle A = 60°$

(b) $b = 22$, $b' = 12$, and $l = 13$ (e) $b = 40$, $b' = 20$, and $m\angle A = 28°$

(c) $b = 16$, $b' = 10$, and $m\angle A = 45°$

24. (a) Find the length of the altitude of a trapezoid if the bases have lengths 13 and 7 and the area is
40.   (10.7)

(b) Find the length of the altitude of a trapezoid if the sum of the lengths of the bases is twice the length
of the altitude and the area is 49.

(c) Find the sum of the lengths of the bases and the median of a trapezoid if the area is 63 and the altitude
has length 7.

(d) Find the lengths of the bases of a trapezoid if the upper base has length 3 less than the lower base, the
altitude has length 4, and the area is 30.

(e) Find the lengths of the bases of a trapezoid if the lower base has length twice that of the upper base,
the altitude has length 6, and the area is 45.

25. In an isosceles trapezoid: (10.7)

(a) Find the lengths of the bases if each leg has length 5, the altitude has length 3, and the area is 39.

(b) Find the lengths of the bases if the altitude has length 5, each base angle measures 45°, and the area
is 90.

(c) Find the lengths of the bases if the area is $42\sqrt{3}$, the altitude has length $3\sqrt{3}$, and each base angle
measures 60°.

(d) Find the length of each leg if the bases have lengths 24 and 32 and the area is 84.

(e)   Find the length of each leg if the area is 300, the median has length 25, and the lower base has length 30.

26.   Find the area of a rhombus if   (10.8)

(a)   The diagonals have lengths 8 and 9.

(b)   The diagonals have lengths 11 and 7.

(c)   The diagonals have lengths 4 and $6\sqrt{3}$.

(d)   The diagonals have lengths $3x$ and $8x$.

(e)   One diagonal has length 10 and a side has length 13.

(f)   The perimeter is 40 and a diagonal has length 12.

(g)   The side has length 6 and an angle measures 30°.

(h)   The perimeter is 28 and an angle measures 45°.

(i)   The perimeter is 32 and the length of the short diagonal equals a side in length.

(j)   A side has length 14 and an angle measures 120°.

27.   Find the area of a rhombus to the nearest integer if (a) the side has length 30 and an angle measures 55°; (b) the perimeter is 20 and an angle measures 33°; (c) the side has length 10 and an angle measures 130°.   (10.8)

28.   In a rhombus, find the length of: (10.8)

(a)   A diagonal if the other diagonal has length 7 and the area is 35   (10.8)

(b)   The diagonals if their ratio is 4:3 and the area is 54

(c)   The diagonals if the longer is twice the shorter and the area is 100

(d)   The side if the area is 24 and one diagonal has length 6

(e)   The side if the area is 6 and one diagonal has length 4 more than the other

29.   A rhombus is equal to a trapezoid whose lower base has length 26 and whose other three sides have length 10. Find the length of the altitude of the rhombus if its perimeter is 36.   (10.8)

30.   Provide the proofs requested in Fig. 10-26.   (10.9)

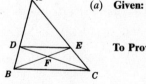   (a)   **Given:**   △ABC
$DB = \frac{1}{3}AB$
$EC = \frac{1}{3}AC$
To Prove: Area(△DFB)
= area(△FEC)

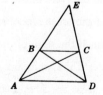   (b)   **Given:**   Trapezoid ABCD
$\overline{AB}$ and $\overline{CD}$ extended
meet at E.
To Prove: Area(△ECA)
= area(△EBD)

Fig. 10-26

31.   Provide the proofs requested in Fig. 10-27.   (10.9)

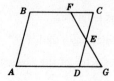   (a)   **Given:**   ▱ABCD
E is midpoint of $\overline{CD}$.
To Prove:
Area (▱ABCD) =
area(trapezoid BFGA)

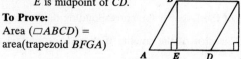   (b)   **Given**   ▱ABCD
$\overline{BE}$ and $\overline{CF} \perp \overline{AF}$.
To Prove:
BCFE is a rectangle and
equal in area to ▱ABCD.

Fig. 10-27

32. Prove each of the following: (10.10)

    (*a*) A median divides a triangle into two triangles having equal areas.

    (*b*) Triangles are equal in area if they have a common base and their vertices lie in a line parallel to the base.

    (*c*) In a triangle, if lines are drawn from a vertex to the trisection points of the opposite sides, the area of the triangle is trisected.

    (*d*) In trapezoid $ABCD$, base $\overline{AD}$ is twice base $\overline{BC}$. If M is the midpoint of $\overline{AD}$, then $ABCM$ and $BCDM$ are parallelograms which are equal in area.

33. (*a*) In $\triangle ABC$, E is a point on $\overline{BM}$, the median to $\overline{AC}$. Prove that area($\triangle BEA$) = area($\triangle BEC$).

    (*b*) In $\triangle ABC$, Q is a point on $\overline{BC}$, M is the midpoint of $\overline{AB}$, and P is the midpoint of $\overline{AC}$. Prove that area($\triangle BQM$) + area($\triangle PQC$) = area(quadrilateral $APQM$).

    (*c*) In quadrilateral $ABCD$, diagonal $\overline{AC}$ bisects diagonal $\overline{BD}$. Prove that area($\triangle ABC$) = area($\triangle ACD$).

    (*d*) Prove that the diagonals of a parallelogram divide the parallelogram into four triangles which are equal in area.   (10.10)

34. Find the ratio of the areas of two similar triangles if the ratio of two corresponding sides is (*a*) 1:7; (*b*) 7:2; (*c*) $1:\sqrt{3}$; (*d*) $a:5a$; (*e*) $9:x$; (*f*) $3:\sqrt{x}$; (*g*) $s:s\sqrt{2}$.   (10.11)

35. Find the ratio of the areas of two similar triangles   (10.11)

    (*a*) If the ratio of the lengths of two corresponding medians is 7:10

    (*b*) If the length of an altitude of the first is two-thirds of a corresponding altitude of the second

    (*c*) If two corresponding angle bisectors have lengths 10 and 12

    (*d*) If the length of each side of the first is one-third the length of each corresponding side of the second

    (*e*) If the radii of their circumscribed circles are $7\frac{1}{2}$ and 5

    (*f*) If their perimeters are 30 and $30\sqrt{2}$

36. Find the ratio of any two corresponding sides of two similar triangles if the ratio of their areas is (*a*) 100:1; (*b*) 1:49; (*c*) 400:81; (*d*) 25:121; (*e*) $4:y^2$; (*f*) $9x^2:1$; (*g*) 3:4; (*h*) 1:2; (*i*) $x^2:5$; (*j*) $x:16$.   (10.11)

37. In two similar triangles, find the ratio of the lengths of   (10.11)

    (*a*) Corresponding sides if the areas are 72 and 50

    (*b*) Corresponding medians if the ratio of the areas is 9:49

    (*c*) Corresponding altitudes if the areas are 18 and 6

    (*d*) The perimeters if the areas are 50 and 40

    (*e*) Radii of the inscribed circles if the ratio of the areas is 1:3

38. The areas of two similar triangles are in the ratio of 25:16. Find   (10.11)

    (*a*) The length of a side of the larger if the corresponding side of the smaller has length 80

    (*b*) The length of a median of the larger if the corresponding median of the smaller has length 10

    (*c*) The length of an angle bisector of the smaller if the corresponding angle bisector of the larger has length 15

    (*d*) The perimeter of the smaller if the perimeter of the larger is 125

(e) The circumference of the inscribed circle of the larger if the circumference of the inscribed circle of the smaller is 84

(f) The diameter of the circumscribed circle of the smaller if the diameter of the circumscribed circle of the larger is 22.5

(g) The length of an altitude of the larger if the corresponding altitude of the smaller has length $16\sqrt{3}$

39. (a) The areas of two similar triangles are 36 and 25. If a median of the smaller triangle has length 10, find the length of the corresponding median of the larger.   (10.12)

(b) Corresponding altitudes of two similar triangles have lengths 3 and 4. If the area of the larger triangle is 112, find the area of the smaller.

(c) Two similar polygons have perimeters of 32 and 24. If the area of the smaller is 27, find the area of the larger.

(d) The areas of two similar pentagons are 88 and 22. If a diagonal of the larger has length 5, find the length of the corresponding diagonal of the smaller.

(e) In two similar polygons, the ratio of the lengths of two corresponding sides is $\sqrt{3}:1$. If the area of the smaller is 15, find the area of the larger.

40. Find the area of $\triangle DEF$, whose vertices are $D(0,0)$ and (a) $E(6,4)$ and $F(8,2)$; (b) $E(3,2)$ and $F(6,-4)$; (c) $E(-2,3)$ and $F(10,7)$.   (10.13)

41. Find the area of a triangle whose vertices are (a) $(0,0)$, $(2,3)$, and $(4,1)$; (b) $(1,1)$, $(7,3)$, and $(3,6)$; (c) $(-1,2)$, $(0,-2)$, and $(3,1)$.   (10.13)

42. The vertices of $\triangle ABC$ are $A(2,1)$, $B(8,9)$, and $C(5,7)$. (a) Find the area of $\triangle ABC$. (b) Find the length of $\overline{AB}$. (c) Find the length of the altitude to $\overline{AB}$.   (10.13)

CHAPTER 11

Regular Polygons and the Circle

11.1 REGULAR POLYGONS

A *regular polygon* is an equilateral and equiangular polygon.

The *center of a regular polygon* is the common center of its inscribed and circumscribed circles.

A *radius of a regular polygon* is a segment joining its center to any vertex. A radius of a regular polygon is also a radius of the circumscribed circle. (Here, as for circles, we may use the word *radius* to mean the number that is "the length of the radius.")

A *central angle of a regular polygon* is an angle included between two radii drawn to successive vertices.

An *apothem of a regular polygon* is a segment from its center perpendicular to one of its sides. An apothem is also a radius of the inscribed circle.

Thus for the regular pentagon shown in Fig. 11-1, $AB = BC = CD = DE = EA$ and $m\angle A = m\angle B = m\angle C = m\angle D = m\angle E$. Also, its center is O, $\overline{OA}$ and $\overline{OB}$ are its radii; $\angle AOB$ is a central angle; and $\overline{OG}$ and $\overline{OF}$ are apothems.

The reader should use a calculator for many of the computations in this chapter. See Chapter 2 for an introduction to the use of calculators.

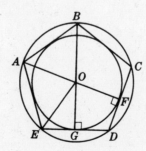

Fig. 11-1

11.1A   Regular-Polygon Principles

PRINCIPLE 1:   *If a regular polygon of n sides has a side of length s, the perimeter is p = ns.*

PRINCIPLE 2: *A circle may be circumscribed about any regular polygon.*

PRINCIPLE 3: *A circle may be inscribed in any regular polygon.*

PRINCIPLE 4: *The center of the circumscribed circle of a regular polygon is also the center of its inscribed circle.*

PRINCIPLE 5: *An equilateral polygon inscribed in a circle is a regular polygon.*

PRINCIPLE 6: *Radii of a regular polygon are congruent.*

PRINCIPLE 7: *A radius of a regular polygon bisects the angle to which it is drawn.*

Thus in Fig. 11-1, $\overline{OB}$ bisects $\angle ABC$.

PRINCIPLE 8: *Apothems of a regular polygon are congruent.*

PRINCIPLE 9: *An apothem of a regular polygon bisects the side to which it is drawn.*

Thus in Fig. 11-1, $\overline{OF}$ bisects $\overline{CD}$, and $\overline{OG}$ bisects $\overline{ED}$.

PRINCIPLE 10: *For a regular polygon of n sides:*

1. *Each central angle c measures* $\dfrac{360°}{n}$.

2. *Each interior angle i measures* $\dfrac{(n-2)180°}{n}$.

3. *Each exterior angle e measures* $\dfrac{360°}{n}$.

Thus for the regular pentagon $ABCDE$ of Fig. 11-2,

$$m\angle AOB = m\angle ABS = \frac{360°}{n} = \frac{360°}{5} = 72° \qquad m\angle ABC = \frac{(n-2)180°}{n} = \frac{(5-2)180°}{5} = 108°$$

and
$$m\angle ABC + m\angle ABS = 180°$$

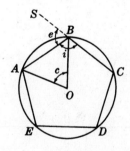

Fig. 11-2

SOLVED PROBLEMS

11.1 FINDING MEASURES OF LINES AND ANGLES IN A REGULAR POLYGON

(*a*) Find the length of a side *s* of a regular pentagon if the perimeter *p* is 35.

(*b*) Find the length of the apothem *a* of a regular pentagon if the radius of the inscribed circle is 21.

(c) In a regular polygon of five sides, find the measures of the central angle c, the exterior angle e, and the interior angle i.

(d) If an interior angle of a regular polygon measures 108°, find the measures of the exterior angle and the central angle and the number of sides.

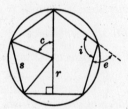

Fig. 11-3

Solutions

(a) $p = 35$. Since $p = 5s$, we have $35 = 5s$ and $s = 7$.

(b) Since an apothem r is a radius of the inscribed circle, it has length 21.

(c) $n = 5$. Then $m\angle c = \dfrac{360°}{n} = \dfrac{360°}{5} = 72°$; $m\angle e = \dfrac{360°}{n} = 72°$; $m\angle i = 180° - m\angle e = 108°$.

(d) $m\angle i = 108°$. Then $m\angle c = 180° - m\angle i = 72°$. Since $m\angle c = \dfrac{360°}{n}$, $n = 5$. (See Fig. 11-3.)

11.2   PROVING A REGULAR-POLYGON PROBLEM STATED IN WORDS

Prove that a vertex angle of a regular pentagon is trisected by diagonals drawn from that vertex.

Solution

Given: Regular pentagon $ABCDE$
          Diagonals $\overline{AC}$ and $\overline{AD}$

To Prove: $\overline{AC}$ and $\overline{AD}$ trisect $\angle A$.

Plan: Circumscribe a circle and show that angles BAC, CAD, and DAE are congruent.

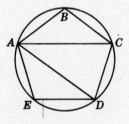

PROOF:

Statements	Reasons
1. $ABCDE$ is a regular pentagon.	1. Given
2. Circumscribe a circle about $ABCDE$.	2. A circle may be circumscribed about any regular polygon
3. $BC = CD = DE$	3. A regular polygon is equilateral.
4. $\overparen{BC} \cong \overparen{CD} \cong \overparen{DE}$	4. In a circle, equal chords have equal arcs.
5. $\angle BAC \cong \angle CAD \cong \angle DAE$	5. In a circle, inscribed angles having congruent arcs are congruent.
6. $\angle A$ is trisected.	6. To divide into three congruent parts is to trisect.

11.2 RELATIONSHIPS OF SEGMENTS IN REGULAR POLYGONS OF 3, 4, AND 6 SIDES

In the regular hexagon, square, and equilateral triangle, special right triangles are formed when the apothem r and a radius R terminating in the same side are drawn. In the case of the square we obtain a 45°-45°-90° triangle, while in the other two cases we obtain a 30°-60°-90° triangle. The formulas in Fig. 11-4 relate the lengths of the sides and radii of these regular polygons.

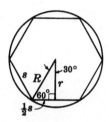

Regular Hexagon
$s = R$
$r = \frac{1}{2}R\sqrt{3}$

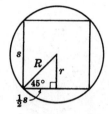

Square
$s = R\sqrt{2}$
$r = \frac{1}{2}s = \frac{1}{2}R\sqrt{2}$

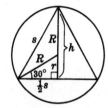

Equilateral Triangle
$s = R\sqrt{3}, h = r + R$
$r = \frac{1}{3}h, R = \frac{2}{3}h, r = \frac{1}{2}R$

Fig. 11-4

SOLVED PROBLEMS

11.3 APPLYING LINE RELATIONSHIPS IN A REGULAR HEXAGON

In a regular hexagon, (a) find the lengths of the side and apothem if the radius is 12; (b) find the radius and length of the apothem if the side has length 8.

Solutions

(a) Since $R = 12$, $s = R = 12$ and $r = \frac{1}{2}R\sqrt{3} = 6\sqrt{3}$.

(b) Since $s = 8$, $R = s = 8$ and $r = \frac{1}{2}R\sqrt{3} = 4\sqrt{3}$.

11.4 APPLYING LINE RELATIONSHIPS IN A SQUARE

In a square, (a) find the lengths of the side and apothem if the radius is 16; (b) find the radius and the length of the apothem if a side has length 10.

Solutions

(a) Since $R = 16$, $s = R\sqrt{2} = 16\sqrt{2}$ and $r = \frac{1}{2}s = 8\sqrt{2}$.

(b) Since $s = 10$, $r = \frac{1}{2}s = 5$ and $R = s/\sqrt{2} = \frac{1}{2}s\sqrt{2} = 5\sqrt{2}$.

11.5 APPLYING LINE RELATIONSHIPS IN AN EQUILATERAL TRIANGLE

In an equilateral triangle, (a) find the lengths of the radius, apothem, and side if the altitude has length 6; (b) find the lengths of the side, apothem, and altitude if the radius is 9.

Solutions

(a) Since $h = 6$, we have $r = \frac{1}{3}h = 2$; $R = \frac{2}{3}h = 4$; and $s = R\sqrt{3} = 4\sqrt{3}$.

(b) Since $R = 9$, $s = R\sqrt{3} = 9\sqrt{3}$; $r = \frac{1}{2}R = 4\frac{1}{2}$; and $h = \frac{3}{2}R = 13\frac{1}{2}$.

11.3   AREA OF A REGULAR POLYGON

The area of a regular polygon equals one-half the product of its perimeter and the length of its apothem.

As shown in Fig. 11-5, by drawing radii we can divide a regular polygon of n sides and perimeter $p = ns$ into n triangles, each of area $\frac{1}{2}rs$. Hence the area of the regular polygon is $n(\frac{1}{2}rs) = \frac{1}{2}nsr = \frac{1}{2}pr$.

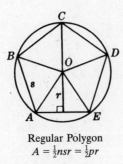

Regular Polygon
$A = \frac{1}{2}nsr = \frac{1}{2}pr$

Fig. 11-5

SOLVED PROBLEMS

11.6   FINDING THE AREA OF A REGULAR POLYGON

(a) Find the area of a regular hexagon if the length of the apothem is $5\sqrt{3}$.

(b) Find the area of a regular pentagon to the nearest integer if the length of the apothem is 20.

Solutions

(a) In a regular hexagon, $r = \frac{1}{2}s\sqrt{3}$. Since $r = 5\sqrt{3}$, $s = 10$ and $p = 6(10) = 60$.
Then $A = \frac{1}{2}pr = \frac{1}{2}(60)(5\sqrt{3}) = 150\sqrt{3}$.

(b) In Fig. 11-6, $m\angle AOE = 360°/5 = 72°$ and $m\angle AOF = \frac{1}{2}m\angle AOE = 36°$. Then $\tan 36° = \frac{1}{2}s/20 = s/40$ or $s = 40\tan 36°$.
Now $A = \frac{1}{2}pr = \frac{1}{2}nsr = \frac{1}{2}(5)(40\tan 36°)(20) = 1453$.

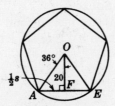

Fig. 11-6

11.4   RATIOS OF SEGMENTS AND AREAS OF REGULAR POLYGONS

PRINCIPLE 1:   *Regular polygons having the same number of sides are similar.*

PRINCIPLE 2:   *Corresponding segments of regular polygons having the same number of sides are in proportion.* "Segments" here includes sides, perimeters, radii or circumferences of circumscribed or inscribed circles, and such.

PRINCIPLE 3: *Areas of regular polygons having the same number of sides are to each other as the squares of the lengths of any two corresponding segments.*

SOLVED PROBLEMS

11.7 RATIOS OF LINES AND AREAS OF REGULAR POLYGONS

   (*a*) In two regular polygons having the same number of sides, find the ratio of the lengths of the apothems if the perimeters are in the ratio 5:3.

   (*b*) In two regular polygons having the same number of sides, find the length of a side of the smaller if the lengths of the apothems are 20 and 50 and a side of the larger has length 32.5.

   (*c*) In two regular polygons having the same number of sides, find the ratio of the areas if the lengths of the sides are in the ratio 1:5.

   (*d*) In two regular polygons having the same number of sides, find the area of the smaller if the sides have lengths 4 and 12 and the area of the larger is 10,260.

Solutions

   (*a*) By Principle 2, $r:r' = p:p' = 5:3$.

   (*b*) By Principle 2, $s:s' = r:r'$; thus $s:32.5 = 20:50$ and $s = 13$.

   (*c*) By Principle 3, $\dfrac{A}{A'} = \left(\dfrac{s}{s'}\right)^2 = \left(\dfrac{1}{5}\right)^2 = \dfrac{1}{25}$.

   (*d*) By Principle 3, $\dfrac{A}{A'} = \left(\dfrac{s}{s'}\right)^2$. Then $\dfrac{A}{10,260} = \left(\dfrac{4}{12}\right)^2$ and $A = 1140$.

11.5 CIRCUMFERENCE AND AREA OF A CIRCLE

π (pi) is the ratio of the circumference C of any circle to its diameter d; that is, $\pi = C/d$. Hence

$$C = \pi d \qquad \text{or} \qquad C = 2\pi r$$

Approximate values for π are 3.1416 or 3.14 or $\frac{22}{7}$. Unless you are told otherwise, we shall use 3.14 for π in solving problems.

   A circle may be regarded as a regular polygon having an infinite number of sides. If a square is inscribed in a circle, and the number of sides is continually doubled (to form an octagon, a 16-gon, and so on), the perimeters of the resulting polygons will get closer and closer to the circumference of the circle (Fig. 11-7).

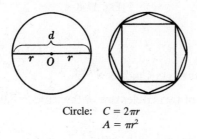

Circle: $C = 2\pi r$
$A = \pi r^2$

Fig. 11-7

Thus to find the area of a circle, the formula $A = \frac{1}{2}pr$ can be used with C substituted for p; doing so, we get

$$A = \tfrac{1}{2}Cr = \tfrac{1}{2}(2\pi r)(r) = \pi r^2$$

All circles are similar figures, since they have the same shape. Because they are similar figures, (1) corresponding segments of circles are in proportion and (2) the areas of two circles are to each other as the squares of their radii or circumferences.

SOLVED PROBLEMS

11.8 FINDING THE CIRCUMFERENCE AND AREA OF A CIRCLE
In a circle, (a) find the circumference and area if the radius is 6; (b) find the radius and area if the circumference is 18π; (c) find the radius and circumference if the area is 144π. (Answer both in terms of π and to the nearest integer.)

Solutions

(a) $r = 6$. Then $C = 2\pi r = 12\pi$ and $A = \pi r^2 = 36\pi \doteq 36(3.14) = 113$.

(b) $C = 18\pi$. Since $C = 2\pi r$, we have $18\pi = 2\pi r$ and $r = 9$. Then $A = \pi r^2 = 81\pi \doteq 254$.

(c) $A = 144\pi$. Since $A = \pi r^2$, we have $144\pi = \pi r^2$ and $r = 12$. Then $C = 2\pi r = 24\pi \doteq 75$.

11.9 CIRCUMFERENCE AND AREA OF CIRCUMSCRIBED AND INSCRIBED CIRCLES
Find the circumference and area of the circumscribed circle and inscribed circle (a) of a regular hexagon whose side has length 8; (b) of an equilateral triangle whose altitude has length $9\sqrt{3}$. (See Fig. 11-8.)

Solutions

(a) Here $R = s = 8$. Then $C = 2\pi R = 16\pi$ and $A = \pi R^2 = 64\pi$.
Also $r = \frac{1}{2}R\sqrt{3} = 4\sqrt{3}$. Then $C = 2\pi r = 8\pi\sqrt{3}$ and $A = \pi r^2 = 48\pi$.

(b) Here $R = \frac{2}{3}h = 6\sqrt{3}$. Then $C = 2\pi R = 12\pi\sqrt{3}$ and $A = \pi R^2 = 108\pi$.
Also $r = \frac{1}{3}h = 3\sqrt{3}$. Then $C = 2\pi r = 6\pi\sqrt{3}$ and $A = \pi r^2 = 27\pi$.

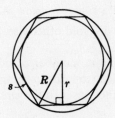
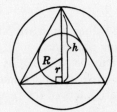

Fig. 11-8

11.10 RATIOS OF SEGMENTS AND AREAS IN CIRCLES

(a) If the circumferences of two circles are in the ratio 2:3, find the ratio of the diameters and the ratio of the areas.

(b) If the areas of two circles are in the ratio 1:25, find the ratio of the diameters and the ratio of the circumferences.

Solutions

(a) $\dfrac{d}{d'} = \dfrac{C}{C'} = \dfrac{2}{3}$ and $\dfrac{A}{A'} = \left(\dfrac{C}{C'}\right)^2 = \left(\dfrac{2}{3}\right)^2 = \dfrac{4}{9}$.

(b) Since $\dfrac{A}{A'} = \left(\dfrac{d}{d'}\right)^2$, $\dfrac{1}{25} = \left(\dfrac{d}{d'}\right)^2$ and $\dfrac{d}{d'} = \dfrac{1}{5}$. Also, $\dfrac{C}{C'} = \dfrac{d}{d'} = \dfrac{1}{5}$.

11.6   LENGTH OF AN ARC; AREA OF A SECTOR AND A SEGMENT

A sector of a circle is a part of a circle bounded by two radii and their intercepted arc. Thus in Fig. 11-9, the shaded section of circle O is sector OAB.

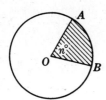

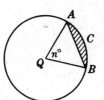

Fig. 11-9   **Fig. 11-10**

A *segment of a circle* is a part of a circle bounded by a chord and its arc. A *minor segment* of a circle is the smaller of the two segments thus formed. Thus in Fig. 11-10, the shaded section of circle Q is minor segment ACB.

PRINCIPLE 1:   *In a circle of radius r, the length l of an arc of measure $n°$ equals $\dfrac{n}{360}$ of the circumference of the circle, or $l = \dfrac{n}{360}2\pi r = \dfrac{\pi n r}{180}$.*

PRINCIPLE 2:   *In a circle of radius r, the area K of a sector of measure $n°$ equals $\dfrac{n}{360}$ of the area of the circle, or $K = \dfrac{n}{360}\pi r^2$.*

PRINCIPLE 3: $\dfrac{\text{Area of a sector of } n°}{\text{Area of the circle}} = \dfrac{\text{length of an arc of measure } n°}{\text{circumference of the circle}} = \dfrac{n}{360}$

PRINCIPLE 4:   *The area of a minor segment of a circle equals the area of its sector less the area of the triangle formed by its radii and chord.*

PRINCIPLE 5:   *If a regular polygon is inscribed in a circle, each segment cut off by the polygon has area equal to the difference between the area of the circle and the area of the polygon divided by the number of sides.*

SOLVED PROBLEMS

11.11   LENGTH OF AN ARC

(a) Find the length of a 36° arc in a circle whose circumference is 45π.

(b) Find the radius of a circle if a 40° arc has a length of 4π.

Solutions

(a) Here $n° = 36°$ and $C = 2\pi r = 45\pi$. Then $l = \dfrac{n}{360}2\pi r = \dfrac{36}{360}45\pi = \dfrac{9}{2}\pi$.

(b) Here $l = 4\pi$ and $n° = 40°$. Then $l = \dfrac{n}{360}2\pi r$ yields $4\pi = \dfrac{40}{360}2\pi r$, and $r = 18$.

11.12 AREA OF A SECTOR

(a) Find the area K of a 300° sector of a circle whose radius is 12.

(b) Find the measure of the central angle of a sector whose area is 6π if the area of the circle is 9π.

(c) Find the radius of a circle if an arc of length 2π has a sector of area 10π.

Solutions

(a) $n° = 300°$ and $r = 12$. Then $K = \dfrac{n}{360}\pi r^2 = \dfrac{300}{360}144\pi = 120\pi$.

(b) $\dfrac{\text{Area of sector}}{\text{Area of circle}} = \dfrac{n}{360}$, so $\dfrac{6\pi}{9\pi} = \dfrac{n}{360}$, and $n = 240$. Thus the central angle measures 240°.

(c) $\dfrac{\text{Length of arc}}{\text{Circumference}} = \dfrac{\text{area of sector}}{\text{area of circle}}$, so $\dfrac{2\pi}{2\pi r} = \dfrac{10\pi}{\pi r^2}$ and $r = 10$.

11.13 AREA OF A SEGMENT OF A CIRCLE

(a) Find the area of a segment if its central angle measures 60° and the radius of the circle is 12.

(b) Find the area of a segment if its central angle measures 90° and the radius of the circle is 8.

(c) Find each segment formed by an inscribed equilateral triangle if the radius of the circle is 8.

Solutions

See Fig. 11-11.

(a) $n° = 60°$ and $r = 12$. Then area of sector $OAB = \dfrac{n}{360}\pi r^2 = \dfrac{60}{360}144\pi = 24\pi$.

Also, area of equilateral $\triangle OAB = \frac{1}{4}s^2\sqrt{3} = \frac{1}{4}(144)\sqrt{3} = 36\sqrt{3}$.
Hence area of segment $ACB = 24\pi - 36\sqrt{3}$.

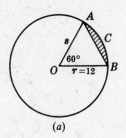   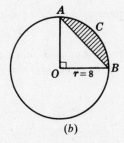   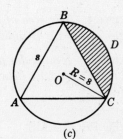

(a)   (b)   (c)

Fig. 11-11

(b) $n° = 90°$ and $r = 8$. Then area of sector $OAB = \dfrac{n}{360}\pi r^2 = \dfrac{90}{360}64\pi = 16\pi$.

Also, area of rt. $\triangle OAB = \frac{1}{2}bh = \frac{1}{2}(8)(8) = 32$.
Hence area of segment $ACB = 16\pi - 32$.

(c) $R = 8$. Since $s = R\sqrt{3} = 8\sqrt{3}$, the area of $\triangle ABC$ is $\frac{1}{4}s^2\sqrt{3} = 48\sqrt{3}$.
Also, area of circle $O = \pi R^2 = 64\pi$.
Hence area of segment $BDC = \frac{1}{3}(64\pi - 48\sqrt{3})$.

11.14 AREA OF A SEGMENT FORMED BY AN INSCRIBED REGULAR POLYGON
Find the area of each segment formed by an inscribed regular polygon of 12 sides (dodecagon) if the radius of the circle is 12. (See Fig. 11-12.)

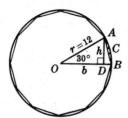

Fig. 11-12

Solution

Area of sector $OAB = \dfrac{n}{360}\,\pi r^2 = \dfrac{30}{360}\,144\pi = 12\pi$.

To find the area $\triangle OAB$, we draw altitude $\overline{AD}$ to base $\overline{OB}$. Since $m\angle AOB = 30°$, $h = AD = \frac{1}{2}r = 6$. Then the area of $\triangle OAB$ is $\frac{1}{2}bh = \frac{1}{2}(12)(6) = 36$.

Hence the area of segment ACB is $12\pi - 36$.

11.7 AREAS OF COMBINATION FIGURES

The areas of combination figures like that in Fig. 11-13 may be found by determining individual areas and then adding or subtracting as required. Thus the shaded area in the figure equals the sum of the areas of the square and the semicircle: $A = 8^2 + \frac{1}{2}(16\pi) = 64 + 8\pi$.

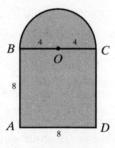

Fig. 11-13

SOLVED PROBLEMS

11.15 FINDING AREAS OF COMBINATION FIGURES
Find the shaded area in each part of Fig. 11-14. In (a), circles A, B, and C are tangent externally and each has radius 3. In (b), each arc is part of a circle of radius 9.

Solutions

(a)   Area of $\triangle ABC = \frac{1}{4}s^2\sqrt{3} = \dfrac{1}{4}(6^2)\sqrt{3} = 9\sqrt{3}$. Area of sector $\mathrm{I} = \dfrac{n°}{360°}(\pi r^2) = \dfrac{300}{360}(9\pi) = \dfrac{15}{2}\pi$.

   Shaded area $= 9\sqrt{3} + 3(\frac{15}{2}\pi) = 9\sqrt{3} + \frac{45}{2}\pi$.

(b)   Area of square $= 18^2 = 324$. Area of sector $\mathrm{I} = \dfrac{n°}{360°}(\pi r^2) = \dfrac{90}{360}(81\pi) = \dfrac{81}{4}\pi$.

   Shaded area $= 324 - 4(\frac{81}{4}\pi) = 324 - 81\pi$.

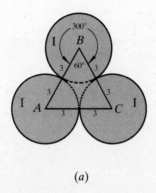

(a)

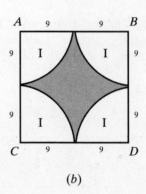

(b)

Fig. 11-14

Supplementary Problems

1. In a regular polygon, find   (11.1)

    (a)   The perimeter if the length of a side is 8 and the number of sides is 25

    (b)   The perimeter if the length of a side is 2.45 and the number of sides is 10

    (c)   The perimeter if the length of a side is $4\frac{2}{3}$ and the number of sides is 24

    (d)   The number of sides if the perimeter is 325 and the length of a side is 25

    (e)   The number of sides if the perimeter is $27\sqrt{3}$ and the length of a side is $3\sqrt{3}$

    (f)   The length of a side if the number of sides is 30 and the perimeter is 100

    (g)   The length of a side if the perimeter is 67.5 and the number of sides is 15

2. In a regular polygon, find   (11.1)

    (a)   The length of the apothem if the diameter of an inscribed circle is 25

    (b)   The length of the apothem if the radius of the inscribed circle is 23.47

    (c)   The radius of the inscribed circle if the length of the apothem is $7\sqrt{3}$

    (d)   The radius of the regular polygon if the diameter of the circumscribed circle is 37

    (e)   The radius of the circumscribed circle if the radius of the regular polygon is $3\sqrt{2}$

3. In a regular polygon of 15 sides, find the measure of (a) the central angle; (b) the exterior angle; (c) the interior angle.   (11.1)

4. If an exterior angle of a regular polygon measures 40°, find (a) the measure of the central angle; (b) the number of sides; (c) the measure of the interior angle.   (11.1)

5. If an interior angle of a regular polygon measures 165°, find (a) the measure of the exterior angle; (b) the measure of the central angle; (c) the number of sides.   (11.1)

6. If a central angle of a regular polygon measures 5°, find (a) the measure of the exterior angle; (b) the number of sides; (c) the measure of the interior angle.   (11.1)

7. Name the regular polygon whose   (11.1)

    (*a*)   Central angle measures 45°   (*d*)   Exterior angle measures 36°

    (*b*)   Central angle measures 60°   (*e*)   Interior angle is congruent to its central angle

    (*c*)   Exterior angle measures 120°   (*f*)   Interior angle measures 150°

8. Prove each of the following: (11.2)

    (*a*)   The diagonals of a regular pentagon are congruent.

    (*b*)   A diagonal of a regular pentagon forms an isosceles trapezoid with three of its sides.

    (*c*)   If two diagonals of a regular pentagon intersect, the longer segment of each diagonal is congruent to a side of the regular pentagon.

9. In a regular hexagon, find   (11.3)

    (*a*)   The length of a side if its radius is 9

    (*b*)   The perimeter if its radius is 5

    (*c*)   The length of the apothem if its radius is 12

    (*d*)   Its radius if the length of a side is 6

    (*e*)   The length of the apothem if the length of a side is 26

    (*f*)   Its radius if the length of the apothem is $3\sqrt{3}$

    (*g*)   The length of a side if the length of the apothem is 30

    (*h*)   The perimeter if the length of the apothem is $5\sqrt{3}$

10. In a square, find   (11.4)

    (*a*)   The length of a side if the radius is 18

    (*b*)   The length of the apothem if the radius is 14

    (*c*)   The perimeter if the radius is $5\sqrt{2}$

    (*d*)   The radius if the length of a side is 16

    (*e*)   The length of a side if the length of the apothem is 1.7

    (*f*)   The perimeter if the length of the apothem is $3\frac{1}{2}$

    (*g*)   The radius if the perimeter is 40

    (*h*)   The length of the apothem if the perimeter is $16\sqrt{2}$

11. In an equilateral triangle, find   (11.5)

    (*a*)   The length of a side if its radius is 30

    (*b*)   The length of the apothem if its radius is 28

    (*c*)   The length of an altitude if its radius is 18

    (*d*)   The perimeter if its radius is $2\sqrt{3}$

    (*e*)   Its radius if the length of a side is 24

    (*f*)   The length of the apothem if the length of a side is 24

    (*g*)   The length of its altitude if the length of a side is 96

    (*h*)   Its radius if the length of the apothem is 21

(*i*)   The length of a side if the length of the apothem is $\sqrt{3}$

(*j*)   The length of the altitude if the length of the apothem is $3\frac{1}{3}$

(*k*)   The length of the altitude if the perimeter is 15

(*l*)   The length of the apothem if the perimeter is 54.

12. (*a*) Find the area of a regular pentagon to the nearest integer if the length of the apothem is 15.

    (*b*) Find the area of a regular decagon to the nearest integer if the length of a side is 20.   (11.6)

13. Find the area of a regular hexagon, in radical form, if (*a*) the length of a side is 6; (*b*) its radius is 8; (*c*) the length of the apothem is $10\sqrt{3}$.   (11.6)

14. Find the area of a square if (*a*) the length of the apothem is 12; (*b*) its radius is $9\sqrt{2}$; (*c*) its perimeter is 40.   (11.6)

15. Find the area of an equilateral triangle, in radical form, if   (11.6)

    (*a*) The length of the apothem is $2\sqrt{3}$.   (*d*) The length of the altitude is $12\sqrt{3}$.

    (*b*) Its radius is 6.   (*e*) The perimeter is $6\sqrt{3}$.

    (*c*) The length of the altitude is 4.   (*f*) The length of the apothem is 4.

16. If the area of a regular hexagon is $150\sqrt{3}$, find (*a*) the length of a side; (*b*) its radius; (*c*) the length of the apothem.   (11.6)

17. If the area of an equilateral triangle is $81\sqrt{3}$, find (*a*) the length of a side; (*b*) the length of the altitude; (*c*) its radius; (*d*) the length of the apothem.   (11.6)

18. Find the ratio of the perimeters of two regular polygons having the same number of sides if   (11.7)

    (*a*) The ratio of the sides is 1:8.

    (*b*) The ratio of their radii is 4:9.

    (*c*) Their radii are 18 and 20.

    (*d*) Their apothems have lengths 16 and 22.

    (*e*) The length of the larger side is triple that of the smaller.

    (*f*) The length of the smaller apothem is two-fifths that of the larger.

    (*g*) The lengths of the apothems are $20\sqrt{2}$ and 15.

    (*h*) The circumference of the larger circumscribed circle is $2\frac{1}{2}$ times that of the smaller.

19. Find the ratio of the perimeters of two equilateral triangles if (*a*) the sides have lengths 20 and 8; (*b*) their radii are 12 and 60; (*c*) their apothems have lengths $2\sqrt{3}$ and $6\sqrt{3}$; (*d*) the circumferences of their inscribed circles are 120 and 160; (*e*) their altitudes have lengths $5x$ and x.   (11.7)

20. Find the ratio of the lengths of the sides of two regular polygons having the same number of sides if the ratio of their areas is (*a*) 25:1; (*b*) 16:49; (*c*) x^2:4; (*d*) 2:1; (*e*) 3:y^2; (*f*) x:18.   (11.7)

21. Find the ratio of the areas of two regular hexagons if (*a*) their sides have lengths 14 and 28; (*b*) their apothems have lengths 3 and 15; (*c*) their radii are $6\sqrt{3}$ and $\sqrt{3}$; (*b*) their perimeters are 75 and 250; (*e*) the circumferences of the circumscribed circles are 28 and 20.   (11.7)

22. Find the circumference of a circle in terms of π if (*a*) the radius is 6; (*b*) the diameter is 14; (*c*) the area is 25π; (*d*) the area is 3π.   (11.8)

23. Find the area of a circle in terms of π if (*a*) the radius is 3; (*b*) the diameter is 10; (*c*) the circumference is 16π; (*d*) the circumference is π; (*e*) the circumference is $6\pi\sqrt{2}$.   (11.8)

24. In a circle, (*a*) find the circumference and area if the radius is 5; (*b*) find the radius and area if the circumference is 16π; (*c*) find the radius and circumference if the area is 16π.   (11.8)

25. In a regular hexagon, find the circumference of the circumscribed circle if (*a*) the length of the apothem is $3\sqrt{3}$; (*b*) the perimeter is 12; (*c*) the length of a side is $3\frac{1}{2}$. Also find the circumference of its inscribed circle if (*d*) the length of the apothem is 13; (*e*) the length of a side is 8; (*f*) the perimeter is $6\sqrt{3}$.   (11.9)

26. For a square, find the area in terms of π of the   (11.9)

    (*a*)   Circumscribed circle if the length of the apothem is 7

    (*b*)   Circumscribed circle if the perimeter is 24

    (*c*)   Circumscribed circle if the length of a side is 8

    (*d*)   Inscribed circle if the length of the apothem is 5

    (*e*)   Inscribed circle if the length of a side is $12\sqrt{2}$

    (*f*)   Inscribed circle if the perimeter is 80

27. Find the circumference and area of the (1) circumscribed circle and (2) inscribed circle of   (11.9)

    (*a*)   A regular hexagon if the length of a side is 4

    (*b*)   A regular hexagon if the length of the apothem is $4\sqrt{3}$

    (*c*)   An equilateral triangle if the length of the altitude is 9

    (*d*)   An equilateral triangle if the length of the apothem is 4

    (*e*)   A square if the length of a side is 20

    (*f*)   A square if the length of the apothem is 3

28. Find the radius of a pipe having the same capacity as two pipes whose radii are (*a*) 6 ft and 8 ft; (*b*) 8 ft and 15 ft; (*c*) 3 ft and 6 ft. (*Hint*: Find the areas of their circular cross sections.)   (11.10)

29. In a circle, find the length of a 90° arc if   (11.11)

    (*a*)   The radius is 4.

    (*b*)   The diameter is 40.

    (*c*)   The circumference is 32.

    (*d*)   The circumference is 44π.

    (*e*)   An inscribed hexagon has a side of length 12.

    (*f*)   An inscribed equilateral triangle has an altitude of length 30.

30. Find the length of   (11.11)

    (*a*)   A 90° arc if the radius of the circle is 6

    (*b*)   A 180° arc if the circumference is 25

(c)   A 30° arc if the circumference is 60π

(d)   A 40° arc if the diameter is 18

(e)   An arc intercepted by the side of a regular hexagon inscribed in a circle of radius 3

(f)   An arc intercepted by a chord of length 12 in a circle of radius 12.

31. In a circle, find the area of a 60° sector if   (11.12)

    (a)   The radius is 6.   (e)   The area of the circle is 27.

    (b)   The diameter is 2.   (f)   The area of a 240° sector is 52.

    (c)   The circumference is 10π.   (g)   An inscribed hexagon has a side of length 12.

    (d)   The area of the circle is 150π. (h)   An inscribed hexagon has an area of $24\sqrt{3}$.

32. Find the area of a   (11.12)

    (a)   60° sector if the radius of the circle is 6   (c)   15° sector if the area of the circle is 72π

    (b)   240° sector if the area of the circle is 30   (d)   90° sector if its arc length is 4π

33. Find the measure of a central angle of an arc whose length is   (11.12)

    (a)   3 m if the circumference is 9 m   (d) 6π if the circumference is 12π

    (b)   2 ft if the circumference is 1 yd   (e)   Three-eighths of the circumference

    (c)   25 if the circumference is 250   (f)   Equal to the radius

34. Find the measure of a central angle of a sector whose area is   (11.12)

    (a)   10, if the area of the circle is 50   (d) 5π if the area of the circle is 12π

    (b)   15 cm², if the area of the circle is 20 cm²   (e)   Eight-ninths of the area of the circle

    (c)   1 ft² if the area of the circle is 1 yd²

35. Find the measure of a central angle of   (11.11, 11.12)

    (a)   An arc whose length is 5π if the area of its sector is 25π

    (b)   An arc whose length is 12π if the area of its sector is 48π

    (c)   A sector whose area is 2π if the length of its arc is π

    (d)   A sector whose area is 10π if the length of its arc is 2π

36. Find the radius of a circle if a   (11.11, 11.12)

    (a)   120° arc has a length of 8π (d)   30° sector has an area of 3π

    (b)   40° arc has a length of 2π (e)   36° sector has an area of $2\frac{1}{2}\pi$

    (c)   270° arc has a length of 15π (f)   120° sector has an area of 6π

37. Find the radius of a circle if a   (11.12)

    (a)   Sector of area 12π has an arc of length 6π.

    (b)   Sector of area 10π has an arc of length 2π.

    (c)   Sector of area 25 cm² has an arc of length 5 cm.

    (d)   Sector of area 162 has an arc of length 36.

38. Find the area of a segment if its central angle is 60° and the radius of the circle is (*a*) 6; (*b*) 12; (*c*) 3; (*d*) *r*;
(*e*) 2*r*. (11.13)

39. Find the area of a segment of a circle if (11.13)

   (*a*) The radius of the circle is 4 and the central angle measures 90°.

   (*b*) The radius of the circle is 30 and the central angle measures 60°.

   (*c*) The radius of the circle and the chord of the segment each have length 12.

   (*d*) The central angle is 90° and the length of the arc is 4π.

   (*e*) Its chord of length 20 units is 10 units from the center of the circle.

40. Find the area of a segment of a circle if the radius of the circle is 8 and the central angle measures (*a*) 120°;
(*b*) 135°; (*c*) 150°. (11.13)

41. If the radius of a circle is 4, find the area of each segment formed by an inscribed (*a*) equilateral triangle;
(*b*) regular hexagon; (*c*) square. (11.13, 11.14)

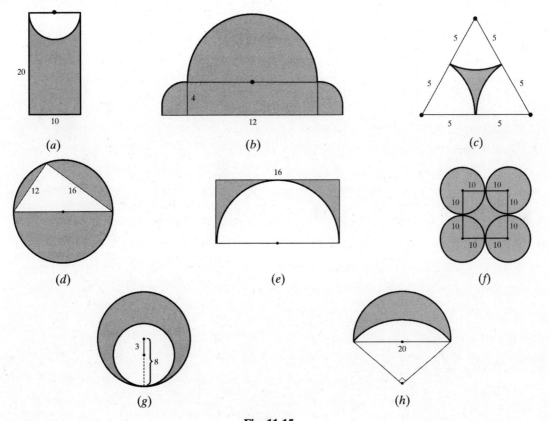

Fig. 11-15

42. Find the area of each segment of a circle if the segments are formed by an inscribed (11.13)

   (*a*) Equilateral triangle and the radius of the circle is 6

   (*b*) Regular hexagon and the radius of the circle is 3

   (*c*) Square and the radius of the circle is 6

43. Find the shaded area in each part of Fig. 11-15. Each heavy dot represents the center of an arc or a circle.   (11.15)

44. Find the shaded area in each part of Fig. 11-16. Each dot represents the center of an arc or circle. (11.15)

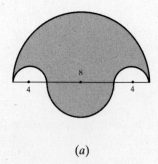

(a)

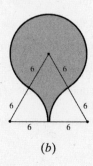

(b)

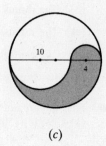

(c)

Fig. 11-16

CHAPTER 12

Locus

12.1 DETERMINING A LOCUS

Locus, in Latin, means *location*. The plural is *loci*. A *locus of points* is the set of points, and only those points, that satisfy given conditions.

Thus the locus of points that are 1 in from a given point P is the set of points 1 in from P. These points lie on a circle with its center at P and a radius of 1 in, and hence this circle is the required locus (Fig. 12-1). Note that we show loci as long-short dashed figures.

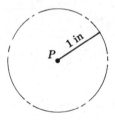

Fig. 12-1

To determine a locus, (1) state what is given and the condition to be satisfied; (2) find several points satisfying the condition which indicate the shape of the locus; then (3) connect the points and describe the locus fully.

All geometric constructions require the use of straightedge and compasses. Hence if a locus is to be *constructed*, such drawing instruments can be used.

12.1A Fundamental Locus Theorems

PRINCIPLE 1: *The locus of points equidistant from two given points is the perpendicular bisector of the line segment joining the two points* (Fig. 12-2).

Fig. 12-2

PRINCIPLE 2:   *The locus of points equidistant from two given parallel lines is a line parallel to the two lines and midway between them* (Fig. 12-3).

PRINCIPLE 3:   *The locus of points equidistant from the sides of a given angle is the bisector of the angle* (Fig. 12-4).

Fig. 12-3   **Fig. 12-4**

PRINCIPLE 4:   *The locus of points equidistant from two given intersecting lines is the bisectors of the angles formed by the lines* (Fig. 12-5).

Fig. 12-5   **Fig. 12-6**

PRINCIPLE 5:   *The locus of points equidistant from two concentric circles is the circle concentric with the given circles and midway between them* (Fig. 12-6).

PRINCIPLE 6:   *The locus of points at a given distance from a given point is a circle whose center is the given point and whose radius is the given distance* (Fig. 12-7).

Fig. 12-7   **Fig. 12-8**

PRINCIPLE 7:   *The locus of points at a given distance from a given line is a pair of lines, parallel to the given line and at the given distance from the given line* (Fig. 12-8).

PRINCIPLE 8:   *The locus of points at a given distance from a given circle whose radius is greater than that distance is a pair of concentric circles, one on either side of the given circle and at the given distance from it* (Fig. 12-9).

Fig. 12-9   **Fig. 12-10**

PRINCIPLE 9: *The locus of points at a given distance from a given circle whose radius is less than the distance is a circle, outside the given circle and concentric with it* (Fig. 12-10). (If $r = d$, the locus also includes the center of the given circle.)

SOLVED PROBLEMS

12.1 DETERMINING LOCI

Determine the locus of (*a*) a runner moving equidistant from the sides of a straight track; (*b*) a plane flying equidistant from two separated aircraft batteries; (*c*) a satellite 100 mi above the earth; (*d*) the furthermost point reached by a gun with a range of 10 mi.

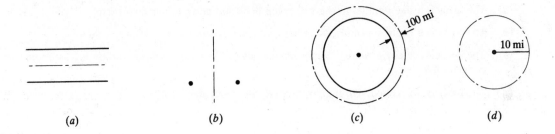

(*a*)   (*b*)   (*c*)   (*d*)

Fig. 12-11

Solutions

See Fig. 12-11.

(*a*) The locus is a line parallel to the two given lines and midway between them.

(*b*) The locus is the perpendicular bisector of the line joining the two points.

(*c*) The locus is a circle concentric with the earth and of radius 100 mi greater than that of the earth.

(*d*) The locus is a circle of radius 10 mi with its center at the gun.

12.2 DETERMINING THE LOCUS OF THE CENTER OF A CIRCLE

Determine the locus of the center of a circular disk (*a*) moving so that it touches each of two parallel lines; (*b*) moving tangentially to two concentric circles; (*c*) moving so that its rim passes through a fixed point; (*d*) rolling along a large fixed circular hoop.

Solutions

See Fig. 12-12.

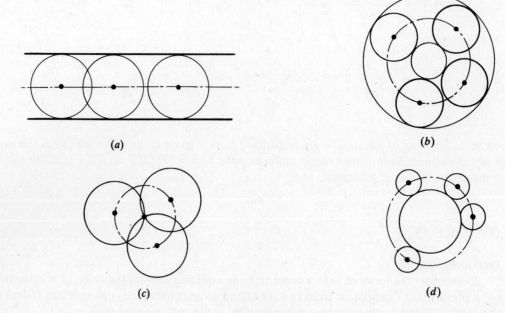

Fig. 12-12

(a)   The locus is a line parallel to the two given lines and midway between them.

(b)   The locus is a circle concentric with the given circles and midway between them.

(c)   The locus is a circle whose center is the given point and whose radius is the radius of the circular disk.

(d)   The locus is a circle outside the given circle and concentric to it.

12.3   CONSTRUCTING LOCI

Construct (a) the locus of points equidistant from two given points; (b) the locus of points equidistant from two given parallel lines; (c) the locus of points at a given distance from a given circle whose radius is less than that distance.

Solutions

See Fig. 12-13.

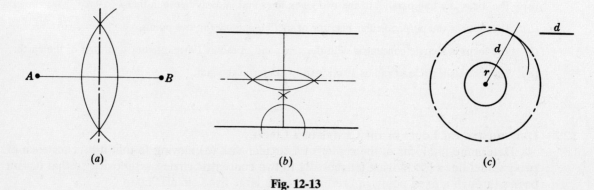

Fig. 12-13

12.2 LOCATING POINTS BY MEANS OF INTERSECTING LOCI

A point or points which satisfy two conditions may be found by drawing the locus for each condition. The required points are the points of intersection of the two loci.

SOLVED PROBLEM

12.4 Locating Points that Satisfy Two Conditions

On a map locate buried treasure that is 3 ft from a tree (T) and equidistant from two points (A and B) in Fig. 12-14.

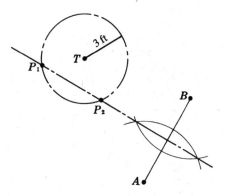

Fig. 12-14

Solution

The required loci are (1) the perpendicular bisector of $\overline{AB}$ and (2) a circle with its center at T and radius 3 ft. As shown, these meet in P_1 and P_2, which are the locations of the treasure.

Note: The diagram shows the two loci intersecting at P_1 and P_2. However, there are three possible kinds of solutions, depending on the location of T with respect to A and B:

1. The solution has two points if the loci intersect.

2. The solution has one point if the perpendicular bisector is tangent to the circle.

3. The solution has no points if the perpendicular bisector does not meet the circle.

12.3 PROVING A LOCUS

To prove that a locus satisfies a given condition, it is necessary to prove the locus theorem *and* its converse or its inverse. Thus to prove that a circle A of radius 2 in is the locus of points 2 in from A, it is necessary to prove either that

1. Any point on circle A is 2 in from A.

2. Any point 2 in from A is on a circle A (converse of statement 1).

or that

1. Any point on circle A is 2 in from A.

2. Any point not on circle A is not 2 in from A (inverse of statement 1).

These statements are easily proved using the principle that a point is outside, on, or inside a circle according as its distance from the center is greater than, equal to, or less than the radius of the circle.

SOLVED PROBLEM

12.5   **PROVING A LOCUS THEOREM**

Prove that the locus of points equidistant from two given points is the perpendicular bisector of the segment joining the two points.

Solution

First prove that any point on the locus satisfies the condition:

Given: Points A and B. $\overline{CD}$ is the $\perp$ bisector of $\overline{AB}$.

To Prove: Any point P on $\overline{CD}$ is equidistant from A and B; that is, $\overline{PA} \cong \overline{PB}$.

Plan: Prove $\triangle PEA \cong \triangle PEB$ to obtain $\overline{PA} \cong \overline{PB}$.

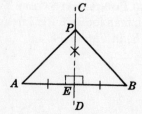

PROOF:

Statements	Reasons
1. $\overline{CD}$ is the $\perp$ bisector of $\overline{AB}$.	1. Given
2. $\angle PEA \cong \angle PEB$	2. Perpendiculars form right angles; all right angles are congruent.
3. $\overline{AE} \cong \overline{EB}$	3. To bisect is to divide into congruent parts.
4. $\overline{PE} \cong \overline{PE}$	4. Reflexive property
5. $\triangle PEA \cong \triangle PEB$	5. s.a.s. $\cong$ s.a.s.
6. $\overline{PA} \cong \overline{PB}$	6. Corresponding parts of $\cong$ triangles are $\cong$.

Then prove that any point satisfying the condition is on the locus:

Given: Any point Q which is equidistant from points A and B ($\overline{QA} \cong \overline{QB}$).

To Prove: Q is on the perpendicular bisector of $\overline{AB}$.

Plan: Draw $\overline{QG}$ perpendicular to $\overline{AB}$ and prove by congruent triangles that $\overline{QG}$ bisects $\overline{AB}$.

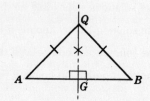

PROOF:

Statements	Reasons
1. Draw $\overline{QG} \perp \overline{AB}$	1. Through an external point, a line can be drawn perpendicular to a given line.
2. $\overline{QA} \cong \overline{QB}$	2. Given
3. $\angle QGA$ and $\angle QGB$ are rt. $\angle$s; $\triangle QGA$ and $\triangle QGB$ are rt. $\triangle$s.	3. Perpendiculars form right angles; $\triangle$s with a rt. $\angle$ are rt. $\triangle$s.
4. $\overline{QG} \cong \overline{QG}$	4. Reflexive property
5. $\triangle QGA \cong \triangle QGB$	5. hy. leg $\cong$ hy. leg
6. $\overline{AG} \cong \overline{GB}$	6. Corresponding parts of $\cong$ triangles are $\cong$.
7. $\overline{QG}$ bisects $\overline{AB}$	7. To bisect is to divide into two congruent parts
8. $\overline{QG}$ is $\perp$ bisector of $\overline{AB}$.	8. A line perpendicular to a segment and bisecting it is its perpendicular bisector.

12.4 LOCUS IN ANALYTIC GEOMETRY

A locus of points is the set of points, and only those points, satisfying a given condition. In geometry, a line or curve (or set of lines or curves) on a graph is the locus of analytic points that satisfy the equation of the line or curve.

Think of the locus as the path of a point moving according to a given condition or as the set of points satisfying a given condition.

PRINCIPLE 1:   *The locus of points whose abscissa is a constant k is a line parallel to the y-axis; its equation is $x = k$.* (See Fig. 12-15.)

PRINCIPLE 2:   *The locus of points whose ordinate is a constant k is a line parallel to the x-axis; its equation is $y = k$.* (See Fig. 12-15.)

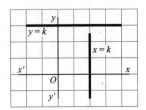

Fig. 12-15

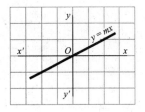

Fig. 12-16

PRINCIPLE 3:   *The locus of points whose ordinate equals the product of a constant m and its abscissa is a straight line passing through the origin; its equation is $y = mx$.*

The constant m is the slope of the line. (See Fig. 12-16.)

PRINCIPLE 4:   *The locus of points whose ordinate and abscissa are related by either of the equations*

$$y = mx + b \qquad or \qquad \frac{y - y_1}{x - x_1} = m$$

where m and b are constants, is a line (Fig. 12-17).

In the equation $y = mx + b$, m is the slope and b is the *y-intercept*. The equation $\frac{y - y_1}{x - x_1} = m$ tells us that the line passes through the fixed point (x_1, y_1) and has a slope of m.

PRINCIPLE 5:   *The locus of points such that the sum of the squares of the coordinates is a constant is a circle whose center is the origin.*

The constant is the square of the radius, and the equation of the circle is

$$x^2 + y^2 = r^2$$

(see Fig. 12-18). Note that for any point $P(x, y)$ on the circle, $x^2 + y^2 = r^2$.

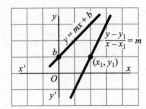

Fig. 12-17

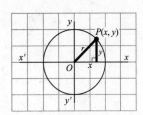

Fig. 12-18

SOLVED PROBLEMS

12.6   APPLYING PRINCIPLES 1 AND 2

Graph and give the equation of the locus of points (*a*) whose ordinate is -2; (*b*) that are 3 units from the *y*-axis; (*c*) that are equidistant from the points $(3,0)$ and $(5,0)$.

Solutions

(*a*)   From Principle 2, the equation is $y = -2$; see Fig. 12-19(*a*).

(*b*)   From Principle 1, the equation is $x = 3$ and $x = -3$; see Fig. 12-19(*b*).

(*c*)   The equation is $x = 4$; see Fig. 12-19(*c*).

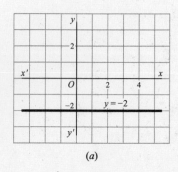   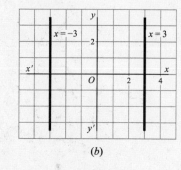   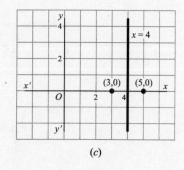

(*a*)   (*b*)   (*c*)

Fig. 12-19

12.7   APPLYING PRINCIPLES 3 AND 4

Graph and describe the locus whose equation is (*a*) $y = \frac{1}{3}x + 1$; (*b*) $y = \frac{3}{2}x$; (*c*) $\dfrac{y-1}{x-1} = \dfrac{3}{4}$.

Solutions

(*a*)   The locus is a line whose *y*-intercept is 1 and whose slope equals $\frac{1}{3}$. See Fig. 12-20(*a*).

(*b*)   The locus is a line which passes through the origin and has slope $\frac{3}{2}$. See Fig. 12-20(*b*).

(*c*)   The locus is a line which passes through the point $(1,1)$ and has slope $\frac{3}{4}$. See Fig. 12-20(*c*).

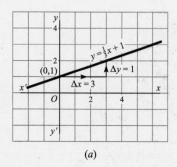   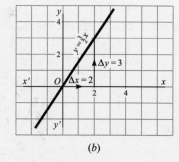   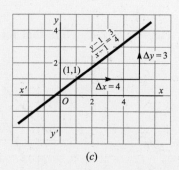

(*a*)   (*b*)   (*c*)

Fig. 12-20

12.8   APPLYING PRINCIPLE 5

Graph and give the equation of the locus of points (*a*) 2 units from the origin; (*b*) 2 units from the locus of $x^2 + y^2 = 9$.

Solutions

(*a*)   The locus is a circle whose equation is $x^2 + y^2 = 4$. See Fig. 12-21(*a*).

(*b*)   The locus is a pair of circles, each 2 units from the circle with center at O and radius 3. Their equations are $x^2 + y^2 = 25$ and $x^2 + y^2 = 1$. See Fig. 12-21(*b*).

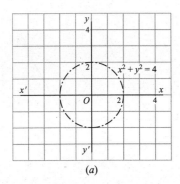   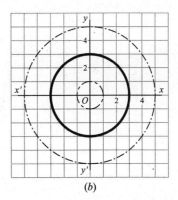

　　　　　　　(*a*)　　　　　　　　　　　　　　　　　(*b*)

Fig. 12-21

Supplementary Problems

1.   Determine the locus of   (12.1)

(*a*)   The midpoints of the radii of a given circle

(*b*)   The midpoints of chords of a given circle parallel to a given line

(*c*)   The midpoints of chords of fixed length in a given circle

(*d*)   The vertex of the right angle of a triangle having a given hypotenuse

(*e*)   The vertex of an isosceles triangle having a given base

(*f*)   The center of a circle which passes through two given points

(*g*)   The center of a circle tangent to a given line at a given point on that line

(*h*)   The center of a circle tangent to the sides of a given angle

2.   Determine the locus of   (12.1)

(*a*)   A boat moving so that it is equidistant from the parallel banks of a stream

(*b*)   A swimmer maintaining the same distance from two floats

(*c*)   A police helicopter in pursuit of a car which has just passed the junction of two straight roads and which may be on either one of them

(*d*)   A treasure buried at the same distance from two intersecting straight roads

3. Determine the locus of (a) a planet moving at a fixed distance from its sun; (b) a boat moving at a fixed distance from the coast of a circular island; (c) plants laid at a distance of 20 ft from a straight row of other plants; (d) the outer extremity of a clock hand.   (12.1)

4. Excluding points lying outside rectangle ABCD in Fig. 12-22, find the locus of points which are   (12.1)

   (a) Equidistant from $\overline{AD}$ and $\overline{BC}$ (e) 5 units from $\overline{BC}$

   (b) Equidistant from $\overline{AB}$ and $\overline{CD}$ (f) 10 units from $\overline{AB}$

   (c) Equidistant from A and B   (g) 20 units from $\overline{CD}$

   (d) Equidistant from B and C   (h) 10 units from B

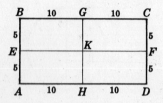

Fig. 12-22

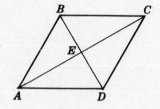

Fig. 12-23

5. Find the locus of points in rhombus ABCD in Fig. 12-23 which are equidistant from (a) $\overline{AB}$ and $\overline{AD}$; (b) $\overline{AB}$ and $\overline{BC}$; (c) A and C; (d) B and D; (e) each of the four sides.   (12.1)

6. In Fig. 12-24, find the locus of points which are on or inside circle C and   (12.1, 12.2)

   (a) 5 units from O   (e) 10 units from circle A

   (b) 15 units from O   (f) 5 units from circle B

   (c) Equidistant from circles A and C   (g) The center of a circle tangent to circles A and C

   (d) 10 units from circle C

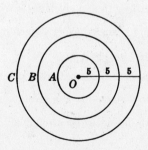

Fig. 12-24

7. Determine the locus of the center of (a) a coin rolling around and touching a smaller coin; (b) a coin rolling around and touching a larger coin; (c) a wheel moving between two parallel bars and touching both of them; (d) a wheel moving along a straight metal bar and touching it.   (12.2)

8. Find the locus of points that are in rectangle $ABCD$ of Fig. 12-25 and the center of a circle   (12.2)

   (a) Tangent to $\overline{AD}$ and $\overline{BC}$ (d) Of radius 10, tangent to $\overline{BC}$

   (b) Tangent to $\overline{AB}$ and $\overline{CD}$ (e) Of radius 20, tangent to $\overline{AD}$

   (c) Tangent to $\overline{AD}$ and $\overline{EF}$ (f) Tangent to $\overline{BC}$ at G

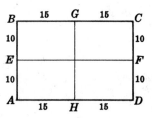

Fig. 12-25

9. Locate each of the following: (12.4)

   (a) Treasure that is buried 5 ft from a straight fence and equidistant from two given points where the fence meets the ground

   (b) Points that are 3 ft from a circle whose radius is 2 ft and are equidistant from two lines which are parallel to each other and tangent to the circle

   (c) A point equidistant from the three vertices of a given triangle

   (d) A point equidistant from two given points and equidistant from two given parallels

   (e) Points equidistant from two given intersecting lines and 5 ft from their intersection

   (f) A point that is equidistant from the sides of an angle and $\frac{1}{2}$ in from their intersection

10. Locate the point or points which satisfy the following conditions with respect to $\triangle ABC$ in Fig. 12-26:
                                                                                                                   (12.4)

   (a) Equidistant from its sides

   (b) Equidistant from its vertices

   (c) Equidistant from A and B and from $\overline{AB}$ and $\overline{BC}$

   (d) Equidistant from $\overline{BC}$ and $\overline{AC}$ and 5 units from C

   (e) 5 units from B and 10 units from A

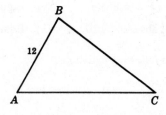

Fig. 12-26

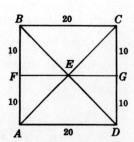

Fig. 12-27

11. Excluding points lying outside square $ABCD$ in Fig. 12-27, how many points are there that are   (12.4)

   (a) Equidistant from its vertices

   (b) Equidistant from its sides

   (c) 5 units from E and on one of the diagonals

   (d) 5 units from E and equidistant from $\overline{AD}$ and $\overline{BC}$

   (e) 5 units from $\overline{FG}$ and equidistant from $\overline{AB}$ and $\overline{CD}$

   (f) 20 units from A and 10 units from B

12. Prove that the locus of points equidistant from the sides of an angle is the bisector of the angle.   (12.5)

13. State the equation of the line or pair of lines which is the locus of points   (12.6)

   (a) Whose abscissa is -5 (f) 3 units from the line $x = 2$

   (b) Whose ordinate is $3\frac{1}{2}$ (g) 6 units above the line $y = -2$

   (c) 3 units from the x-axis   (h) 1 unit to the right of the y-axis

   (d) 5 units below the x-axis   (i) Equidistant from the lines $x = 5$ and $x = 13$

   (e) 4 units from the y-axis

14. State the equation of the locus of the center of a circle that   (12.8)

   (a) Is tangent to the x-axis at $(6,0)$ (d) Passes through the origin and $(10,0)$

   (b) Is tangent to the y-axis at $(0,5)$ (e) Passes through $(3,7)$ and $(9,7)$

   (c) Is tangent to the lines $x = 4$ and $x = 8$ (f) Passes through $(3,-2)$ and $(3,8)$

15. State the equation of the line or pair of lines which is the locus of points   (12.6)

   (a) Whose coordinates are equal   (e) The sum of whose coordinates is 12

   (b) Whose ordinate is 5 more than the abscissa (f) The difference of whose coordinates is 2

   (c) Whose abscissa is 4 less than the ordinate (g) Equidistant from the x-axis and y-axis

   (d) Whose ordinate exceeds the abscissa by 10   (h) Equidistant from $x + y = 3$ and $x + y = 7$

16. Describe the locus of each of the following equations: (12.7)

   (a) $y = 2x + 5$ (c) $\dfrac{y+3}{x+2} = \dfrac{5}{4}$ (e) $x + y = 7$

   (b) $\dfrac{y-3}{x-2} = 4$ (d) $y = \frac{1}{2}x$ (f) $3y = x$

17. State the equation of a line which passes through the origin and has a slope of (*a*) 4; (*b*) −2; (*c*) $\frac{3}{2}$; (*d*) −2/5; (*e*) 0.   (12.7)

18. State the equation of a line which has a *y*-intercept of   (12.7)

(*a*) 5 and a slope of 4   (*d*) 8 and is parallel to $y = 3x - 2$

(*b*) 2 and a slope of −3   (*e*) −3 and is parallel to $y = 7 - 4x$

(*c*) −1 and a slope of $\frac{1}{3}$ (*f*) 0 and is parallel to $y - 2x = 8$

19. State the equation of a line which has a slope of 2 and passes through (*a*) (1,4); (*b*) (−2,3); (*c*) (−4,0); (*d*) (0,−7).   (12.7)

20. State the equation of a line   (12.7)

(*a*) Which passes through the origin and has a slope of 4

(*b*) Which passes through (0,3) and has a slope of $\frac{1}{2}$

(*c*) Which passes through (1,2) and has a slope of 3

(*d*) Which passes through (−1,−2) and has a slope of $\frac{1}{3}$

(*e*) Which passes through the origin and is parallel to a line that has a slope of 2

21. (*a*) Describe the locus of the equation $x^2 + y^2 = 49$.   (12.8)

(*b*) State the equation of the locus of points 4 units from the origin.

(*c*) State the equations of the locus of points 3 units from the locus of $x^2 + y^2 = 25$.

22. State the equation of the locus of points 5 units from (*a*) the origin; (*b*) the circle $x^2 + y^2 = 16$; (*c*) the circle $x^2 + y^2 = 49$.   (12.8)

23. What is the radius of the circle whose equation is (*a*) $x^2 + y^2 = 9$; (*b*) $x^2 + y^2 = \frac{16}{9}$; (*c*) $9x^2 + 9y^2 = 36$; (*d*) $x^2 + y^2 = 3$?   (12.8)

24. What is the equation of a circle whose center is the origin and whose radius is (*a*) 4; (*b*) 11; (*c*) $\frac{2}{3}$; (*d*) $1\frac{1}{2}$; (*e*) $\sqrt{5}$; (*f*) $\frac{1}{2}\sqrt{3}$?   (12.8)

CHAPTER 13

Inequalities and Indirect Reasoning

13.1 INEQUALITIES

An inequality is a statement that quantities are not equal. If two quantities are unequal, the first is either greater than or less than the other. The inequality symbols are: $\neq$, meaning unequal to; $>$, meaning greater than; and $<$, meaning less than. Thus $4 \neq 3$ is read "four is unequal to three"; $7 > 2$ is read "seven is greater than two"; and $1 < 5$ is read "one is less than five."

Two inequalities may be of the same order or of opposite order. In inequalities of the same order, the same inequality symbol is used; in inequalities of the opposite order, opposite inequality symbols are used. Thus $5 > 3$ and $10 > 7$ are inequalities of the same order; $5 > 3$ and $7 < 10$ are inequalities of opposite order.

Inequalities of the same order may be combined, as follows. The inequalities $x < y$ and $y < z$ may be combined into $x < y < z$, which states that y is greater than x and less than z. The inequalities $a > b$ and $b > c$ may be combined into $a > b > c$, which states b is less than a and greater than c.

13.1A Inequality Axioms

Axioms are statements that are accepted as true without proof and are used in the same way as theorems.

AXIOM 1: *A quantity may be substituted for its equal in any inequality.*

Thus if $x > y$ and $y = 10$, then $x > 10$.

AXIOM 2: *If the first of three quantities is greater than the second, and the second is greater than the third, then the first is greater than the third.*

Thus if $x > y$ and $y > z$, then $x > z$.

AXIOM 3: *The whole is greater than any of its parts.*

Thus $AB > AM$ and $m\angle BAD > m\angle BAC$ in Fig. 13-1.

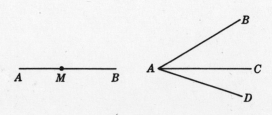

Fig. 13-1

13.1B   Inequality Axioms of Operation

AXIOM 4:   *If equals are added to unequals, the sums are unequal in the same order.*

Since $5 > 4$ and $4 = 4$, we know that $5 + 4 > 4 + 4$ (or $9 > 8$). If $x - 4 < 5$, then $x - 4 + 4 < 5 + 4$ or $x < 9$.

AXIOM 5:   *If unequals are added to unequals of the same order, the sums are unequal in the same order.*

Since $5 > 3$ and $4 > 1$, we have $5 + 4 > 3 + 1$ (or $9 > 4$). If $2x - 4 < 5$ and $x + 4 < 8$, then $2x - 4 + x + 4 < 5 + 8$ or $3x < 13$.

AXIOM 6:   *If equals are subtracted from unequals, the differences are unequal in the same order.*

Since $10 > 5$ and $3 = 3$, we have $10 - 3 > 5 - 3$ (or $7 > 2$). If $x + 6 < 9$ and $6 = 6$, then $x + 6 - 6 < 9 - 6$ or $x < 3$.

AXIOM 7:   *If unequals are subtracted from equals, the differences are unequal in the opposite order.*

Since $10 = 10$ and $5 > 3$, we have $10 - 5 < 10 - 3$ (or $5 < 7$). If $x + y = 12$ and $y > 5$, then $x + y - y < 12 - 5$ or $x < 7$.

AXIOM 8:   *If unequals are multiplied by the same positive number, the products are unequal in the same order.*

Thus if $\frac{1}{4}x < 5$, then $4(\frac{1}{4}x) < 4(5)$ or $x < 20$.

AXIOM 9:   *If unequals are multiplied by the same negative number, the results are unequal in the opposite order.*

Thus if $\frac{1}{2}x < 5$, then $(-2)(\frac{1}{2}x) > (-2)(5)$ or $-x > -10$ or $x < 10$.

AXIOM 10:   *If unequals are divided by the same positive number, the results are unequal in the same order.*

Thus if $4x > 20$, then $\frac{4x}{4} > \frac{20}{4}$ or $x > 5$.

AXIOM 11:   *If unequals are divided by the same negative number, the results are unequal in the opposite order.*

Thus if $-7x < 42$, then $\frac{-7x}{-7} > \frac{42}{-7}$ or $x > -6$.

13.1C   Inequality Postulate

POSTULATE 1:   *The length of a line segment is the shortest distance between two points.*

13.1D   Triangle Inequality Theorems

PRINCIPLE 1:   *The sum of the lengths of two sides of a triangle is greater than the length of the third side.* (Corollary: *The length of the longest side of a triangle is less than the sum of the lengths of the other two sides and greater than their difference.*)

Thus in Fig. 13-2, $BC + CA > AB$.

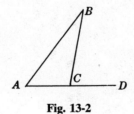

Fig. 13-2

PRINCIPLE 2: *In a triangle, the measure of an exterior angle is larger than the measure of either nonadjacent interior angle.*

Thus in Fig. 13-2, $m\angle BCD > m\angle BAC$ and $m\angle BCD > m\angle ABC$.

PRINCIPLE 3: *If the lengths of two sides of a triangle are unequal, the measures of the angles opposite these sides are unequal, the larger angle being opposite the longer side.* (Corollary: *The largest angle of a triangle is opposite the longest side.*)

Thus in Fig. 13-2, if $BC > AC$, then $m\angle A > m\angle B$.

PRINCIPLE 4: *If the measures of two angles of a triangle are unequal, the lengths of the sides opposite these angles are unequal, the longer side being opposite the larger angle.* (Corollary: *The longest side of a triangle is opposite the largest angle.*)

Thus in Fig. 13-2, if $m\angle A > m\angle B$, then $BC > AC$.

PRINCIPLE 5: *The perpendicular from a point to a line is the shortest segment from the point to the line.*

Thus in Fig. 13-3, if $\overline{PC} \perp \overline{AB}$ and $\overline{PD}$ is any other line from P to $\overline{AB}$, then $PC < PD$.

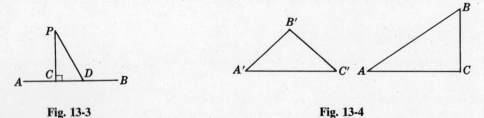

Fig. 13-3   Fig. 13-4

PRINCIPLE 6: *If two sides of a triangle are congruent to two sides of another triangle, the triangle having the greater included angle has the greater third side.*

Thus in Fig. 13-4, if $BC = B'C'$, $AC = A'C'$, and $m\angle C > m\angle C'$, then $AB > A'B'$.

PRINCIPLE 7: *If two sides of a triangle are congruent to two sides of another triangle, the triangle having the greater third side has the greater angle opposite this side.*

Thus in Fig. 13-4, if $BC = B'C'$, $AC = A'C'$, and $AB > A'B'$, then $m\angle C > m\angle C'$.

13.1E Circle Inequality Theorems

PRINCIPLE 8: *In the same or equal circles, the greater central angle has the greater arc.*

Thus in Fig. 13-5, if $m\angle AOB > m\angle COD$, then $m\overgroup{AB} > m\overgroup{CD}$.

PRINCIPLE 9: *In the same or equal circles, the greater arc has the greater central angle.* (This is the converse of Principle 8.)

Thus in Fig. 13-5, if $m\overgroup{AB} > m\overgroup{CD}$, then $m\angle AOB > m\angle COD$.

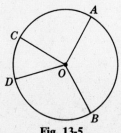

Fig. 13-5

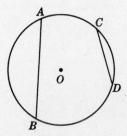

Fig. 13-6

PRINCIPLE 10: *In the same or equal circles, the greater chord has the greater minor arc.*

Thus in Fig. 13-6, if $AB > CD$, then $m\widehat{AB} > m\widehat{CD}$.

PRINCIPLE 11: *In the same or equal circles, the greater minor arc has the greater chord.* (This is the converse of Principle 10.)

Thus in Fig. 13-6, if $m\widehat{AB} > m\widehat{CD}$, then $AB > CD$.

PRINCIPLE 12: *In the same or equal circles, the greater chord is at a smaller distance from the center.*

Thus in Fig. 13-7, if $AB > CD$, then $OE < OF$.

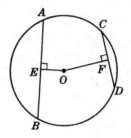

Fig. 13-7

PRINCIPLE 13: *In the same or equal circles, the chord at the smaller distance from the center is the greater chord.* (This is the converse of Principle 12.)

Thus in Fig. 13-7, if $OE < OF$, then $AB > CD$.

SOLVED PROBLEMS

13.1 SELECTING INEQUALITY SYMBOLS

Determine which inequality symbol, $>$ or $<$, makes each of the following true:

(*a*) 5 _?_ 3   (*c*) -5 _?_ 3   (*e*) If $x = 3$, then x^2 _?_ x.

(*b*) 6 _?_ 9   (*d*) -5 _?_ -3 (*f*) If $x > 10$, then 10 _?_ x.

Solutions

(*a*) >   (*b*) <   (*c*) <   (*d*) <   (*e*) >   (*f*) <

13.2 APPLYING INEQUALITY AXIOMS

Complete each of the following statements:

(*a*) If $a > b$ and $b > 8$, then a _?_ 8.

(*b*) If $x > y$ and $y = 15$, then x _?_ 15.

(*c*) If $c < 20$ and $d < 5$, then $c + d$ _?_ 25.

(*d*) If $x > y$ and $y > 6$, then x _?_ y _?_ 6.

(*e*) If $x > y$, then $\frac{1}{2}x$ _?_ $\frac{1}{2}y$.

(*f*) If $e < \frac{1}{4}f$, then $4e$ _?_ f.

(*g*) If $-y < z$ then y _?_ $-z$.

(*h*) If $-4x > p$, then x _?_ $-\frac{1}{4}p$.

(*i*) If Paul and Jack have equal amounts of money and Paul spends more than Jack, then Paul will have ___?___ than Jack.

(*j*) If Anne is now older than Helen, then 10 years ago, Anne was ___?___ than Helen.

Solutions

(*a*) >	(*c*) <	(*e*) >	(*g*) >	(*i*) less
(*b*) >	(*d*) >, >	(*f*) <	(*h*) <	(*j*) older

13.3 APPLYING TRIANGLE INEQUALITY THEOREMS (Fig. 13-8)

(*a*) Determine the integer values that the length of side *a* of the triangle can have if the other two sides have lengths 3 and 7.

(*b*) Determine which is the longest side of the triangle if two angles have measures 59° and 60°.

(*c*) Determine which is the longest side of parallelogram *ABCD* if *E* is the midpoint of the diagonals and $m\angle AEB > m\angle AED$.

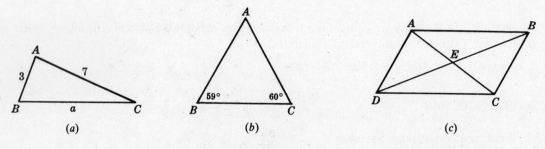

| (*a*) | (*b*) | (*c*) |

Fig. 13-8

Solutions

(*a*) Since *a* must be less than $3 + 7 = 10$ and greater than $7 - 3 = 4$, *a* can have the integer values of 5, 6, 7, 8, 9.

(*b*) Since $m\angle B = 59°$ and $m\angle C = 60°$, $m\angle A = 180° - (59° + 60°) = 61°$. Then the longest side is opposite the largest angle, $\angle A$, so the longest side is *BC*.

(*c*) In $\square ABCD$, $AE = CE$ and $DE = EB$. Since $m\angle AEB > m\angle AED$, $AB > AD$ or $AB (= DC)$ is the longest side. (Principle 6)

13.4 APPLYING CIRCLE INEQUALITY THEOREMS

In Fig. 13-9, compare

(*a*) *OD* and *OF* if $\angle C$ is the largest angle of $\triangle ABC$

(*b*) *AC* and *BC* if $m\widehat{AC} > m\widehat{BC}$

(*c*) $m\widehat{BC}$ and $m\widehat{AC}$ if $OF > OE$

(*d*) $m\angle AOB$ and $m\angle BOC$ if $m\widehat{AB} > m\widehat{BC}$

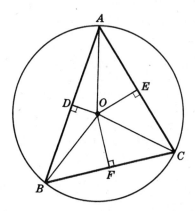

Fig. 13-9

Solutions

(a) Since $\angle C$ is the largest angle of the triangle, $\overline{AB}$ is the longest side, or $AB > BC$; hence $OD < OF$ by Principle 12.

(b) Since $m\widehat{AC} > m\widehat{BC}$, $AC > BC$, the greater arc having the greater chord.

(c) Since $OF > OE$, $BC < AC$ by Principle 13; hence $m\widehat{BC} < m\widehat{AC}$ by Principle 10.

(d) Since $m\widehat{AB} > m\widehat{BC}$, $m\angle AOB > m\angle BOC$, the greater arc having the greater central angle.

13.5 PROVING AN INEQUALITY PROBLEM

Prove that in $\triangle ABC$, if M is the midpoint of $\overline{AC}$ and $BM > AM$, then $m\angle A + m\angle C > m\angle B$.

Given: $\triangle ABC$, M is midpoint of AC.
   $BM > AM$

To Prove: $m\angle A + m\angle C > m\angle B$

Plan: Prove $m\angle A > m\angle 1$ and $m\angle C > m\angle 2$ and then add unequals.

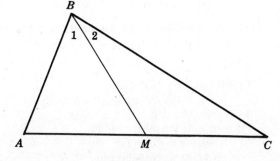

PROOF:

Statements	Reasons
1. M is the midpoint of $\overline{AC}$	1. Given
2. $\overline{AM} \cong \overline{MC}$	2. A midpoint divides a line into two congruent parts.
3. $BM > AM$	3. Given
4. $BM > MC$	4. A quantity may be substituted for its equal in any inequality. Definition of congruent segments.
	5. In a triangle, the larger angle lies opposite the longer side.
5. In $\triangle AMB$, $m\angle A > \angle 1$ In $\triangle BMC$, $m\angle C > \angle 2$	6. If unequals are added to unequals, the sums are unequal in the same order.
6. $m\angle A + m\angle C > m\angle B$	

13.2 INDIRECT REASONING

We often arrive at a correct conclusion by *indirect* reasoning. In this form of reasoning, the correct conclusion is reached by eliminating all possible conclusions except one. The remaining possibility must be the correct one. Suppose we are given the years 1492, 1809, and 1960 and are assured that one of these years is the year in which a president of the United States was born. By eliminating 1492 and 1960 as impossibilities, we know by indirect reasoning that 1809 is the correct answer. (Had we known that 1809 was the year in which Lincoln was born, the reasoning would have been direct.)

In proving a theorem by indirect reasoning, a possible conclusion may be eliminated if we assume it is true and that assumption results in a contradiction of some given or known fact.

SOLVED PROBLEMS

13.6 APPLYING INDIRECT REASONING IN LIFE SITUATIONS

Explain how indirect reasoning is used in each of the following situations:

(*a*) A detective determines the murderer of a slain person.

(*b*) A librarian determines which volume of a set of books is in use.

Solutions

(*a*) The detective, using a list of all those who could have been a murderer in the case, eliminates all except one. He or she concludes that the remaining one is the murderer.

(*b*) The librarian finds all the books of the set except one by looking on the shelf and checking the records. He or she concludes that the missing one is the one in use.

13.7 PROVING AN INEQUALITY THEOREM BY THE INDIRECT METHOD

Prove that in the same or equal circles, unequal chords are unequally distant from the center.

Given: Circle O, $AB \neq CD$
$\overline{OE} \perp \overline{AB}, \overline{OF} \perp \overline{CD}$
To Prove: $OE \neq OF$
Plan: Assume the other possible
conclusion, $OE = OF$, and
arrive at a contradiction.

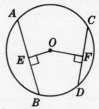

PROOF:

Statements	Reasons
1. Either $OE = OF$ or $OE \neq OF$.	1. Two quantities are either equal or unequal.
2. Assume $OE = OF$.	2. This is one of the possible conclusions.
3. If $OE = OF$, then $AB = CD$.	3. In the same or equal circles, chords equally distant from the center are equal.
4. But $AB \neq CD$	4. Given
5. The assumption $OE = OF$ is not valid.	5. It leads to a contradiction.
6. Hence $OE \neq OF$.	6. This is the only remaining possibility.

Supplementary Problems

1. Determine which inequality symbol, $>$ or $<$, makes each of the following true: (13.1)

(a) If $y > 15$, then 15 __?__ y.

(b) If $x = 2$, then $3x - 1$ __?__ 4.

(c) If $x = 2$ and $y = 3$, then xy __?__ 5.

(d) If $a = 4$ and $b = \frac{1}{4}$, then a/b __?__ 15.

(e) If $a = 5$, then a^2 __?__ $4a$.

(f) If $b = \frac{1}{2}$, then b^2 __?__ b.

2. Complete each of the following statements: (13.2)

(a) If $y > x$ and $x = z$, then y __?__ z.

(b) If $a + b > c$ and $b = d$, then $a + d$ __?__ c.

(c) If $a < b$ and $b < 15$, then a __?__ 15.

(d) If $z > y$, $y > x$, and $x = 10$, then z __?__ 10.

3. Complete each of the following statements about Fig. 13-10:   (13.2)

(a) BC __?__ BD

(b) $m\angle BAD$ __?__ $m\angle BAC$

(c) $\triangle ADC$ __?__ $\triangle ABC$

(d) If $m\angle A = m\angle C$, then AB __?__ BD.

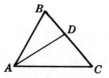

Fig. 13-10

4. Complete each of the following statements.   (13.2)

(a) If Mary and Ann earn the same weekly wage and Mary is to receive a larger increase than Ann, then Mary will earn __?__ Ann earns.

(b) If Bernice, who is the same weight as Helen, reduces more than Helen, then Bernice will weigh __?__ Helen weighs.

5. Complete each of the following statements: (13.2)

(a) If $a > 3$, then $4a$ __?__ 12.

(b) If $x - 3 > 15$, then x __?__ 18.

(c) If $3x < 18$, then x __?__ 6.

(d) If $f > 8$, then $f + 7$ __?__ 15.

(e) If $x = y$, then $x + 5$ __?__ $y + 6$.

(f) If $g = h$, then $g - 10$ __?__ $h - 9$.

6. Which of the following sets of numbers can be the lengths of the sides of a triangle?   (13.3)

(a) 3, 4, 8   (b) 5, 7, 12   (c) 3, 4, 6   (d) 2, 7, 8   (e) 50, 50, 5

7. What integer values can the length of the third side of a triangle have if the two sides have lengths (a) 2 and 6; (b) 3 and 8; (c) 4 and 7; (d) 4 and 6; (e) 4 and 5; (f) 7 and 7?   (13.3)

8. In Fig. 13-11, arrange, in descending order of size, (*a*) the angles of △*ABC*; (*b*) the sides of △*DEF*; (*c*) the angles 1, 2, and 3. (13.3)

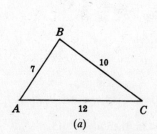
(*a*)

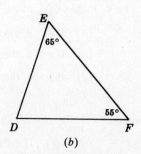
(*b*)

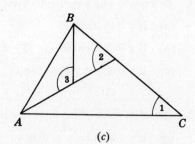
(*c*)

Fig. 13-11

9. (*a*) In quadrilateral *ABCD* of Fig. 13-12, compare *m*∠*BAC* and *m*∠*ACD* if *AB* = *CD* and *BC* > *AD*.

(*b*) In △*ABC* of Fig. 13-13, compare *AB* and *BC* if $\overline{BM}$ is the median to $\overline{AC}$ and *m*∠*AMB* > *m*∠*BMC*. (13.3)

10. Arrange, in descending order of magnitude, (13.4)

(*a*) The sides of △*ABC* in Fig. 13-14

(*b*) The central angles *AOB*, *BOC*, and *AOC* in Fig. 13-14

(*c*) The sides of trapezoid *ABCD* in Fig. 13-15

(*d*) The distances of the sides of △*DEF* from the center in Fig. 13-16

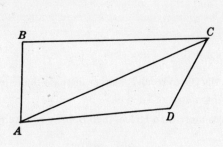
Fig. 13-12

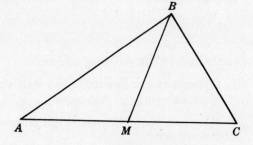
Fig. 13-13

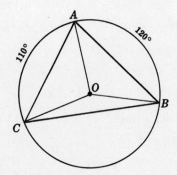
Fig. 13-14

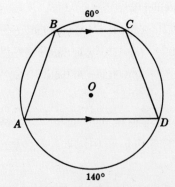
Fig. 13-15

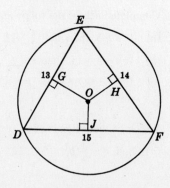
Fig. 13-16

11. Provide the proofs requested in Fig. 13-17.   (13.5)

(a) **Given:** Parallelogram $ABCD$ (b) **Given:** Rhombus $FGHJ$ (c) **Given:** $\overline{AD}$ bisects $\angle A$
$AC > BD$ $m\angle G > m\angle F$ $\overline{CD}$ bisects $\angle C$, $AB > BC$
To Prove: $m\angle BDA > m\angle CAD$ **To Prove:** $FL > GL$ **To Prove:** $AD > CD$

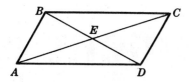 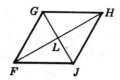 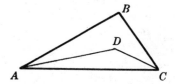

(d) **Given:** $\overrightarrow{AE} \| \overrightarrow{BC}$ (e) **Given:** Quad. $ABCD$ (f) **Given:** $AB = AC$
$m\angle DAE > m\angle EAC$ $AB > BC$, $AD > CD$ **To Prove:** $BD > CD$
To Prove: $AC > AB$ **To Prove:** $m\angle C > m\angle A$

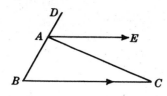 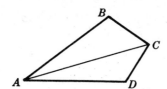 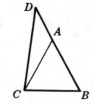

Fig. 13-17

12. Explain how indirect reasoning is used in each of the following situations: (13.6)

(a) A person determines which of his ties has been borrowed by his roommate.

(b) A girl determines that the electric motor in her train set is not defective even though her toy trains do not run.

(c) A teacher finds which of his students did not do their assigned homework.

(d) A mechanic finds the reason why the battery in a car does not work.

(e) A person accused of a crime proves her innocence by means of an alibi.

13. Prove each of the following: (13.7)

(a) The base angles of an isosceles triangle cannot be right angles.

(b) A scalene triangle cannot have two congruent angles.

(c) The median to the base of a scalene triangle cannot be perpendicular to the base.

(d) If the diagonals of a parallelogram are not congruent, then it is not a rectangle.

(e) If a diagonal of a parallelogram does not bisect a vertex angle, then the parallelogram is not a rhombus.

(f) If two angles of a triangle are unequal, the sides opposite are unequal, the longer side being opposite the larger angle.

CHAPTER 14

Improvement of Reasoning

14.1 DEFINITIONS

"Was Lincoln an educated man?" is a question that cannot be properly answered unless we agree upon the meaning of "an educated man." Understanding cannot exist and progress cannot be made in any discussion or problem unless the terms involved are properly defined or, by agreement, are to be undefined.

14.1A Requirements of a Good Definition

PRINCIPLE 1: *All terms in a definition must have been previously defined (or be those that, by agreement, are left undefined).*

Thus, if we are to define a regular polygon as an equilateral and equiangular polygon, it is necessary that equilateral, equiangular, and polygon be previously defined.

PRINCIPLE 2: *The term being defined should be placed in the next larger set or class to which it belongs.*

Thus the terms polygon, quadrilateral, parallelogram and rectangle should be defined in that order. Once the term polygon has been defined, the term quadrilateral is then defined as a kind of polygon. Then the term parallelogram is defined as a kind of quadrilateral and, lastly, the term rectangle is defined as a kind of parallelogram.

Proper sequence in definition can be understood by using a circle to represent a set of objects. In Fig. 14-1, the set of rectangles is in the next larger set of parallelograms. In turn, the set of parallelograms is in the next larger set of quadrilaterals, and, finally, the set of quadrilaterals is in the next larger set of polygons.

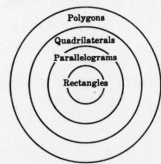

Fig. 14-1

PRINCIPLE 3: *The term being defined should be distinguished from all other members of its class.*

Thus the definition of a triangle as a polygon with three sides is a good one since it shows how the triangle differs from all other polygons.

Principle 4: *The distinguishing characteristics of a defined term should be as few as possible.*

Thus a right triangle should be defined as a triangle having a right angle, and not as a triangle having a right angle and two acute angles.

SOLVED PROBLEMS

14.1 Observing Proper Sequence in Definition

In which order should the terms in each of the following sets be defined: (*a*) Englishman, European, Londoner; (*b*) quadrilateral, square, rectangle, parallelogram

Answers

(*a*) European, Englishman, Londoner

(*b*) Quadrilateral, parallelogram, rectangle, square

14.2 Correcting Faulty Definitions

Correct the following definition: A trapezoid is a quadrilateral having two parallel sides.

Solution

The given definition is incomplete. The correct definition is "a trapezoid is a quadrilateral having only two parallel sides." This definition distinguishes a trapezoid from a parallelogram.

14.2 DEDUCTIVE REASONING IN GEOMETRY

The kinds of terms and statements discussed in this section comprise the deductive structure of geometry, which can be visualized as in Fig. 14-2.

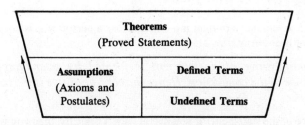

DEDUCTIVE STRUCTURE OF GEOMETRY

Fig. 14-2

14.2A Undefined and Defined Terms

Point, line, and surface are the terms in geometry which are, by agreement, not defined. These undefined terms begin the process of definition in geometry and underlie the definitions of all other geometric terms.

Thus we can define a triangle in terms of a polygon, a polygon in terms of a geometric figure, and a geometric figure as a figure composed of line segments, or parts of lines. However, the process of definition cannot be continued further because the term "line" is undefined.

14.2B Assumptions

Postulates and axioms are the statements which are not proved in geometry. They are called assumptions because we willingly accept them as true. These assumptions enable us to begin the process of proof in the same way that undefined terms enable us to begin the process of definition.

Thus when we draw a line segment between two points, we justify this by using as a reason the postulate "two points determine one and only one straight line." This reason is an assumption since we assume it to be true without requiring further justification.

14.2C Theorems

Theorems are the statements which are proved in geometry. By using definitions and assumptions as reasons, we deduce or prove the basic theorems. As we use each new theorem to prove still more theorems, the process of deduction grows. However, if a new theorem is used to prove a previous one, the logical sequence is violated.

For example, the theorem "the sum of the measures of the angles of a triangle equals 180°" is used to prove that "the sum of the measures of the angles of a pentagon is 540°." This, in turn, enables us to prove that "each angle of a regular pentagon measures 108°." However, it would be violating logical sequence if we tried to use the last theorem to prove either of the first two.

14.3 CONVERSE, INVERSE, AND CONTRAPOSITIVE OF A STATEMENT

DEFINITION 1: The *converse of a statement* is the statement that is formed by interchanging the hypothesis and conclusion.

Thus the converse of the statement "lions are wild animals" is "wild animals are lions." Note that the converse is not necessarily true.

DEFINITION 2: *The negative of a statement is the denial of the statement.*

Thus the negative of the statement "a burglar is a criminal" is "a burglar is not a criminal."

DEFINITION 3: *The inverse of a statement is formed by denying both the hypothesis and the conclusion.*

Thus the inverse of the statement "a burglar is a criminal" is "a person who is not a burglar is not a criminal." Note that the inverse is not necessarily true.

DEFINITION 4: *The contrapositive of a statement is formed by interchanging the negative of the hypothesis with the negative of the conclusion. Hence the contrapositive is the converse of the inverse and the inverse of the converse.*

Thus the contrapositive of the statement "if you live in New York City, then you will live in New York State" is "if you do not live in New York State, then you do not live in New York City." Note that both statements are true.

14.3A Converse, Inverse, and Contrapositive Principles

PRINCIPLE 1: *A statement is considered false if one false instance of the statement exists.*

PRINCIPLE 2: *The converse of a definition is true.*

Thus the definition "a quadrilateral is a four-sided polygon" and its converse "a four-sided polygon is a quadrilateral" are both true.

PRINCIPLE 3: *The converse of a true statement other than a definition is not necessarily true.*

The statement "vertical angles are congruent angles" is true, but its converse, "congruent angles are vertical angles" is not necessarily true.

PRINCIPLE 4:   *The inverse of a true statement is not necessarily true.*

The statement "a square is a quadrilateral" is true, but its inverse, "a non-square is not a quadrilateral," is not necessarily true.

PRINCIPLE 5:   *The contrapositive of a true statement is true, and the contrapositive of a false statement is false.*

The statement "a triangle is a square" is false, and its contrapositive, "a non-square is not a triangle," is also false.

The statement "right angles are congruent angles" is true, and its contrapositive, "angles that are not congruent are not right angles," is also true.

14.3B   Logically Equivalent Statements

Logically equivalent statements are pairs of related statements that are either both true or both false. Thus according to Principle 5, a statement and its contrapositive are logically equivalent statements. Also, the converse and inverse of a statement are logically equivalent, since each is the contrapositive of the other.

The relationships among a statement and its inverse, converse, and contrapositive are summed up in the rectangle of logical equivalency in Fig. 14-3:

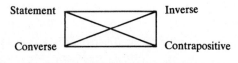

Rectangle of Logical Equivalency

Fig. 14-3

1. Logically equivalent statements are at diagonally opposite vertices. Thus the logically equivalent pairs of statements are (*a*) a statement and its contrapositive, and (*b*) the inverse and converse of the same statement.

2. Statements that are not logically equivalent are at adjacent vertices. Thus pairs of statements that are not logically equivalent are (*a*) a statement and its inverse, (*b*) a statement and its converse, (*c*) the converse and contrapositive of the same statement, and (*d*) the inverse and contrapositive of the same statement.

SOLVED PROBLEMS

14.3   **CONVERSE OF A STATEMENT**

State the converse of each of the following statements, and indicate whether or not it is true.

(*a*)   Supplementary angles are two angles the sum of whose measures is 180°.

(*b*)   A square is a parallelogram with a right angle.

(*c*)   A regular polygon is an equilateral and equiangular polygon.

Solutions

(*a*)   Two angles the sum of whose measures is 180° are supplementary. (True)

(*b*)   A parallelogram with a right angle is a square. (False)

(*c*)   An equilateral and equiangular polygon is a regular polygon. (True)

14.4   NEGATIVE OF A STATEMENT

State the negative of (*a*) $a = b$; (*b*) $m\angle B \neq m\angle C$; (*c*) $\angle C$ is the complement of $\angle D$; (*d*) "the point does not lie on the line."

Solutions

(*a*) $a \neq b$

(*b*) $m\angle B = m\angle C$

(*c*) $\angle C$ is not the complement of $\angle D$.

(*d*)   The point lies on the line.

14.5   INVERSE OF A STATEMENT

State the inverse of each of the following statements, and indicate whether or not it is true.

(*a*)   A person born in the United States is a citizen of the United States.

(*b*)   A sculptor is a talented person.

(*c*)   A triangle is a polygon.

Solutions

(*a*)   A person who is not born in the United States is not a citizen of the United States. (False, since there are naturalized citizens)

(*b*)   One who is not a sculptor is not a talented person. (False, since one may be a fine musician, etc.)

(*c*)   A figure that is not a triangle is not a polygon. (False, since the figure may be a quadrilateral, etc.)

14.6   FORMING THE CONVERSE, INVERSE, AND CONTRAPOSITIVE

State the converse, inverse, and contrapositive of the statement "a square is a rectangle." Determine the truth or falsity of each, and check the logical equivalence of the statement and its contrapositive, and of the converse and inverse.

Solutions

Statement: A square is a rectangle. (True)
Converse: A rectangle is a square. (False)
Inverse: A figure that is not a square is not a rectangle. (False)
Contrapositive: A figure that is not a rectangle is not a square. (True)
Thus the statement and its contrapositive are true, and the converse and inverse are false.

14.4   PARTIAL CONVERSE AND PARTIAL INVERSE OF A THEOREM

A partial converse of a theorem is formed by interchanging any one condition in the hypothesis with one consequence in the conclusion.

A partial inverse of a theorem is formed by denying one condition in the hypothesis and one consequence in the conclusion.

Thus from the theorem "if a line bisects the vertex angle of an isosceles triangle, then it is an altitude to the base," we can form a partial inverse or partial converse as shown in Fig. 14-4.

In forming a partial converse or inverse, the basic figure, such as the triangle in Fig. 14-4, is kept and not interchanged or denied.

(*a*) Theorem

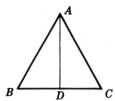

(*b*) Partial Converse

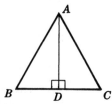

(*c*) Partial Inverse

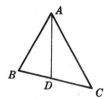

Given: △*ABC*
    (*1*) $\overline{AB} \cong \overline{AC}$
    (*2*) $\overline{AB}$ bisects ∠*A*.
To Prove: (*3*) $\overline{AD}$ is the altitude
         to $\overline{BC}$.

Given: △*ABC*
    (*2*) $\overline{AD}$ bisects ∠*A*.
    (*3*) $\overline{AD}$ is the altitude to $\overline{BC}$.
To Prove: (*1*) $\overline{AB} \cong \overline{AC}$.

Given: △*ABC*
    (*1'*) $\overline{AB} \not\cong AC$
    (*2*) $\overline{AD}$ bisects ∠*A*.
To Prove: (*3'*) $\overline{AD}$ is not the
         altitude to $\overline{BC}$.

Fig. 14-4

In Fig. 14-4(*b*), the partial converse is formed by interchanging statements (*1*) and (*3*). Stated in words, the partial converse is: "If the bisector of an angle of a triangle is an altitude, then the triangle is isosceles." Another partial converse may be formed by interchanging (*2*) and (*3*).

In Fig. 14-4(*c*), the partial inverse is formed by replacing statements (*1*) and (*3*) with their negatives, (*1'*) and (*3'*). Stated in words, the partial inverse is: "If two sides of a triangle are not congruent, the line segment that bisects their included angle is not an altitude to the third side." Another partial inverse may be formed by negating (*2*) and (*3*).

SOLVED PROBLEMS

14.7　**FORMING PARTIAL CONVERSES WITH PARTIAL INVERSES OF A THEOREM**
    Form (*a*) partial converses and (*b*) partial inverses of the statement "equal supplementary angles are right angles."

Solutions

(*a*) Partial converses:　(1) Congruent right angles are supplementary.
                   (2) Supplementary right angles are congruent.
(*b*) Partial inverses:　(1) Congruent angles that are not supplementary are not right angles.
                   (2) Supplementary angles that are not congruent are not right angles.

14.5 NECESSARY AND SUFFICIENT CONDITIONS

In logic and in geometry, it is often important to determine whether the conditions in the hypothesis of a statement are necessary or sufficient to justify its conclusion. This is done by ascertaining the truth or falsity of the statement and its converse, and then applying the following principles.

PRINCIPLE 1:　*If a statement and its converse are both true, then the conditions in the hypothesis of the statement are necessary and sufficient for its conclusion.*

For example, the statement "if angles are right angles, then they are congruent and supplementary" is true, and its converse, "if angles are congruent and supplementary, then they are right angles" is also true. Hence being right angles is necessary and sufficient for the angles to be congruent and supplementary.

PRINCIPLE 2:　*If a statement is true and its converse is false, then the conditions in the hypothesis of the statement are sufficient but not necessary for its conclusion.*

The statement "if angles are right angles, then they are congruent" is true, and its converse, "if angles are congruent, then they are right angles," is false. Hence being right angles is sufficient for the angles to be congruent. However, the angles need not be right angles to be congruent.

PRINCIPLE 3: *If a statement is false and its converse is true, then the conditions in the hypothesis are necessary but not sufficient for its conclusion.*

The statement "if angles are supplementary, then they are right angles" is false, and its converse, "if angles are right angles, then they are supplementary," is true. Hence angles need to be supplementary to be right angles, but being supplementary is not sufficient for angles to be right angles.

PRINCIPLE 4: *If a statement and its converse are both false, then the conditions in the hypothesis are neither necessary nor sufficient for its conclusion.*

Thus the statement "if angles are supplementary, then they are congruent" is false, and its converse, "if angles are congruent, then they are supplementary," is false. Hence being supplementary is neither necessary nor sufficient for the angles to be congruent.

These principles are summarized in the table that follows.

When the Conditions in the Hypothesis of a Statement are Necessary or Sufficient to Justify its Conclusion

Principle	Statement	Converse	Sufficient?	Necessary?
1	True	True	Yes	Yes
2	True	False	Yes	No
3	False	True	No	Yes
4	False	False	No	No

SOLVED PROBLEMS

14.8 DETERMINING NECESSARY AND SUFFICIENT CONDITIONS

For each of the following statements, determine whether the conditions in the hypothesis are necessary or sufficient to justify the conclusion.

(*a*) A regular polygon is equilateral and equiangular.

(*b*) An equiangular polygon is regular.

(*c*) A regular polygon is equilateral.

(*d*) An equilateral polygon is equiangular.

Solutions

(*a*) Since the statement and its converse are both true, the conditions are necessary and sufficient.

(*b*) Since the statement is false and its converse is true, the conditions are necessary but not sufficient.

(*c*) Since the statement is true and its converse is false, the conditions are sufficient but not necessary.

(*d*) Since both the statement and its converse are false, the conditions are neither necessary nor sufficient.

Supplementary Problems

1. State the order in which the terms in each of the following sets should be defined: (14.1)

   (a) Jewelry, wedding ring, ornament, ring

   (b) Automobile, vehicle, commercial automobile, truck

   (c) Quadrilateral, rhombus, polygon, parallelogram

   (d) Obtuse triangle, obtuse angle, angle, isosceles obtuse triangle

2. Correct each of the following definitions: (14.2)

   (a) A regular polygon is an equilateral polygon.

   (b) An isosceles triangle is a triangle having at least two congruent sides and angles.

   (c) A pentagon is a geometric figure having five sides.

   (d) A rectangle is a parallelogram whose angles are right angles.

   (e) An inscribed angle is an angle formed by two chords.

   (f) A parallelogram is a quadrilateral whose opposite sides are congruent and parallel.

   (g) An obtuse angle is an angle larger than a right angle.

3. State the negative of each of the following statements: (14.4)

   (a) $x + 2 = 4$ (d) His mark was more than 65.

   (b) $3y \neq 15$ (e) Joe is heavier than Dick.

   (c) She loves you.   (f) $a + b \neq c$

4. State the inverse of each of the following statements, and indicate whether or not it is true.   (14.5)

   (a) A square has congruent diagonals.   (c) A bachelor is an unmarried person.

   (b) An equiangular triangle is equilateral.   (d) Zero is not a positive number.

5. State the converse, inverse, and contrapositive of each of the following statements. Indicate the truth or falsity of each, and check the logical equivalence of the statement and its contrapositive, and of the converse and inverse.   (14.6)

   (a) If two sides of a triangle are congruent, the angles opposite these sides are congruent.

   (b) Congruent triangles are similar triangles.

   (c) If two lines intersect, then they are not parallel.

   (d) A senator of the United States is a member of its Congress.

6. Form partial converses and partial inverses of the theorems given in Fig. 14-5.   (14.7)

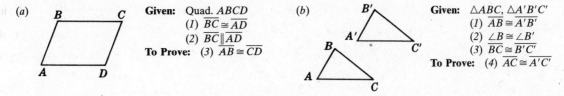

(a) **Given:** Quad. $ABCD$
      (1) $\overline{BC} \cong \overline{AD}$
      (2) $\overline{BC} \| \overline{AD}$
   To Prove: (3) $\overline{AB} \cong \overline{CD}$

(b) **Given:** $\triangle ABC$, $\triangle A'B'C'$
      (1) $\overline{AB} \cong \overline{A'B'}$
      (2) $\angle B \cong \angle B'$
      (3) $\overline{BC} \cong \overline{B'C'}$
   To Prove: (4) $\overline{AC} \cong \overline{A'C'}$

Fig. 14-5

7. For each of the following statements, determine whether the conditions in the hypothesis are necessary or sufficient to justify the conclusion.   (14.8)

   (a) Senators of the United States are elected members of Congress, two from each state.

   (b) Elected members of Congress are senators of the United States.

   (c) Elected persons are government officials.

   (d) If a woman lives in New York City, then she lives in New York State.

   (e) A bachelor is an unmarried man.

   (f) A bachelor is an unmarried person.

   (g) A quadrilateral having two pairs of congruent sides is a parallelogram.

CHAPTER 15

Constructions

15.1 INTRODUCTION

Geometric figures are constructed with straightedge and compass. Since constructions are based on deductive reasoning, measuring instruments such as the ruler and protractor are not permitted. However, a ruler may be used as a straightedge if its markings are disregarded.

In constructions, it is advisable to plan ahead by making a sketch of the situation; such a sketch will usually reveal the needed construction steps. Construction lines should be made light to distinguish them from the required figure.

The following constructions are detailed in this chapter:

1. To construct a line segment congruent to a given line segment

2. To construct an angle congruent to a given angle

3. To bisect a given angle

4. To construct a line perpendicular to a given line through a given point on the line

5. To bisect a given line segment

6. To construct a line perpendicular to a given line through a given external point

7. To construct a triangle given its three sides

8. To construct an angle of measure 60°

9. To construct a triangle given two sides and the included angle

10. To construct a triangle given two angles and the included side

11. To construct a triangle given two angles and a side not included

12. To construct a right triangle given its hypotenuse and a leg

13. To construct a line parallel to a given line through a given external point

14. To construct a tangent to a given circle through a given point on the circle

15. To construct a tangent to a given circle through a given point outside the circle

16. To circumscribe a circle about a triangle

17. To locate the center of a given circle

18. To inscribe a circle in a given triangle

19. To inscribe a square in a given circle

20. To inscribe a regular octagon in a given circle

21. To inscribe a regular hexagon in a given circle

22. To inscribe an equilateral triangle in a given circle

23. To construct a triangle similar to a given triangle on a given line segment as base

15.2   DUPLICATING SEGMENTS AND ANGLES

CONSTRUCTION 1:   *To construct a line segment congruent to a given line segment*
Given: Line segment $\overline{AB}$ (Fig. 15-1)
To construct: A line segment congruent to $\overline{AB}$
Construction: On a working line w, with any point C as a center and a radius equal to AB, construct an arc intersecting w at D. Then $\overline{CD}$ is the required line segment.

<table>
<tr>
<td>

Fig. 15-1
</td>
<td>

Fig. 15-2
</td>
</tr>
</table>

CONSTRUCTION 2:   *To construct an angle congruent to a given angle*
Given: $\angle A$ (Fig. 15-2)
To construct: An angle congruent to $\angle A$
Construction: With A as center and a convenient radius, construct an arc (1) intersecting the sides of $\angle A$ at B and C. With A', a point on a working line w, as center and the same radius, construct arc (2) intersecting w at B'. With B' as center and a radius equal to BC, construct arc (3) intersecting arc (2) at C'. Draw $A'C'$. Then $\angle A'$ is the required angle. ($\triangle ABC \cong A'B'C'$ by s.s.s. $\cong$ s.s.s.; hence $\angle A \cong \angle A'$.)

SOLVED PROBLEMS

15.1   COMBINING LINE SEGMENTS
   Given line segments with lengths a and b (Fig. 15-3), construct line segments with lengths equal to (*a*) $a + 2b$; (*b*) $2(a + b)$; (*c*) $b - a$.

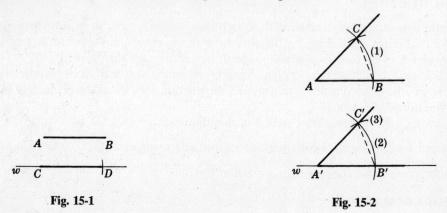

(*a*)   (*b*)   (*c*)

Fig. 15-3

Solutions

Use construction 1.

(*a*)   On a working line *w*, construct a line segment $\overline{AB}$ with length *a*. From *B*, construct a line segment with length equal to *b*, to point *C*; and from *C* construct a line segment with length *b*, to point *D*. Then $\overline{AD}$ is the required line segment.

(*b*)   Similar to (*a*). $AD = a + b + (a + b)$.

(*c*)   Similar to (*a*). First construct $\overline{AB}$ with length *b*, then $\overline{BC}$ with length *a*. $AC = b - a$.

15.2   COMBINING ANGLES

Given $\triangle ABC$ in Fig. 15-4, construct angles whose measures are equal to (*a*) 2*A*; (*b*) $A + B + C$; (*c*) $B - A$.

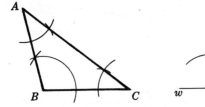 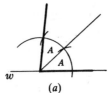 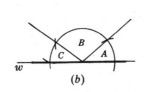 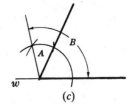

Fig. 15-4

Solutions

Use construction 2.

(*a*)   Using a working line *w* as one side, duplicate $\angle A$. Construct another duplicate of $\angle A$ adjacent to $\angle A$, as shown. The exterior sides of the copied angles form the required angle.

(*b*)   Using a working line *w* as one side, duplicate $\angle A$. Construct $\angle B$ adjacent to $\angle A$. Then construct $\angle C$ adjacent to $\angle B$. The exterior sides of the copied angles *A* and *C* form the required angle. Note that the angle is a straight angle.

(*c*)   Using a working line *w* as one side, duplicate $\angle B$. Then duplicate $\angle A$ from the new side of $\angle B$ as shown. The difference is the required angle.

15.3   CONSTRUCTING BISECTORS AND PERPENDICULARS

CONSTRUCTION 3:   *To bisect a given angle*
Given: $\angle A$ (Fig. 15-5)
To construct: The bisector of $\angle A$
Construction: With *A* as center and a convenient radius, construct an arc intersecting the sides of $\angle A$ at *B* and *C*. With *B* and *C* as centers and equal radii, construct arcs intersecting in *D*. Draw $\overrightarrow{A}D$. Then $\overrightarrow{AD}$ is the required bisector. ($\triangle ABD \cong \triangle ADC$ by s.s.s. $\cong$ s.s.s.; hence $\angle 1 \cong \angle 2$.)

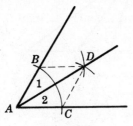 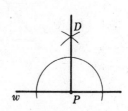

Fig. 15-5   **Fig. 15-6**

CONSTRUCTION 4: *To construct a line perpendicular to a given line through a given point on the line*
Given: Line *w* and point *P* on *w* (Fig. 15-6)
To construct: A perpendicular to *w* at *P*
Construction: Using construction 3, bisect the straight angle at *P*. Then $\overrightarrow{DP}$ is the required perpendicular; $\overrightarrow{DP}$ is the required line.

CONSTRUCTION 5: *To bisect a given line segment* (to construct the perpendicular bisector of a given line segment)
Given: Line segment $\overline{AB}$ (Fig. 15-7)
To construct: The perpendicular bisector of $\overline{AB}$
Construction: With *A* as center and a radius of more than half $\overline{AB}$, construct arc (1). With *B* as center and the same radius, construct arc (2) intersecting arc (1) at *C* and *D*. Draw $\overleftrightarrow{CD}$. $\overleftrightarrow{CD}$ is the required perpendicular bisector of $\overline{AB}$. (Two points each equidistant from the ends of a segment determine the perpendicular bisector of the segment.)

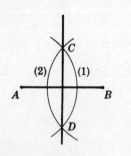

Fig. 15-7

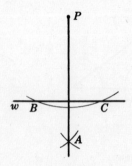

Fig. 15-8

CONSTRUCTION 6: *To construct a line perpendicular to a given line through a given external point*
Given: Line *w* and point *P* outside of *w* (Fig. 15-8)
To construct: A perpendicular to *w* through *P*
Construction: With *P* as center and a sufficiently long radius, construct an arc intersecting *w* at *B* and *C*. With *B* and *C* as centers and equal radii of more than half $\overline{BC}$, construct arcs intersecting at *A*. Draw $\overleftrightarrow{PA}$. Then $\overleftrightarrow{PA}$ is the required perpendicular. (Points *P* and *A* are each equidistant from *B* and *C*.)

SOLVED PROBLEMS

15.3 CONSTRUCTING SPECIAL LINES IN A TRIANGLE

In scalene $\triangle ABC$ [Fig. 15-9(*a*)], construct (*a*) a perpendicular bisector of $\overline{AB}$ and (*b*) a median to $\overline{AB}$. In $\triangle DEF$ [Fig. 15-9(*b*)], *D* is an obtuse angle; construct (*c*) the altitude to $\overline{DF}$ and (*d*) the bisector of $\angle E$.

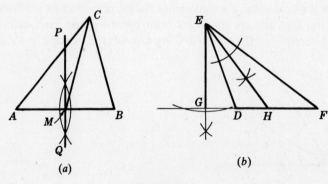

(*a*)   (*b*)

Fig. 15-9

Solutions

(a)   Use construction 5 to obtain $\overrightarrow{PQ}$ the perpendicular bisector of $\overline{AB}$.

(b)   Point M is the midpoint of $\overline{AB}$. Draw $\overline{CM}$, the median to $\overline{AB}$.

(c)   Use construction 6 to obtain $\overline{EG}$, the altitude to $\overline{DF}$ (extended).

(d)   Use construction 3 to bisect $\angle E$. $\overrightarrow{EH}$ is the required bisector.

15.4   CONSTRUCTING BISECTORS AND PERPENDICULARS TO OBTAIN REQUIRED ANGLES

(a)   Construct angles measuring 90°, 45°, and 135°.

(b)   Given an angle with measure A (Fig. 15-10), construct an angle whose measure is $90° + A$.

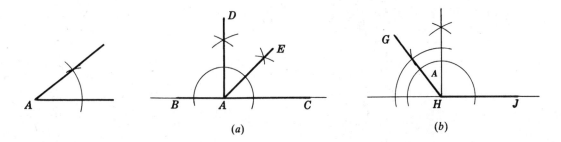

Fig. 15-10

Solutions

(a)   In Fig. 15-10(a), $m\angle DAB = 90°$, $m\angle CAE = 45°$, $m\angle BAE = 135°$

(b)   In Fig. 15-10(b), $m\angle GHJ = 90° + A$.

15.4   CONSTRUCTING A TRIANGLE

15.4A   Determining a Triangle

A triangle is determined when a set of given data fix its size and shape. Since the parts needed to prove congruent triangles fix the size and shape of the triangles, a triangle is determined when the given data consist of three sides, or two sides and the angle included by those sides, or two angles and a side included by those angles, or two angles and a side not included by those angles, or the hypotenuse and either leg of a right triangle.

15.4B   Sketching Triangles to be Constructed

Before doing the actual construction, it is very helpful to make a preliminary sketch of the required triangle. In this sketch:

1.   Show the position of each of the given parts of the triangle.

2.   Draw the given parts heavy, the remaining parts light.

3.   Approximate the sizes of the given parts.

4.   Use small letters for sides to agree with the capital letters for the angles opposite them.

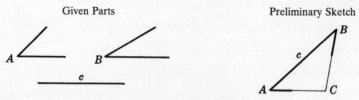

Fig. 15-11

As an example, you might make a sketch like that in Fig. 15-11 before constructing a triangle given two angles and an included side.

15.4C   Triangle Constructions

CONSTRUCTION 7:   *To construct a triangle given its three sides*
Given: Sides of lengths a, b, and c (Fig. 15-12)
To construct: $\triangle ABC$
Construction: On a working line w, construct $\overline{AC}$ such that $AC = b$. With A as center and c as radius, construct arc (1). Then with C as center and a as radius, construct arc (2) intersecting arc (1) at B. Draw $\overline{BC}$ and $\overline{AB}$. $\triangle ABC$ is the required triangle.

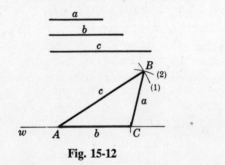

Fig. 15-12

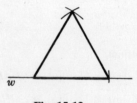

Fig. 15-13

CONSTRUCTION 8:   *To construct an angle of measure 60°*
Given: Line w (Fig. 15-13)
To construct: An angle of measure 60°
Construction: Using a convenient length as a side, construct an equilateral triangle using construction 7. Then any angle of the equilateral triangle is the required angle.

CONSTRUCTION 9:   *To construct a triangle given two sides and the included angle*
Given: $\angle A$, segments of lengths b and c (Fig. 15-14)
To construct: $\triangle ABC$

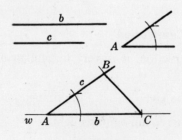

Fig. 15-14

Construction: On a working line *w*, construct $\overline{AC}$ such that $AC = b$. At *A*, construct $\angle A$ with one side $\overline{AC}$. On the other side of $\angle A$, construct $\overline{AB}$ such that $AB = c$. Draw $\overline{BC}$. Then the required triangle is $\triangle ABC$.

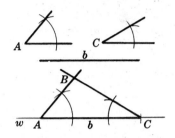

Fig. 15-15

CONSTRUCTION 10:   *To construct a triangle given two angles and the included side*
Given: $\angle A$, $\angle C$, and a segment of length *b* (Fig. 15-15)
To construct: $\triangle ABC$
Construction: On a working line *w*, construct $\overline{AC}$ such that $AC = b$. At *A* construct $\angle A$ with one side on $\overline{AC}$, and at *C* construct $\angle C$ with one side on $\overline{AC}$. Extend the new sides of the angles until they meet, at *B*.

CONSTRUCTION 11:   *To construct a triangle given two angles and a side not included*
Given: $\angle A$, $\angle B$, and a segment of length *b* (Fig. 15-16)
To construct: $\triangle ABC$
Construction: On a working line *w*, construct $\overline{AC}$ such that $AC = b$. At *C* construct an angle with measure equal to $m\angle A + m\angle B$ so that the extension of $\overline{AC}$ will be one side of the angle. The remainder of the straight angle at *C* will be $\angle C$. At *A* construct $\angle A$ with one side on $\overline{AC}$. The intersection of the new sides of the angles is *B*.

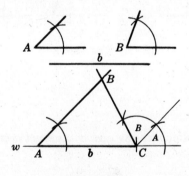

Fig. 15-16

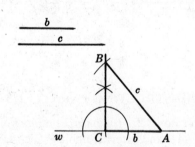

Fig. 15-17

CONSTRUCTION 12:   *To construct a right triangle given its hypotenuse and a leg*
Given: Hypotenuse with length *c* and leg with length *b* of right triangle *ABC* (Fig. 15-17)
To construct: Right triangle *ABC*
Construction: On a working line *w*, construct $\overline{AC}$ such that $AC = b$. At *C* construct a perpendicular to $\overline{AC}$. With *A* as center and a radius of *c*, construct an arc intersecting the perpendicular at *B*.

SOLVED PROBLEMS

15.5   CONSTRUCTING A TRIANGLE

Construct an isosceles triangle, given the lengths of the base and an arm (Fig. 15-18).

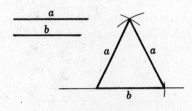

Fig. 15-18

Solution

Use construction 7, since all three sides of the triangle are known.

15.6   CONSTRUCTING ANGLES BASED ON THE CONSTRUCTION OF THE 60° ANGLE
Construct an angle of measure (*a*) 120°; (*b*) 30°; (*c*) 150°; (*d*) 105°; (*e*) 75°.

Solutions

(*a*)   Use construction 8 [Fig. 15-19(*a*)] to construct 120° as 180° − 60°.

(*b*)   Use constructions 8 and 3 to construct 30° as $\frac{1}{2}(60°)$ [Fig. 15-19(*b*)].

(*c*)   Use (*b*) to construct 150° as 180° − 30° [Fig. 15-19(*b*)].

(*d*)   Use constructions 3, 4, and 8 to construct 105° as $60° + \frac{1}{2}(90°)$ [Fig. 15-19(*c*)].

(*e*)   Use (*d*) to construct 75° as 180° − 105° [Fig. 15-19(*c*)].

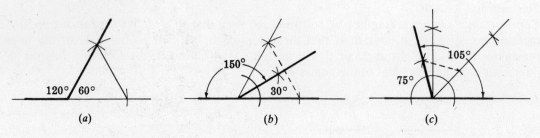

Fig. 15-19

15.5   CONSTRUCTING PARALLEL LINES

CONSTRUCTION 13:   *To construct a line parallel to a given line through a given external point*
Given: $\overleftrightarrow{AB}$ and external point P (Fig. 15-20)
To construct: A line through P parallel to $\overleftrightarrow{AB}$
Construction: Draw a line $\overleftrightarrow{RS}$ through P intersecting $\overleftrightarrow{AB}$ in Q. Construct $\angle SPD \cong \angle PQB$. Then $\overleftrightarrow{CD}$ is the required parallel. (If two corresponding angles are congruent the lines cut by the transversal are parallel.)

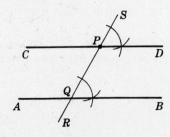

Fig. 15-20

SOLVED PROBLEM

15.7   CONSTRUCTING A PARALLELOGRAM
Construct a parallelogram given the lengths of two adjacent sides a and b and of a diagonal d (Fig. 15-21).

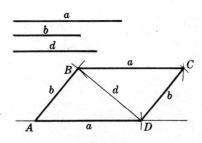

Fig. 15-21

Solution
Three vertices of the parallelogram are obtained by constructing $\triangle ABD$ by construction 7. The fourth vertex, C, is obtained by constructing $\triangle BCD$ upon diagonal $\overline{BD}$ by construction 7. Vertex C may also be obtained by constructing $\overline{BC} \| \overline{AD}$ and $\overline{DC} \| \overline{AB}$.

15.6   CIRCLE CONSTRUCTIONS

CONSTRUCTION 14:   *To construct a tangent to a given circle through a given point on the circle*
Given: Circle O and point P on the circle (Fig. 15-22)
To construct: A tangent to circle O at P
Construction: Draw radius $\overline{OP}$ and extend it outside the circle. Construct $\overleftrightarrow{AB} \perp \overleftrightarrow{OP}$ at P. $\overleftrightarrow{AB}$ is the required tangent. (A line perpendicular to a radius at its outer extremity is a tangent to the circle.)

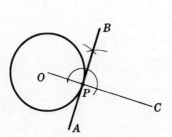

Fig. 15-22

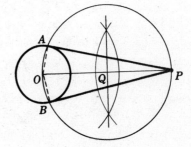

Fig. 15-23

CONSTRUCTION 15:   *To construct a tangent to a given circle through a given point outside the circle*
Given: Circle O and point P outside the circle (Fig. 15-23)
To construct: A tangent to circle O from P
Construction: Draw $\overline{OP}$, and make $\overline{OP}$ the diameter of a new circle Q. Connect P to A and B, the intersections of circles O and Q. Then $\overline{PA}$ and $\overline{PB}$ are tangents. ($\angle OAP$ and $\angle OBP$ are right angles, since angles inscribed in semicircles are right angles.)

CONSTRUCTION 16: *To circumscribe a circle about a triangle*

Given: △ABC (Fig. 15-24)

To construct: The circumscribed circle of △ABC

Construction: Construct the perpendicular bisectors of two sides of the triangle. Their intersection is the center of the required circle, and the distance to any vertex is the radius. (Any point on the perpendicular bisector of a segment is equidistant from the ends of the segment.)

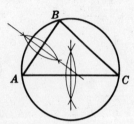   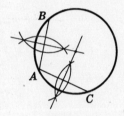

Fig. 15-24   **Fig. 15-25**

CONSTRUCTION 17: *To locate the center of a given circle*

Given: A circle (Fig. 15-25)

To construct: The center of the given circle

Construction: Select any three points A, B, and C on the circle. Construct the perpendicular bisectors of line segments $\overline{AB}$ and $\overline{AC}$. The intersection of these perpendicular bisectors is the center of the circle.

CONSTRUCTION 18: *To inscribe a circle in a given triangle*

Given: △ABC (Fig. 15-6)

To construct: The circle inscribed in △ABC

Construction: Construct the bisectors of two of the angles of △ABC. Their intersection is the center of the required circle, and the distance (perpendicular) to any side is the radius. (Any point on the bisector of an angle is equidistant from the sides of the angle.)

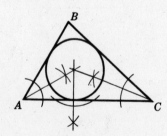

Fig. 15-26

SOLVED PROBLEMS

15.8 CONSTRUCTING TANGENTS

A secant from a point P outside circle O in Fig. 15-27 meets the circle in B and A. Construct a triangle circumscribed about the circle so that two of its sides meet in P and the third side is tangent to the circle at A.

Solution

Use constructions 14 and 15: At *A* construct a tangent to circle *O*. From *P* construct tangents to circle *O* intersecting the first tangent in *C* and *D*. The required triangle is △*PCD*.

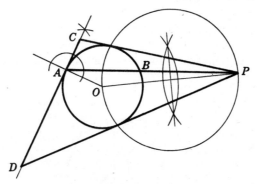

Fig. 15-27

15.9   CONSTRUCTING CIRCLES

Construct the circumscribed and inscribed circles of isosceles triangle *DEF* in Fig. 15-28.

Solution

Use constructions 16 and 18. In doing so, note that the bisector of ∠*E* is also the perpendicular bisector of $\overline{DF}$. Then the center of each circle is on $\overline{EG}$. *I*, the center of the inscribed circle, is found by constructing the bisector of ∠*D* or ∠*F*. *C*, the center of the circumscribed circle, is found by constructing the perpendicular bisector of $\overline{DE}$ or $\overline{EF}$.

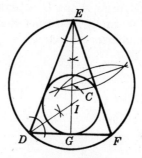

Fig. 15-28

15.7   INSCRIBING AND CIRCUMSCRIBING REGULAR POLYGONS

CONSTRUCTION 19:   *To inscribe a square in a given circle*
Given: Circle *O* (Fig. 15-29)
To construct: A square inscribed in circle *O*

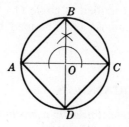

Fig. 15-29

Construction: Draw a diameter, and construct another diameter perpendicular to it. Join the end points of the diameters to form the required square.

CONSTRUCTION 20:   *To inscribe a regular octagon in a given circle*
Given: Circle O (Fig. 15-30)
To construct: A regular octagon inscribed in circle O
Construction: As in construction 19, construct perpendicular diameters. Then bisect the angles formed by these diameters, dividing the circle into eight congruent arcs. The chords of these arcs are the sides of the required regular octagon.

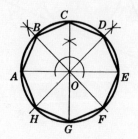

Fig. 15-30

CONSTRUCTION 21:   *To inscribe a regular hexagon in a given circle*
Given: Circle O (Fig. 15-31)
To construct: A regular hexagon inscribed in circle O
Construction: Draw diameter $\overline{AD}$ and, using A and D as centers, construct four arcs having the same radius as circle O and intersecting the circle. Construct the required regular hexagon by joining consecutive points in which these arcs intersect the circle.

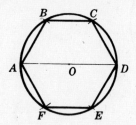

Fig. 15-31

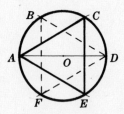

Fig. 15-32

CONSTRUCTION 22:   *To inscribe an equilateral triangle in a given circle*
Given: Circle O (Fig. 15-32)
To construct: An equilateral triangle inscribed in circle O
Construction: Inscribed equilateral triangles are obtained by joining alternately the six points of division obtained in construction 21.

15.8   CONSTRUCTING SIMILAR TRIANGLES

CONSTRUCTION 23:   *To construct a triangle similar to a given triangle on a given line segment as base*
Given: $\triangle ABC$ and line segment $\overline{A'C'}$ (Fig. 15-33)

To construct: △*A'B'C'* ~ △*ABC* on $\overline{A'C'}$ as base
Construction: On $\overline{A'C'}$, construct ∠*A'* ≅ ∠*A* and ∠*C'* ≅ ∠*C* using construction 2. Extend the other sides until they meet, at *B*. (If two angles of one triangle are congruent to two angles of another triangle, the triangles are similar.)

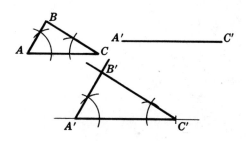

Fig. 15-33

SOLVED PROBLEM

15.10 Constructing Similar Triangles
            Construct a triangle similar to triangle *ABC* in Fig. 15-34, with a base twice as long as the base of the given triangle.

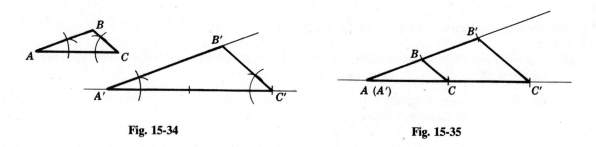

Fig. 15-34   **Fig. 15-35**

Solution
      Construct $\overline{A'C'}$ twice as long as $\overline{AC}$, and then use construction 23.
      Alternative method (Fig. 15-35): Extend two sides of △*ABC* to twice their lengths and join the endpoints.

Supplementary Problems

1. Given line segments with lengths *a* and *b* as follows: ——————*a*—————— ————*b*————. Construct a line segment whose length equals (*a*) *a* + *b*; (*b*) *a* − *b*; (*c*) 2*a* + *b*; (*d*) *a* + 3*b*; (*e*) 2(*a* + *b*); (*f*) 2(3*b* − *a*).   (15.1)

2. Given line segments with lengths *a*, *b*, and *c*: ——*a*—— ————*b*———— ———*c*———. Construct a line segment whose length equals (*a*) *a* + *b* + *c*; (*b*) *a* + *c* − *b*; (*c*) *a* + 2(*b* + *c*); (*d*) *b* + 2(*a* − *c*); (*e*) 3(*b* + *c* − *a*).   (15.1)

3. Given angles with measures *A* and *B* (Fig. 15-36). Construct an angle with measure (*a*) *A* + *B*; (*b*) *A* − *B*; (*c*) 2*B* − *A*; (*d*) 2*A* − *B*; (*e*) 2(*A* − *B*).   (15.2)

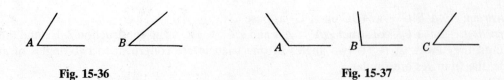

Fig. 15-36   **Fig. 15-37**

4. Given angles with measures A, B, and C (Fig. 15-37). Construct an angle with measure (*a*) $A + C$; (*b*) $B + C - A$; (*c*) $2C$; (*d*) $B - C$; (*e*) $2(A - B)$.   (15.2)

5. In a right triangle, construct (*a*) the bisector of the right angle; (*b*) the perpendicular bisector of the hypotenuse; (*c*) the median to the hypotenuse.   (15.3)

6. For each kind of triangle (acute, right, and obtuse), show that the following sets of rays and segments are concurrent, that is, they intersect in one point: (*a*) the angle bisectors; (*b*) the medians; (*c*) the altitudes; (*d*) the perpendicular bisectors.   (15.3)

7. Given $\triangle ABC$ in Fig. 15-38, construct (*a*) the supplement of $\angle A$; (*b*) the complement of $\angle B$; (*c*) the complement of $\frac{1}{2}\angle C$.   (15.4)

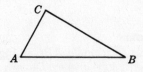

Fig. 15-38

8. Construct an angle with measure equal to (*a*) $22\frac{1}{2}°$; (*b*) $67\frac{1}{2}°$; (*c*) $112\frac{1}{2}°$.   (15.4)

9. Given an acute angle, construct (*a*) its supplement; (*b*) its complement; (*c*) half its supplement; (*d*) half its complement.   (15.4)

10. By actual construction, illustrate that the difference between the measures of the supplement and complement of an acute angle equals 90°.   (15.4)

11. Construct a right triangle given its (*a*) legs; (*b*) hypotenuse and a leg; (*c*) leg and an acute angle adjacent to the leg; (*d*) leg and an acute angle opposite the leg; (*e*) hypotenuse and an acute angle.   (15.5)

12. Construct an isosceles triangle given (*a*) an arm and a vertex angle; (*b*) an arm and a base angle; (*c*) an arm and the altitude to the base; (*d*) the base and the altitude to the base.   (15.5)

13. Construct an isosceles right triangle given (*a*) a leg; (*b*) the hypotenuse; (*c*) the altitude to the hypotenuse.   (15.5)

14. Construct a triangle given (*a*) two sides and the median to one of them; (*b*) two sides and the altitude to one of them; (*c*) an angle, the angle bisector of the given angle, and a side adjacent to the given angle.   (15.5)

15. Construct angles of measure 15° and 165°.   (15.6)

16. Given an angle with measure A, construct angles with measure (*a*) $A + 60°$; (*b*) $A + 30°$; (*c*) $A + 120°$.   (15.6)

17. Construct a parallelogram, given (*a*) two adjacent sides and an angle; (*b*) the diagonals and the acute angle at their intersection; (*c*) the diagonals and a side; (*d*) two adjacent sides and the altitude to one of them; (*e*) a side, an angle, and the altitude to the given side.   (15.7)

18. Circumscribe a triangle about a given circle, if the points of tangency are given.   (15.8)

19. Secant $\overleftrightarrow{AB}$ passes through the center of circle O in Fig. 15-39. Circumscribe a quadrilateral about the circle so that A and B are opposite vertices.   (15.8)

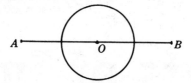

Fig. 15-39

20. Circumscribe and inscribe circles about (a) an acute triangle; (b) an obtuse triangle.   (15.9)

21. Circumscribe a circle about (a) a right triangle; (b) a rectangle; (c) a square.   (15.9)

22. Construct the inscribed and circumscribed circles of an equilateral triangle.   (15.9)

23. Locate the center of a circle drawn around the outside of a half-dollar piece.   (15.9)

24. In a given circle, inscribe (a) a square; (b) a regular octagon; (c) a regular 16-gon; (d) a regular hexagon; (e) an equilateral triangle; (f) a regular dodecagon.

25. Construct a triangle similar to a given triangle with a base (a) three times as long; (b) half as long; (c) one and one-half times as long.   (15.10)

CHAPTER 16

Proofs of Important Theorems

16.1 INTRODUCTION

The theorems proved in this chapter are considered the most important in the logical sequence of geometry. They are as follows:

1. If two sides of a triangle are congruent, the angles opposite these sides are congruent. (Base angles of an isosceles triangle are congruent.)

2. The sum of the measures of the angles in a triangle equals 180°.

3. If two angles of a triangle are congruent, the sides opposite these angles are congruent.

4. Two right triangles are congruent if the hypotenuse and a leg of one are congruent to the corresponding parts of the other.

5. A diameter perpendicular to a chord bisects the chord and its arcs.

6. An angle inscribed in a circle is measured by one-half its intercepted arc.

7. An angle formed by two chords intersecting inside a circle is measured by one-half the sum of the intercepted arcs.

8a. An angle formed by two secants intersecting outside a circle is measured by one-half the difference of its intercepted arcs.

8b. An angle formed by a tangent and a secant intersecting outside a circle is measured by one-half the difference of its intercepted arcs.

8c. An angle formed by two tangents intersecting outside a circle is measured by one-half the difference of its intercepted arcs.

9. If three angles of one triangle are congruent to three angles of another triangle, the triangles are similar.

10. If the altitude is drawn to the hypotenuse of a right triangle, then (a) the two triangles thus formed are similar to the given triangle and to each other, and (b) each leg of the given triangle is the mean proportional between the hypotenuse and the projection of that leg upon the hypotenuse.

11. The square of the length of the hypotenuse of a right triangle equals the sum of the squares of the lengths of the other two sides.

12. The area of a parallelogram equals the product of the length of one side and the length of the altitude to that side.

13. The area of a triangle is equal to one-half the product of the length of one side and the length of the altitude to that side.

14. The area of a trapezoid is equal to one-half the product of the length of the altitude and the sum of the lengths of the bases.

15. The area of a regular polygon is equal to one-half the product of its perimeter and the length of its apothem.

16.2 THE PROOFS

1. If two sides of a triangle are congruent, the angles opposite these sides are congruent. (Base angles of an isosceles triangle are congruent.)

Given: $\triangle ABC, \overline{AB} \cong \overline{BC}$
To Prove: $\angle A \cong \angle C$
Plan: When the bisector of the vertex angle is drawn, the angles to be proved congruent become corresponding angles of congruent triangles.

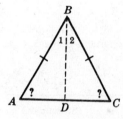

PROOF:

Statements	Reasons
1. Draw $\overline{BD}$ bisecting $\angle B$.	1. An angle may be bisected.
2. $\angle 1 \cong \angle 2$	2. To bisect is to divide into two congruent parts.
3. $\overline{AB} \cong \overline{BC}$	3. Given
4. $\overline{BD} \cong \overline{BD}$	4. Reflexive property
5. $\triangle ADB \cong \triangle BDC$	5. s.a.s. $\cong$ s.a.s.
6. $\angle A \cong \angle C$	6. Corresponding parts of congruent triangles are congruent.

2. The sum of the measures of the angles in a triangle equals 180°.

Given: $\triangle ABC$
To Prove: $m\angle A + m\angle B + m\angle C = 180°$
Plan: When a line is drawn through one vertex parallel to the opposite side, a straight angle is formed whose parts can be proved congruent to the angles of the triangle.

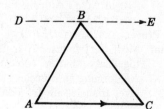

PROOF:

Statements	Reasons
1. Through B, draw $\overleftrightarrow{DE} \parallel \overleftrightarrow{AC}$.	1. Through an external point, a line can be drawn parallel to a given line.
2. $m\angle DBE = 180°$	2. A straight angle is an angle whose measure is 180°
3. $m\angle DBA + m\angle ABC + mCBE = 180°$	3. The whole equals the sum of its parts.
4. $\angle A \cong \angle DBA, \ \angle C \cong CBE$	4. Alternate interior angles of parallel lines are congruent.
5. $m\angle A + m\angle B + m\angle C = 180°$	5. Substitution Postulate

3. If two angles of a triangle are congruent, the sides opposite these angles are congruent.

Given: $\triangle ABC, \ \angle A \cong \angle C$
To Prove: $\overline{AB} \cong \overline{BC}$
Plan: When the bisector of $\angle B$ is drawn, the sides to be proved congruent become corresponding sides of congruent triangles.

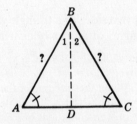

PROOF:

Statements	Reasons
1. Draw $\overline{BD}$ bisecting $\angle B$.	1. An angle may be bisected.
2. $\angle 1 \cong \angle 2$	2. To bisect is to divide into two congruent parts.
3. $\angle A \cong \angle C$	3. Given
4. $\overline{BD} \cong \overline{BD}$	4. Reflexive property
5. $\triangle BDA \cong \triangle BDC$	5. s.a.a. $\cong$ s.a.a.
6. $\overline{AB} \cong \overline{BC}$	6. Corresponding parts of congruent triangles are congruent.

4. Two right triangles are congruent if the hypotenuse and a leg of one are congruent to the corresponding parts of the other.

Given: Right $\triangle ABC$ with right angle at C
        Right $\triangle DEF$ with right angle at F
        $\overline{AB} \cong \overline{DE}, \ \overline{BC} \cong \overline{EF}$
To Prove: $\triangle ABC \cong \triangle DEF$
Plan: Move the two given triangles together so that $\overline{BC}$ coincides with $\overline{EF}$, forming an isosceles triangle. The given triangles are proved congruent by using Theorem 1 and s.a.a. $\cong$ s.a.a.

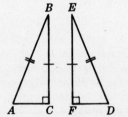

PROOF:

Statements	Reasons
1. $\overline{BC} \cong \overline{EF}$	1. Given
2. Move triangles ABC and DEF together so that $\overline{BC}$ coincides with $\overline{EF}$, and A and D are on opposite sides of $\overline{BC}$.	2. A geometric figure may be moved without changing its size or shape. Equal lines may be made to coincide.
3. $\angle C$ and $\angle F$ are right angles.	3. Given
4. $\angle ACD$ is a straight angle.	4. The whole equals the sum of its parts.
5. $\overline{AD}$ is a straight line segment.	5. The sides of a straight angle lie in a straight line.
6. $\overline{AB} \cong \overline{DE}$	6. Given
7. $\angle A \cong \angle D$	7. If two sides of a triangle are congruent, the angles opposite these sides are congruent.
8. $\triangle ABC \cong \triangle DEF$	8. s.a.a. $\cong$ s.a.a.

5. A diameter perpendicular to a chord bisects the chord and its arcs.

Given: Circle O, diameter $\overline{AB} \perp \overline{CD}$
To Prove: $\overline{CE} \cong \overline{ED}$, $\overparen{BC} \cong \overparen{BD}$, $\overparen{AC} \cong \overparen{AD}$
Plan: Congruent triangles are formed when radii are drawn to C and D, proving $\overline{CE} \cong \overline{ED}$. Equal central angles are used to prove $\overparen{BC} \cong \overparen{BD}$; then the Subtraction Postulate is used to prove $\overparen{AC} \cong \overparen{AD}$.

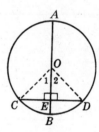

PROOF:

Statements	Reasons
1. Draw $\overline{OC}$ and $\overline{OD}$.	1. A straight line may be drawn between two points.
2. $\overline{OC} \cong \overline{OD}$	2. Radii of a circle are congruent.
3. $\overline{AB} \perp \overline{CD}$	3. Given
4. $\angle OEC$ and $\angle OED$ are right angles.	4. Perpendiculars form right angles.
5. $\overline{OE} \cong \overline{OE}$	5. Reflexive property
6. $\triangle OEC \cong \triangle OED$	6. hy. leg $\cong$ hy. leg
7. $\overline{CE} \cong \overline{ED}$, $\angle 1 \cong \angle 2$	7. Corresponding parts of congruent triangles are congruent.
8. $\overparen{CB} \cong \overparen{BD}$	8. In a circle, congruent central angles have congruent arcs.
9. $\overparen{ACB} \cong \overparen{ADB}$	9. A diameter bisects a circle.
10. $\overparen{AC} \cong \overparen{AD}$	10. In a circle, congruent arcs are equal arcs; Subtraction Postulate

6. An angle inscribed in a circle is measured by one-half its intercepted arc.

Case I: The center of the circle is on one side of the angle.

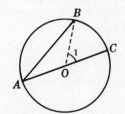

Given: $\angle A$ is inscribed in circle O. O is on side AC.

To Prove: $\angle A \stackrel{\circ}{=} \frac{1}{2}\widehat{BC}$

Plan: When radius $\overline{OB}$ is drawn, isosceles $\triangle AOB$ is formed. $\angle A$ is proved to be equal in measure to one-half central $\angle 1$, which is measured by $\widehat{BC}$.

PROOF:

Statements	Reasons
1. Draw $\overline{OB}$.	1. A straight line can be drawn between two points.
2. $\overline{AO} \cong \overline{OB}$	2. Radii of a circle are congruent.
3. $\angle A \cong \angle B$	3. If two sides of a triangle are congruent, the angles opposite these sides are congruent.
4. $m\angle A + m\angle B = m\angle 1$	4. In a triangle the measure of an exterior angle equals the sum of the measures of the two adjacent interior angles.
5. $m\angle A + m\angle A = 2m\angle A = m\angle 1$	5. Substitution Postulate
6. $m\angle A = \frac{1}{2}m\angle 1$	6. Halves of equals are equal.
7. $\angle 1 \stackrel{\circ}{=} \widehat{BC}$	7. A central angle is measured by its intercepted arc.
8. $\angle A \stackrel{\circ}{=} \frac{1}{2}\widehat{BC}$	8. Substitution Postulate

Case II: The center is inside the angle.

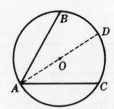

Given: $\angle BAC$ is inscribed in circle O. O is inside $\angle BAC$.

To Prove: $\angle BAC \stackrel{\circ}{=} \frac{1}{2}\widehat{BC}$

Plan: When a diameter is drawn, $\angle BAC$ is divided into two angles which can be measured by applying Case I.

PROOF:

Statements	Reasons
1. Draw diameter $\overline{AD}$.	1. A straight line may be drawn between two points.
2. $\angle BAD \stackrel{\circ}{=} \frac{1}{2}\widehat{BD}$, $\angle DAC \stackrel{\circ}{=} \frac{1}{2}\widehat{DC}$	2. An inscribed angle is measured by one-half its intercepted arc if the center of the circle is on one side.
3. $\angle BAC \stackrel{\circ}{=} \frac{1}{2}\widehat{BD} + \frac{1}{2}\widehat{DC}$ or $\angle BAC \stackrel{\circ}{=} \frac{1}{2}(\widehat{BD} + \widehat{DC})$	3. If equals are added to equals, the sums are equal.
4. $\angle BAC \stackrel{\circ}{=} \frac{1}{2}\widehat{BC}$	4. Substitution Postulate

Case III: The center is outside the angle.

Given: $\angle BAC$ is inscribed in circle O. O is
outside $\angle BAC$.
To Prove: $\angle BAC \stackrel{\cdot}{=} \frac{1}{2}\widehat{BC}$
Plan: When a diameter is drawn, $\angle BAC$
becomes the difference of two angles
which can be measured by applying
Case I.

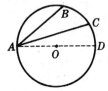

PROOF:

Statements	Reasons
1. Draw diameter $\overline{AD}$.	1. A straight line may be drawn between two points.
2. $\angle BAD \stackrel{\cdot}{=} \frac{1}{2}\widehat{BD}$, $\angle CAD \stackrel{\cdot}{=} \frac{1}{2}\widehat{CD}$	2. An inscribed angle is measured by one-half its intercepted arc if the center of the circle is on one side.
3. $\angle BAC \stackrel{\cdot}{=} \frac{1}{2}\widehat{BD} - \frac{1}{2}\widehat{CD}$ or $\angle BAC \stackrel{\cdot}{=} \frac{1}{2}(\widehat{BD} - \widehat{CD})$	3. If equals are subtracted from equals, the differences are equal.
4. $\angle BAC \stackrel{\cdot}{=} \frac{1}{2}\widehat{BC}$	4. Substitution Postulate

7. An angle formed by two chords intersecting inside a circle is measured by one-half the sum of the intercepted arcs.

Given: $\angle 1$ formed by chords $\overline{AB}$ and $\overline{CD}$
intersecting at point E inside circle O
To Prove: $\angle 1 \stackrel{\cdot}{=} \frac{1}{2}(\widehat{AC} + \widehat{BD})$
Plan: When chord $\overline{AD}$ is drawn, $\angle 1$ becomes
an exterior angle of a triangle whose
nonadjacent interior angles are inscribed
angles measured by $\frac{1}{2}\widehat{AC}$ and $\frac{1}{2}\widehat{BD}$.

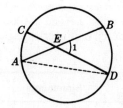

PROOF:

Statements	Reasons
1. Draw $\overline{AD}$.	1. A straight line may be drawn between two points.
2. $m\angle 1 = m\angle A + m\angle D$	2. The measure of an exterior angle of a triangle equals the sum of the measures of the nonadjacent interior angles.
3. $\angle A \stackrel{\cdot}{=} \frac{1}{2}\widehat{BD}$, $\angle D \stackrel{\cdot}{=} \frac{1}{2}\widehat{AC}$	3. An angle inscribed in a circle is measured by one-half its intercepted arc.
4. $\angle 1 \stackrel{\cdot}{=} \frac{1}{2}\widehat{BD} + \frac{1}{2}\widehat{AC} \stackrel{\cdot}{=} \frac{1}{2}(\widehat{BD} + \widehat{AC})$	4. Substitution Postulate

8a. An angle formed by two secants intersecting outside a circle is measured by one-half the difference of its intercepted arcs.

Given: $\angle P$ formed by secants $\overline{PBA}$ and $\overline{PDC}$ intersecting at P, a point outside circle O.

To Prove: $\angle P \doteq \frac{1}{2}(\widehat{AC} - \widehat{BD})$

Plan: When $\overline{AD}$ is drawn, $\angle 1$ becomes an exterior angle of $\triangle ADP$, of which $\angle P$ is a nonadjacent interior angle.

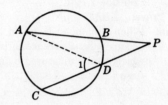

PROOF:

Statements	Reasons
1. Draw $\overline{AD}$.	1. A straight line may be drawn between two points.
2. $m\angle P + m\angle A = m\angle 1$	2. The measure of an exterior angle of a triangle equals the sum of the measures of the nonadjacent interior angles.
3. $m\angle P = m\angle 1 - m\angle A$	3. Subtraction Postulate
4. $\angle 1 \doteq \frac{1}{2}\widehat{AC}$, $\angle A \doteq \frac{1}{2}\widehat{BD}$	4. An angle inscribed in a circle is measured by one-half its intercepted arc.
5. $\angle P \doteq \frac{1}{2}\widehat{AC} - \frac{1}{2}\widehat{BD}$ or $\angle P \doteq \frac{1}{2}(\widehat{AC} - \widehat{BD})$	5. Substitution Postulate

8b. An angle formed by a secant and a tangent intersecting outside a circle is measured by one-half the difference of its intercepted arcs.

Given: $\angle P$ formed by secant $\overline{PBA}$ and tangent $\overline{PDC}$ intersecting P, a point outside circle O.

To Prove: $\angle P \doteq \frac{1}{2}(\widehat{AD} - \widehat{BD})$

Plan: When chord $\overline{AD}$ is drawn, $\angle 1$ becomes an exterior angle of $\triangle ADP$, of which $\angle P$ and $\angle A$ are nonadjacent interior angles.

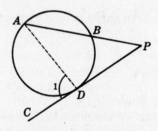

PROOF:

Statements	Reasons
1. Draw $\overline{AD}$.	1. A straight line may be drawn between two points.
2. $m\angle P + m\angle A = m\angle 1$	2. The measure of an exterior angle of a triangle equals the sum of the measures of the nonadjacent interior angles.
3. $m\angle P = m\angle 1 - m\angle A$	3. Subtraction Postulate
4. $\angle 1 \doteq \frac{1}{2}\widehat{AD}$	4. An angle formed by a tangent and a chord is measured by one-half its intercepted arc.
5. $\angle A \doteq \frac{1}{2}\widehat{BD}$	5. An inscribed angle is measured by one-half its intercepted arc.
6. $\angle P \doteq \frac{1}{2}\widehat{AD} - \frac{1}{2}\widehat{BD}$ or $\angle P \doteq \frac{1}{2}(\widehat{AD} - \widehat{BD})$	6. Substitution Postulate

8c. An angle formed by two tangents intersecting outside a circle is measured by one-half the difference of its intercepted arcs.

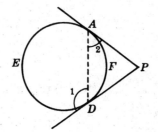

Given: $\angle P$ formed by tangents $\overline{PA}$ and $\overline{PD}$ intersecting at P, a point outside circle O.

To Prove: $\angle P \stackrel{.}{=} \frac{1}{2}(\widehat{AED} - \widehat{AFD})$

Plan: When chord $\overline{AD}$ is drawn, $\angle 1$ becomes an exterior angle of $\triangle ADP$, of which $\angle P$ and $\angle 2$ are nonadjacent interior angles.

PROOF:

Statements	Reasons
1. Draw $\overline{AD}$.	1. A straight line may be drawn between two points.
2. $m\angle P + m\angle 2 = m\angle 1$	2. The measure of an exterior angle of a triangle equals the sum of the measures of the nonadjacent interior angles.
3. $m\angle P = m\angle 1 - m\angle 2$	3. Subtraction Postulate.
4. $\angle 1 \stackrel{.}{=} \frac{1}{2}\widehat{AED},\ \angle 2 \stackrel{.}{=} \frac{1}{2}\widehat{AFD}$	4. An angle formed by a tangent and a chord is measured by one-half its intercepted arc.
5. $\angle P \stackrel{.}{=} \frac{1}{2}\widehat{AED} - \frac{1}{2}\widehat{AFD}$ or $\angle P \stackrel{.}{=} \frac{1}{2}(\widehat{AED} - \widehat{AFD})$	5. Substitution Postulate

9. If three angles of one triangle are congruent to three angles of another triangle, the triangles are similar.

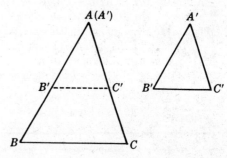

Given: $\triangle ABC$ and $\triangle A'B'C' \angle A \cong \angle A'$, $\angle B \cong \angle B', \angle C \cong \angle C'$

To Prove: $\triangle ABC \sim \triangle A'B'C'$

Plan: To prove the triangles similar, it must be shown that corresponding sides are in proportion. This is done by placing the triangles so that a pair of congruent angles coincide, and then repeating this so that another pair of congruent angles coincide.

PROOF:

Statements	Reasons
1. $\angle A \cong \angle A'$	1. Given
2. Place $\triangle A'B'C'$ on $\triangle ABC$ so that $\angle A'$ coincides with $\angle A$.	2. A geometric figure may be moved without change of its size or shape. Equal angles may be made to coincide.
3. $\angle B \cong \angle B'$	3. Given
4. $\overline{B'C'} \parallel \overline{BC}$	4. Two lines are parallel if their corresponding angles are congruent.
5. $\dfrac{A'B'}{AB} = \dfrac{A'C'}{AC}$	5. A line parallel to one side of a triangle divides the other two sides proportionately

(contd.)

PROOF:

Statements	Reasons
6. In like manner, by placing $\triangle A'B'C'$ on $\triangle ABC$ so that $\angle B'$ coincides with $\angle B$, show that $\dfrac{A'B'}{AB} = \dfrac{B'C'}{BC}$	6. Reasons 1 to 5.
7. $\dfrac{A'B'}{AB} = \dfrac{A'C'}{AC} = \dfrac{B'C'}{BC}$	7. Things (ratios) equal to the same thing are equal to each other.
8. $\triangle A'B'C' \sim \triangle ABC$	8. Two polygons are similar if their corresponding angles are congruent and their corresponding sides are in proportion.

10. If the altitude is drawn to the hypotenuse of a right triangle, then (*a*) the two triangles thus formed are similar to the given triangle and to each other, and (*b*) each leg of the given triangle is the mean proportional between the hypotenuse and the projection of that leg upon the hypotenuse.

Given: $\triangle ABC$ with a right angle at C, altitude $\overline{CD}$ to hypotenuse $\overline{AB}$
To Prove: (*a*) $\triangle ADC \sim \angle BDC \sim \triangle ABC$
　　　　　　(*b*) $c:a = a:p,\ c:b = b:q$
Plan: The triangles are similar since they have a right angle and a pair of congruent acute angles. The proportions follow from the similar triangles.

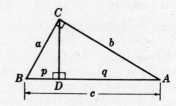

PROOF:

Statements	Reasons
1. $\angle C$ is a right angle.	1. Given
2. $\overline{CD}$ is the altitude to $\overline{AB}$.	2. Given
3. $\overline{CD} \perp \overline{AB}$	3. An altitude to a side of a triangle is perpendicular to that side.
4. $\angle CDB$ and $\angle CDA$ are right angles.	4. Perpendiculars form right angles.
5. $\angle A \cong \angle A,\ \angle B \cong \angle B$	5. Reflexive property
6. $\triangle ADC \sim \triangle ABC,\ \triangle BDC \sim \triangle ABC$	6. Right triangles are similar if an acute angle of one is congruent to an acute angle of the other.
7. $\triangle ADC \sim \triangle BDC$	7. Triangles similar to the same triangle are similar to each other.
8. $c:a = a:p,\ c:b = b:q$	8. Corresponding sides of similar triangles are in proportion.

11.   The square of the length of the hypotenuse of a right triangle equals the sum of the squares of the lengths of the other two sides.

Given: Right $\triangle ABC$, with a right angle at
C. Legs have lengths a and b, and
hypotenuse has length c.
To Prove: $c^2 = a^2 + b^2$
Plan: Draw $\overline{CD} \perp \overline{AB}$ and apply Theorem
10.

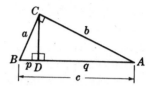

PROOF:

Statements	Reasons
1. Draw $\overline{CD} \perp \overline{AB}$.	1. Through an external point, a line may be drawn perpendicular to a given line.
2. $\dfrac{c}{a} = \dfrac{a}{p}, \dfrac{c}{b} = \dfrac{b}{q}$	2. If the altitude is drawn to the hypotenuse of a right triangle, either leg is the mean proportional between the hypotenuse and the projection of that leg upon the hypotenuse.
3. $a^2 = cp, \ b^2 = cq$	3. In a proportion, the product of the means equals the product of the extremes.
4. $a^2 + b^2 = cp + cq = c(p + q)$	4. If equals are added to equals, the sums are equal.
5. $c = p + q$	5. The whole equals the sum of its parts.
6. $a^2 + b^2 = c(c) = c^2$	6. Substitution Postulate

12.   The area of a parallelogram equals the product of the length of one side and the length of the altitude to that side.

Given: $\square ABCD$, length of base $\overline{AD} = b$,
length of altitude $\overline{BE} = h$
To Prove: Area of $ABCD = bh$
Plan: When a perpendicular is dropped to the
base, extended, a rectangle is formed
having the same base and altitude as the
parallelogram. By adding congruent
triangles to a common area, the
rectangle and parallelogram are proved
equal in area.

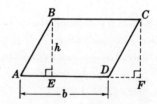

PROOF:

Statements	Reasons
1. Draw $\overrightarrow{CF} \perp \overrightarrow{AD}$ (extended).	1. Through an external point, a line may be drawn perpendicular to a given line.
2. $\overline{CF} \| \overline{BE}$	2. Segments perpendicular to the same line are parallel.
3. $\overline{BC} \| \overline{AD}$	3. Opposite sides of a parallelogram are parallel.
4. $\angle CFD$ and $\angle BEA$ are right angles.	4. Perpendiculars form right angles.
5. $BCFE$ is a rectangle.	5. A parallelogram having a right angle is a rectangle.

(contd.)

PROOF:

Statements	Reasons
6. $\overline{AB} \cong \overline{CD}$, $\overline{CF} \cong \overline{BE}$	6. Opposite sides of a parallelogram are equal.
7. $\triangle ABE \cong \triangle CFD$	7. Hy. leg $\cong$ hy. leg
8. Area(quadrilateral $BCDE$) = area(quadrilateral $BCDE$)	8. Reflexive property
9. Area ($\triangle ABE$) + area (quadrilateral $BCDE$) = area($\triangle CFD$) + area(quadrilateral $BCDE$) or area(rectangle $BCFE$) = area($\square ABCD$)	9. If equals are added to equals, the sums are equal.
10. Area of rectangle $BCFE = bh$	10. The area of a rectangle equals the product of the lengths of its base and altitude.
11. Area of $\square ABCD = bh$	11. Substitution Postulate

13. The area of a triangle is equal to one-half the product of the length of one side and the length of the altitude to that side.

Given: $\triangle ABC$, length of base $\overline{AC} = b$, length of altitude $\overline{BD} = h$

To Prove: Area of $\triangle ABC = \frac{1}{2}bh$

Plan: Drawing $\overline{BE} \parallel \overline{AC}$ and $\overline{EC} \parallel \overline{AB}$ forms a parallelogram having the same base and altitude as the triangle. Then the area of the triangle is half the area of the parallelogram.

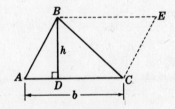

PROOF:

Statements	Reasons
1. Draw $\overline{BE} \parallel \overline{AC}$, $\overline{CE} \parallel \overline{AB}$.	1. Through an external point, a line may be drawn parallel to a given line.
2. $ABEC$ is a parallelogram with base b and altitude h.	2. A quadrilateral is a parallelogram if its opposite sides are parallel.
3. Area($\triangle ABC$) = $\frac{1}{2}$area ($\square ABEC$)	3. A diagonal divides a parallelogram into two congruent triangles.
4. Area($\square ABEC$) = bh	4. The area of a parallelogram equals the product of the lengths of its base and altitude.
5. Area of $\triangle ABC = \frac{1}{2}bh$	5. Substitution Postulate

14. The area of a trapezoid is equal to one-half the product of the length of the altitude and the sum of the lengths of the bases.

Given: Trapezoid $ABCD$, altitude $\overline{BE}$ with length h, base $\overline{AD}$ with length b, base $\overline{BC}$ with length b'.

To Prove: Area of $ABCD = \frac{1}{2}h(b + b')$

Plan: When a diagonal is drawn, the trapezoid is divided into two triangles having common altitude h and bases b and b'.

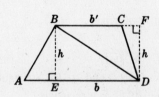

PROOF:

Statements	Reasons
1. Draw $\overline{BD}$.	1. A straight line may be drawn between two points.
2. Draw $\overline{DF} \perp \overrightarrow{BC}$ (extended).	2. Through an external point, a line may be drawn perpendicular to a given line.
3. $DF = BE = h$	3. Parallel lines are everywhere equidistant.
4. Area($\triangle ABD$) $= \frac{1}{2}bh$, Area($\triangle BCD$) $= \frac{1}{2}b'h$	4. The area of a triangle equals one-half the product of the lengths of its base and altitude.
5. Area of $ABCD = \frac{1}{2}bh + \frac{1}{2}b'h$ $= \frac{1}{2}h(b + b')$	5. If equals are added to equals, the sums are equal.

15. The area of a regular polygon is equal to one-half the product of its perimeter and apothem.

Given: Regular polygon $ABCDE \ldots$ having center O, apothem of length r, perimeter p

To Prove: Area of $ABCDE \ldots = \frac{1}{2}rp$

Plan: By joining each vertex to the center, congruent triangles are obtained, the sum of whose areas equals the area of the regular polygon.

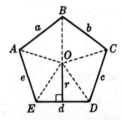

PROOF:

Statements	Reasons
1. Draw $\overline{OA}, \overline{OB}, \overline{OC}, \overline{OD}, \overline{OE}, \ldots$	1. A straight line segment may be drawn between two points.
2. r is the altitude of each triangle formed.	2. Apothems of a regular polygon are congruent.
3. Area of $\triangle AOB = \frac{1}{2}ar$ $\triangle BOC = \frac{1}{2}br$ $\triangle COD = \frac{1}{2}cr$ $\cdots\cdots\cdots\cdots$	3. The area of a triangle equals one-half the product of the length of its base and altitude.
4. Area of regular polygon $ABCDE \ldots$ $= \frac{1}{2}ar + \frac{1}{2}br + \frac{1}{2}cr + \cdots$ $= \frac{1}{2}r(a + b + c + \cdots)$	4. If equals are added to equals, the sums are equal.
5. $p = a + b + c + \cdots$	5. The whole equals the sum of its parts.
6. Area of $ABCDE \ldots = \frac{1}{2}rp$	6. Substitution Postulate

CHAPTER 17

Transformational Geometry

17.1 INTRODUCTION TO TRANSFORMATIONS

If you look back at the previous 16 chapters, you will notice that, while we have concentrated on different topics from chapter to chapter, all the material had one very important thing in common: The *positions* of all the geometric figures were *fixed*. In other words, when we considered a triangle such as △*ABC* in Fig. 17-1, we did not move it. Actually, there is an exception. See page 313, proof #9. In this chapter, we consider objects in geometry as they change position. These objects (such as triangles, lines, points, and circles) will move as a result of transformations of the plane:

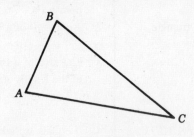

Fig. 17-1

DEFINITION 1:   *By a transformation of the plane, we mean a rule that assigns to each point in the plane a different point or the point itself.*

Note that *each* point in the plane is assigned to *exactly one point*. Points that are assigned to themselves are called *fixed* points. If point *P* is assigned to point *Q*, then we say the *image* of *P* is *Q*, and the image of *Q* is *P*.

17.2 REFLECTIONS

Imagine that a mirror is placed along line *m* in Fig. 17-2. What would be the image of point *S* in the mirror? How would you describe *S'*, the image of *S*? If we actually placed a mirror along *m*, we would see that the image of S lies on l, on the other side of m, and that the distance from S to O is equal to the distance from O to S' (see Fig. 17-3). We say that S' is the image of S under a reflection in line m. Notice that, under this reflection, O is the image of O.

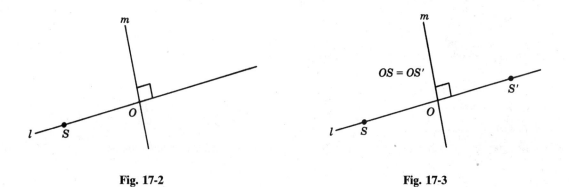

Fig. 17-2   Fig. 17-3

DEFINITION 2:   *A reflection in line m is a transformation of the plane having the property that the image of any point S not on m is S', where m is the perpendicular bisector of $\overline{SS'}$; the image of any point O on m is O itself.*

We write $R_m(S) = S'$ to mean S' is the image of S under the reflection in line m.

SOLVED PROBLEMS

17.1   IMAGE OF A POINT

Find the image of (*a*) A, (*b*) B, (*c*) C, (*d*) $\overline{AC}$, and (*e*) $\angle DAC$ under the reflection in line t indicated in Fig. 17-4.

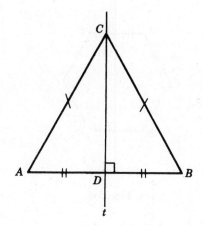

Fig. 17-4

Solutions

(a) *B*, because *t* is the perpendicular bisector of $\overline{AB}$

(b) *A*

(c) *C*, because *C* is on *t*

(d) $\overline{BC}$ (Why?)

(e) $\angle DBC$, because *D* and *C* are fixed, and $R_t(B) = A$

17.2 IMAGE OF A TRIANGLE

What is the image of $\triangle ABC$ in Fig. 17-4 under a reflection in line *t*?

Solution

We saw that $R_t(A) = B$, $R_t(B) = A$, and $R_t(C) = C$; thus, $\triangle ABC$ is its own image.

17.2A Line Symmetry

Notice that the images of angles are angles and the images of segments are segments under a reflection in a line. When a figure is its own image under a reflection in a line (like $\triangle ABC$ in Fig. 17-4), we say the figure has *line symmetry*.

DEFINITION 3: *A figure F exhibits line symmetry if there exists a line l such that the image of F under a reflection in line l is F itself. In this case, l is called a line of symmetry or an axis of symmetry.*

Notice that when a figure exhibits line symmetry, all its points are not necessarily fixed. In Fig. 17-4, only points *C* and *D* are fixed in triangle *ABC*.

SOLVED PROBLEMS

17.3 FINDING THE AXIS OF SYMMETRY

In Fig. 17-5, find all axes of symmetry for regular hexagon *ABCDEF*.

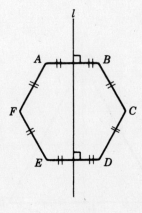

Fig. 17-5

Solution

$\overline{AD}$, $\overline{FC}$, $\overline{BE}$, and the indicated line *l* are all axes of symmetry. Find two others.

17.4   DISCOVERING LINE SYMMETRY
Which of the objects in Fig. 17-6 exhibit line symmetry?

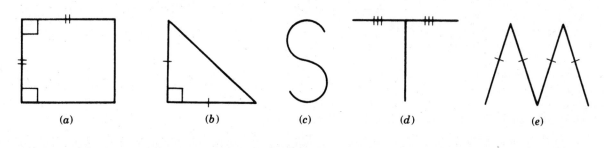

(a)   (b)   (c)   (d)   (e)

Fig. 17-6

Solution

All except (c).

17.2B   Point Symmetry

Not only can we transform the plane by reflections in a line, but we can also reflect in a point P. In Fig. 17-7, for example, we can reflect Q in the point P by finding the point Q' such that $QP = PQ'$.

DEFINITION 4:   *A reflection in the point P is a transformation of the plane such that the image of any point Q except P is Q', where $QP = PQ'$, and the image of P is P (i.e., P is fixed). If figure F is its own image under such a transformation, then we say F exhibits point symmetry.*

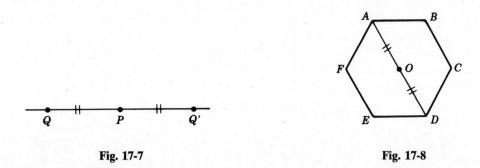

Fig. 17-7   **Fig. 17-8**

Figure 17-8 shows a regular hexagon $ABCDEF$, with $AO = OD$. Notice that A is the image of D under the reflection in O. We use the notation $R_O(A) = D$ and $R_O(D) = A$ to indicate that A and D are each other's images under a reflection in point O.

SOLVED PROBLEMS

17.5   FINDING IMAGES UNDER A REFLECTION IN A POINT
Referring to Fig. 17-8, find (a) $R_O(B)$; (b) $R_O(C)$; (c) $R_O(\overline{AD})$; (d) $R_O(\angle AOB)$; and (e) $R_O(ABCDEF)$.

Solutions

(a) E (b) F (c) $\overline{DA}$ (d) $\angle DOE$

(e)   Hexagon $DEFABC$. (Thus, $ABCDEF$ exhibits point symmetry.)

17.6   FINDING POINT SYMMETRY

Which of the following exhibit point symmetry?

(*a*)   Squares   (*c*)   Scalene triangles

(*b*)   Rhombuses   (*d*)   *S*

Solution

All except (*c*).

17.3   REFLECTIONS AND ANALYTIC GEOMETRY

Since points can change position in transformational geometry, analytic geometry is a particularly useful tool for these transformations. Recall that in analytic geometry, we deal extensively with the positions of points; being able to locate points and determine distances is of great help in exploring the properties of transformations.

SOLVED PROBLEMS

17.7   IMAGES UNDER REFLECTIONS (Fig. 17-9)

(*a*)   What is the image of point *A* under a reflection in the *x*-axis. The *y*-axis?

(*b*)   What is the image of *B* under a reflection in the *y*-axis?

(*c*)   What is the image of *O* under a reflection in the point *O*?

(*d*)   What is the image of *B* under a reflection in the line *y* = *x*?

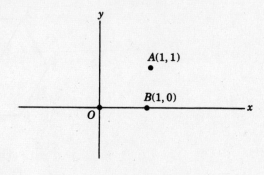

Fig. 17-9

(*e*)   What is the image of *A* under a reflection in the line *x* = −1?

(*f*)   What is the image of △*AOB* under a reflection in the *y*-axis? Under a reflection in *O*?

Solutions

(*a*)   Point *A′* in Fig. 17-10 is the image of *A* under a reflection in the *x*-axis; the coordinates of *A′* are (1, −1). Point A″ is the image of *A* under a reflection in the *y*-axis; *A″* = (−1, 1).

(*b*)   Point *B′* in Fig. 17-11 is the image of *B* under a reflection in the *y*-axis. Its coordinates are (−1, 0).

(*c*)   Point *O* is a fixed point. The point in which we reflect is always fixed.

(*d*) $R_l(B) = B'(0, 1)$ in Fig. 17-12. Notice that line *l* is the perpendicular bisector of $\overline{BB'}$.

(*e*) $R_m(A) = A'(-3, 1)$ in Fig. 17-13. Note that *m* is the perpendicular bisector of $\overline{AA'}$.

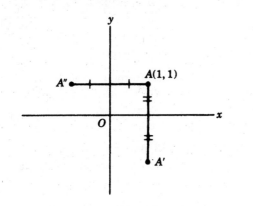

Fig. 17-10

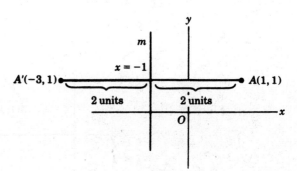

Fig. 17-11

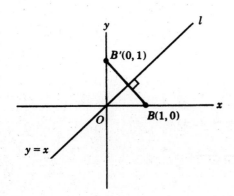

Fig. 17-12

Fig. 17-13

(f) The image of $\triangle AOB$ under a reflection in the y-axis is $\triangle A'B'O$ in Fig. 17-14(a), where $A' = (-1, 1)$, $B' = (-1, 0)$, and $O = (0, 0)$. The image under a reflection in the origin is $\triangle A''B''O$ in Fig. 17-14(b), where $A'' = (-1, -1)$, $B'' = (-1, 0)$, and $O = (0, 0)$.

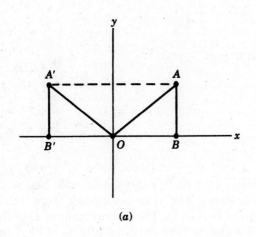

(a)

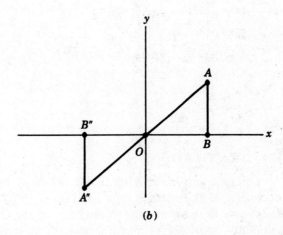

(b)

Fig. 17-14

17.3A   Patterns in Reflections

We can observe several patterns in the results of the last solved problem:

1. The distance from A' to B' in Fig. 17-14(a) equals the distance from A to B. In other words, distance is *preserved* under a reflection. Observe that measures of angles are also preserved. In other words, $m\angle BAO = m\angle B'A'O$ in Fig. 17-14(a), and that property appears to hold for other reflections. As you will see, other properties are preserved as well.

2. Under a reflection in the x-axis, the point (a, b) moves to $(a, -b)$; under a reflection in the y-axis, (a, b) moves to $(-a, b)$; and under a reflection in the origin, (a, b) moves to $(-a, -b)$. These patterns hold only for these reflections.

SOLVED PROBLEMS

17.8   MORE IMAGES UNDER REFLECTIONS
In Fig. 17-15, find:

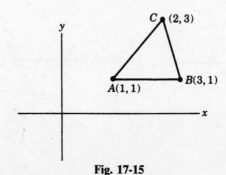

Fig. 17-15

(a) The reflection of C in the y-axis

(b) The reflection of B in the origin

(c) The reflection of $\angle CAB$ in the x-axis

Solutions

(a) $(-2, 3)$

(b) $(-3, -1)$

(c) $\angle C'A'B'$, where $C' = (2, -3)$, $A' = (1, -1)$, and $B' = (3, -1)$

17.4   TRANSLATIONS

Let us transform $\triangle ABC$ in Fig. 17-16(a) by adding 1 to each x-coordinate and 2 to each y-coordinate. The result is shown in Fig. 17-16(b). Notice that $\triangle ABC$ does not change shape, but it does move in the plane, in the direction of ray $\vec{OD}$, where $D = (1, 2)$. The x-coordinate of D is the "amount" by which the x-coordinates of the triangle are shifted, and the y-coordinate of D is the "amount" by which the y-coordinates are shifted. We call this kind of transformation a *translation*.

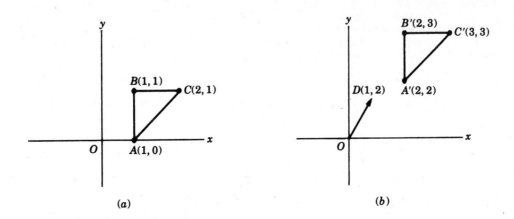

Fig. 17-16

DEFINITION 5: *A translation is a transformation of the plane such that the image of every point* (a, b) *is the point* $(a + h, b + k)$, *where h and k are given.*

A translation has the effect of moving every point the *same distance* in the *same direction*. We use the notation $T_{(h,k)}(a, b)$ to mean the image of (a, b) under a translation of h units in the x-direction and k units in the y-direction.

As in a reflection, distance and angle measure are preserved in a translation.

SOLVED PROBLEMS

17.9   FINDING THE IMAGE OF A POINT

Find $T_{(-1,1)}(1, 4)$ and $T_{(-1,1)}(-1, 2)$.

Solutions

$$T_{(-1,1)}(1, 4) = (1 + (-1), 4 + 1) = (0, 5)$$
$$T_{(-1,1)}(-1, 2) = (-1 + (-1), 2 + 1) = (-2, 3)$$

Notice in Fig. 17-17 that $(1, 4)$ and $(-1, 2)$ are translated the same number of units in the same direction by the same translation T.

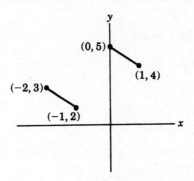

Fig. 17-17

17.10 FINDING THE IMAGE OF A TRIANGLE

Find the image of $\triangle ABC$ under the translation $T_{(1,2)}$, where $A = (0,0)$, $B = (1,1)$, and $C = (1,0)$

Solutions

$T_{(1,2)}(0,0) = (1,2)$, $T_{(1,2)}(1,1) = (2,3)$, and $T_{(1,2)}(1,0) = (2,2)$. Hence the image of $\triangle ABC$ is $\triangle A'B'C'$ in Fig. 17-18. All points are translated along ray $\overrightarrow{AA'} = \overrightarrow{OA'}$, where $A'(1,2)$ has the coordinates of the translation.

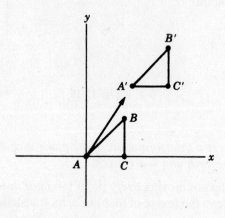

Fig. 17-18

17.11 FINDING THE IMAGES FROM ANOTHER IMAGE

Under a certain translation, $T(5,2) = (7,1)$. Find $T(-3,6)$ under the same translation.

Solutions

We have $T_{(h,k)}(5,2) = (7,1)$. Thus $5 + h = 7$, or $h = 2$; and $2 + k = 1$, or $k = -1$.
Then $T_{(2,-1)}(-3,6) = (2 + (-3), -1 + 6) = (-1,5)$.

17.12 FINDING VARIOUS IMAGES UNDER TRANSLATION

(a) Find $T_{(-1,0)}(6,2)$.

(b) Find h and k if $T_{(h,k)}(1,7) = (0,0)$.

(c) Find the image of square $ABCD$ under the translation $T_{(1,1)}$, where $A = (0,0)$, $B = (1,0)$, $C = (0,1)$, and $D = (1,1)$.

(d) Find $T_{(h,k)}(1,6)$ if $T_{(h,k)}(4,1) = (0,-7)$.

(e) Find all fixed points under $T_{(-1,4)}$.

Solutions

(a) $T(6,2) = (6 + (-1), 2 + 0) = (5,2)$

(b) $h = 0 - 1 = -1; k = 0 - 7 = -7$

(c) $A'B'C'D'$, where $A' = (1,1)$, $B' = (2,1)$, $C' = (1,2)$, and $D' = (2,2)$

(d) $h = 0 - 4 = -4$ and $k = -7 - 1 = -8$, so $T(1,6) = (-3,-2)$

(e) Only $T_{(0,0)}$ has fixed points. Any other translation, including $T_{(-1,4)}$, has none.

17.13 **FINDING IMAGES OF FIGURES**

Let $A = (1,1)$, $B = (2,2)$, and $C = (3,1)$. Find the image under $T_{(2,-1)}$, of (*a*) $\overline{AB}$, (*b*) $\triangle ABC$, and (*c*) $\angle CBA$.

Solutions

(*a*) $\overline{A'B'}$, where $A' = (3,0)$ and $B' = (4,1)$

(*b*) $\triangle A'B'C'$, with $C' = (5,0)$

(*c*) $\angle C'B'A'$

17.5 ROTATIONS

Consider square *ABCD* in Fig. 17-19(*a*). Suppose we were to rotate that square counterclockwise 90° about *P*, as shown by the arrow. (Imagine that the square is separate from the page, but held to it by a pin through point *P*.) Then:

The image of *B* would be *A*.   The image of *C* would be *B*.
The image of *D* would be *C*.   The image of *A* would be *D*.

Now consider point *S* in Fig. 17-19(*b*). We can rotate it counterclockwise by, say, 50° about *P*, as if it were one end of a ruler that was nailed to the page at *P*. The image of *S'* is *S'* in the diagram.

In both these rotations, the segment from *P* to the point being rotated is congruent to the segment from *P* to the image of that point.

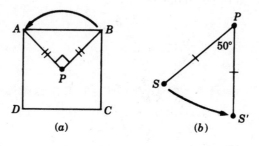

(*a*)				(*b*)

Fig. 17-19

DEFINITION 6:   *A rotation through an angle of measure* θ *degrees about a point P is a transformation of the plane such that the image of P is P and, for any point B ≠ P, the image of B is B', where* $m\angle BPB' = \theta$ *and* $\overline{BP} \cong B'P$.

Figure 17-20 shows *P*, *B*, *B'*, and θ. If θ > 0, the rotation is counterclockwise. If θ < 0, the rotation is clockwise.

We use the notation $\text{Rot}_{(P,\theta)}(B)$ to mean the rotation of point *B* about point *P* through θ°. Segment lengths and measures of angles are preserved under rotations.

Let us rotate point *B* in Fig. 17-21 through 180° about *O*. The image of *B* is $B'(-2, -2)$. Notice that this is also the image of *B* under the reflection in *O*.

Now let us rotate point *B* through 90° about *O*. The image here is $B''(-2, 2)$. Notice that this is the image of *B* under the reflection in the *y*-axis.

Be careful: These two similarities of transformations are coincidences. They come about because we are rotating 90° and 180° *about the origin*. Do not generalize beyond these cases! Note, though, that these coincidences yield the following formulas:

$$\text{Rot}_{(O,90°)}(a,b) = (-b,a) \quad \text{and} \quad \text{Rot}_{(O,180°)}(a,b) = (-a,-b)$$

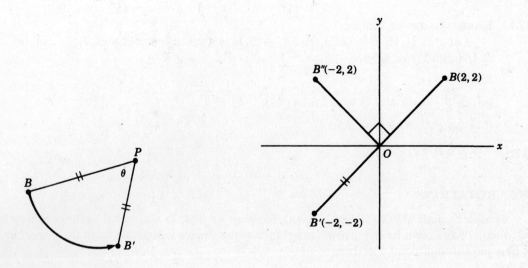

Fig. 17-20   Fig. 17-21

SOLVED PROBLEMS

17.14 FINDING THE ROTATION OF A POINT

Let $A = (1, 3)$ and $B = (2, 1)$, and find (a) $\text{Rot}_{(O, 90°)}(A)$, (b) $\text{Rot}_{(O, 90°)}(B)$, and (c) the image of $\overline{AB}$ under a rotation of 90° about O.

Solutions

(a) $\text{Rot}_{(O, 90°)}(a, b) = (-b, a) = (-3, 1)$

(b) $\text{Rot}_{(O, 90°)}(2, 1) = (-1, 2)$

(c) The image is $\overline{A'B'}$, where $A' = (-3, 1)$ and $B' = (-1, 2)$, as shown in Fig. 17-22.

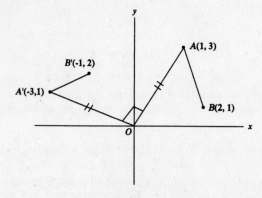

Fig. 17-22

17.15 FINDING THE IMAGE OF A TRIANGLE

(a) Find the image of $\triangle ABC$ under a rotation of 180° about O, if $A = (1, 3)$, $B = (2, 1)$, and $C = (1, 1)$.

(b) Find the image of $\angle BAC$.

Solutions

(a) $\text{Rot}_{(O, 180°)}(A) = (-1, -3) = A'$
$\text{Rot}_{(O, 180°)}(B) = (-2, -1) = B'$
$\text{Rot}_{(O, 180°)}(C) = (-1, -1) = C'$
The image of $\triangle ABC$ is $\triangle A'B'C'$.

(b) The image of $\angle BAC$ is $\angle B'A'C'$.

17.5A Symmetry of Rotation

The image of the square in Fig. 17-19 under a rotation of 90° is the square itself. This is also true for a rotation of −90°, or 180°, and so on.

SOLVED PROBLEMS

17.16 DETERMINING ROTATIONAL SYMMETRY (Fig. 17-23)

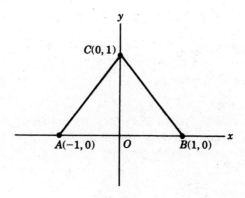

Fig. 17-23

(a) Find $\text{Rot}_{(O, 45°)}(O)$.

(b) Find $\text{Rot}_{(O, 90°)}$ of A, B, and C.

(c) Find $\text{Rot}_{(O, -90°)}$ of A, B, and C.

(d) Find the image of $\triangle ABC$ under $\text{Rot}_{(O, -90°)}$.

(e) Does $\triangle ABC$ exhibit rotational symmetry?

Solutions

(a) $\text{Rot}(O) = O$

(b) $\text{Rot}(A) = A'(0, -1)$; $\text{Rot}(B) = B'(0, 1)$; $\text{Rot}(C) = C'(-1, 0)$

(c) $\text{Rot}(A) = A''(0, 1)$; $\text{Rot}(B) = B''(0, -1)$; $\text{Rot}(C) = C''(1, 0)$

(d) $\text{Rot}(\triangle ABC) = A''B''C''$

(e) No, because it is not its own image for any rotation except one of 360°.

17.6 DILATIONS

Suppose we blew a balloon up most of the way and traced its outline and then blew it up all the way and traced it again. The outlines might look like those in Fig. 17-24. Although the balloon has changed size from (a) to (b) its shape has not changed. Notice that if C is on $\overset{\frown}{AB}$, then its image C' is on $\overset{\frown}{A'B'}$. Such a transformation in the plane is called a *dilation* (or dilatation). The "reverse" transformation is also a dilation: The balloon could be reduced in size in a transformation from that in (b) to that in (a).

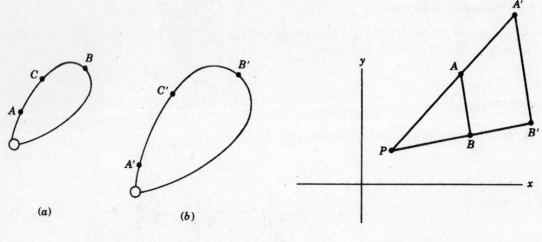

(a)   (b)

Fig. 17-24   **Fig. 17-25**

DEFINITION 7: *Given a point P in the plane and a positive number n, a transformation of the plane having the following properties is called a dilation of n, and P is called the center of dilation: Point P is fixed, and for any point Q, the image of Q is the point Q' such that $PQ' = (n)(PQ)$ and $\vec{PQ}$ and $\vec{PQ'}$ are identical rays.*

The point Q' is usually denoted $D_n(Q)$.

Figure 17-25 shows a dilation in which $n = 2$ and the center of dilation is P. Hence, $D_2(A) = A'$, $D_2(B) = B'$, and $D_2(P) = P$. In addition, because $n = 2$, we know that $PA' = 2PA$ and $PB' = 2PB$.

Several properties of dilations are evident in Fig. 17-25:

1. Dilations *do not* preserve distance.

2. The image of a figure is similar to the figure under a dilation. In Fig. 17-25, $\triangle PAB \sim PA'B'$.

3. Angles *are* preserved under dilations (because of item 2 above).

When the center of a dilation is $O = (0,0)$, we can find the images of points very easily: $D_n(x, y) = (nx, ny)$.

SOLVED PROBLEMS

17.17 FINDING THE DILATION OF A TRIANGLE

Find the image of $\triangle ABC$ in Fig. 17-26 under a dilation of $n = \frac{1}{2}$ with center of dilation at $(0,0)$.

Solution

$D_{1/2}(1,1) = (\frac{1}{2},\frac{1}{2}) = B'$ as shown in Fig. 17-26. Also, $D_{1/2}(1,0) = (\frac{1}{2},0) = A'$ and $D_{1/2}(2,1) = (1,\frac{1}{2}) = C'$. Then $\triangle B'A'C'$ is the image of $\triangle BAC$, and $\triangle B'A'C' \sim \triangle BAC$. Note that the image here is smaller than the original triangle because $n = \frac{1}{2}$.

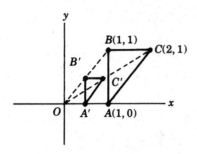

Fig. 17-26

17.18 FINDING AN UNKNOWN n

Given that $D_n(8,0) = (1,0)$, find n for a dilation in which $(0,0)$ is the center of dilation.

Solution

Since the origin is the center of dilation, $(1,0) = (8n, 0n)$. Therefore, $8n = 1$ and $n = \frac{1}{8}$.

17.19 DILATING A SQUARE

Draw a square $ABCD$ in the coordinate plane such that $A = (1,1)$, $B = (1,2)$, $C = (2,1)$, and $D = (2,2)$. Then,

(a) With O as the center of dilation, find the image of $ABCD$ under a dilation with $n = \frac{1}{3}$.

(b) Find the midpoint M of $\overline{AB}$ and the midpoint M' of $\overline{A'B'}$.

(c) Find $D_{1/3}(M)$.

Solutions

(a) For a dilation with center $(0,0)$ and $n = \frac{1}{3}$, we have $D(x,y) = (\frac{1}{3}x, \frac{1}{3}y)$. The image of $ABCD$ is $A'B'C'D'$, where $A' = (\frac{1}{3}, \frac{1}{3})$, $B' = (\frac{1}{3}, \frac{2}{3})$, $C' = (\frac{2}{3}, \frac{1}{3})$, and $D' = (\frac{2}{3}, \frac{2}{3})$.

(b) $M = (\frac{1}{2}(1+1), \frac{1}{2}(1+2)) = (1, \frac{3}{2})$, and $M' = (\frac{1}{3}(1), \frac{1}{3}(\frac{3}{2})) = (\frac{1}{3}, \frac{1}{2})$.

(c) $D(M) = M'$.

17.7 PROPERTIES OF TRANSFORMATIONS

We are now in a position to summarize the properties of transformations. In particular, we are interested in what is preserved under each kind of transformation.

1. Reflections preserve (a) distance, (b) angle measure, (c) midpoints, (d) parallelism, and (e) collinearity.

2. Translations preserve these same five properties, (a) through (e).

3. Rotations preserve all five properties as well.

4. Dilations preserve all except distance, that is, (b) through (e).

Supplementary Problems

1. Find the image of each of the following under the reflection in line t in Fig. 17-27: (a) point D; (b) point C; (c) point B; (d) $\overline{AC}$. (17.1)

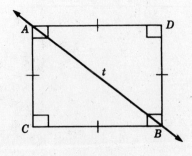

Fig. 17-27

2. Find the image of square $ACBD$ under the reflection in line t in Fig. 17-27. (17.2)

3. Is it true or false that every circle is its own image under a reflection in a diameter? (17.2)

4. Find all axes of symmetry for the square in Fig. 17-27. (17.2)

5. Give (or draw) an example of a five-sided polygon that does not exhibit line symmetry. (17.4)

6. Explain why each figure in Fig. 17-28 exhibits line symmetry. (17.4)

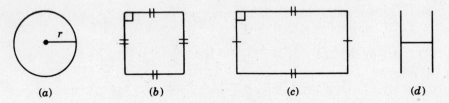

(a)   (b)   (c)   (d)

Fig. 17-28

7. In Fig. 17-29, find (a) $R_O(B)$; (b) $R_O(A)$; (c) $R_O(O)$; (d) $R_O(\triangle AOB)$. (17.5)

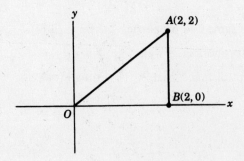

Fig. 17-29

8. In Fig. 17-30, find   (17.5, 17.8)

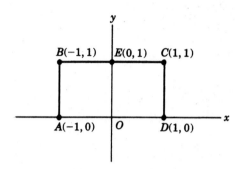

Fig. 17-30

(*a*)   The image of E under a reflection in the y-axis

(*b*)   The image of B under a reflection in the y-axis

(*c*)   The image of $\overline{AB}$ under a reflection in the y-axis

(*d*)   The image of $\overline{BC}$ under a reflection in the x-axis

(*e*)   The image of $ABCD$ under a reflection in the x-axis

9.   Find (*a*) $T_{(1,3)}(2,8)$; (*b*) $T_{(1,3)}(6,5)$; (*c*) $T_{(1,3)}(0,0)$; (*d*) $T_{(1,3)}(1,1)$.   (17.9)

10.   Find the image of rectangle $ABCD$ in Fig. 17-30 under the translation $T_{(3,6)}$.   (17.10)

11.   Under a particular translation, $T(3,4) = (0,0)$. Find $T(-8, -6)$ under that same translation.   (17.11)

12.   Find (*a*) $T_{(4,3)}(0, -6)$; (*b*) $T_{(h,k)}(3,7)$; (*c*) $T_{(h,k)}(e,f)$; (*d*) $T_{(h,k)}(4,1)$ if $T_{(h,k)}(1,1) = (2,2)$.   (17.12)

13.   In Fig. 17-31, find (*a*) $T_{(0,0)}(\overline{EF})$; (*b*) $T_{(1,0)}(\overline{EF})$; (*c*) $T_{(0,1)}(\overline{EF})$; (*d*) $T_{(0,0)}(\triangle OEF)$.   (17.13)

14.   Let $x = (4,1)$ and $y = (0,3)$, and find (*a*) $\text{Rot}_{(O,90°)}(x)$; (*b*) $\text{Rot}_{(O,90°)}(y)$; (*c*) the image of $\overline{yx}$ under a rotation of 90° about O.   (17.14)

15.   Find the image of $\triangle EOF$ in Fig. 17-31 under a rotation of 180° about O.   (17.15)

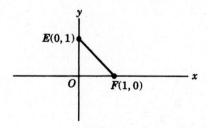

Fig. 17-31

16. In Fig. 17-32, find (17.16)

    (*a*) $\text{Rot}_{(O,90°)}(A)$ and $\text{Rot}_{(O,90°)}(B)$

    (*b*) $\text{Rot}_{(O,-90°)}(A)$ and $\text{Rot}_{(O,90°)}(C)$

    (*c*) The image of $OABC$ under $\text{Rot}_{(O,-90°)}$

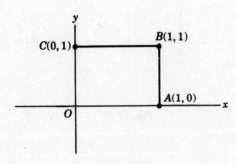

Fig. 17-32

17. Find (*a*) $D_{1/3}(-1,3)$, (*b*) $D_{1/2}(5,-3)$, (*c*) $D_4(0,0)$, and (*d*) $D_5(1,6)$, where the center of dilation in each case is the origin. (17.17)

18. If the center of a dilation D_n is $(0,0)$ and $D_n(3,6) = (5,10)$, find n and $D_n(0,-7)$. (17.18)

19. For A and B as given in Fig. 17-33 and dilations with centers at $(0,0)$, find (18.19)

    (*a*) The image of $\triangle OAB$ under a dilation with $n = \frac{1}{2}$

    (*b*) The image of the midpoint of $\overline{AB}$ under a dilation with $n = 3$

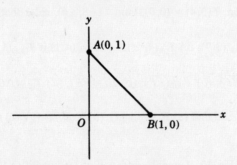

Fig. 17-33

Formulas for Reference

Angle Formulas

1.	Complement of $a°$	1.	$c = 90° - a°$
2.	Supplement of $a°$	2.	$s = 180° - a°$
3.	Sum of measures of angles of a triangle	3.	$S = 180°$
4.	Sum of measures of angles of a quadrilateral	4.	$S = 360°$
5.	Sum of measures of exterior angles of an n-gon	5.	$S = 360°$
6.	Sum of measures of interior angles of an n-gon	6.	$S = 180°(n - 2)$
7.	Measure of each interior angle of an equiangular or regular n-gon	7.	$S = \dfrac{180°(n - 2)}{n}$
8.	Measure of each exterior angle of an equiangular or regular n-gon	8.	$S = \dfrac{360°}{n}$
9.	Measure of central $\angle O$ intercepting an arc of $a°$	9.	$m\angle O = a°$
10.	Measure of inscribed $\angle A$ intercepting an arc of $a°$	10.	$m\angle A = \frac{1}{2}a°$
11.	Measure of $\angle A$ formed by a tangent and a chord and intercepting an arc of $a°$	11.	$m\angle A = \frac{1}{2}a°$
12	Measure of $\angle A$ formed by two intersecting chords and intercepting arcs of $a°$ and $b°$	12	$m\angle A = \frac{1}{2}(a° + b°)$
13.	Measure of $\angle A$ formed by two intersecting tangents, two intersecting secants, or an intersecting tangent and secant and intercepting arcs of $a°$ and $b°$	13.	$m\angle A = \frac{1}{2}(a° - b°)$
14.	Measure of $\angle A$ inscribed in a semicircle	14.	$m\angle A = 90°$
15.	Opposite $\angle s A$ and B of an inscribed quadrilateral	15.	$m\angle A = 180° - m\angle B$

Area Formulas

1.	Area of a rectangle	1.	$K = bh$
2.	Area of a square	2.	$K = s^2$, $\quad K = \frac{1}{2}d^2$
3.	Area of a parallelogram	3.	$K = bh$, $\quad K = ab \sin C$
4.	Area of a triangle	4.	$K = \frac{1}{2}bh$, $\quad K = \frac{1}{2}ab \sin C$
5.	Area of a trapezoid	5.	$K = \frac{1}{2}h(b + b')$, $\quad K = hm$
6.	Area of an equilateral triangle	6.	$K = \frac{1}{4}s^2\sqrt{3}$, $\quad K = \frac{1}{3}h^2\sqrt{3}$
7.	Area of a rhombus	7.	$K = \frac{1}{2}dd'$
8.	Area of a regular polygon	8.	$K = \frac{1}{2}pr$
9.	Area of a circle	9.	$K = \pi r^2$, $\quad K = \frac{1}{4}\pi d^2$
10.	Area of a sector	10.	$K = \dfrac{n}{360}(\pi r^2)$
11.	Area of a minor segment	11.	$K = $ area of sector $-$ area of triangle

Circle Intersection Formulas

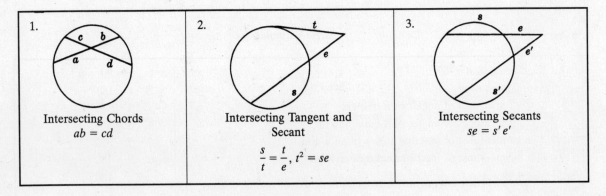

1. Intersecting Chords
$ab = cd$

2. Intersecting Tangent and Secant
$\dfrac{s}{t} = \dfrac{t}{e}, t^2 = se$

3. Intersecting Secants
$se = s'e'$

Right-Triangle Formulas

1.		Pythagorean Theorem	1. $c^2 = a^2 + b^2$
2.		Leg opposite 30° angle Leg opposite 45° angle Leg opposite 60° angle	2. $b = \frac{1}{2}c$ $b = \frac{1}{2}c\sqrt{2}, b = a$ $a = \frac{1}{2}c\sqrt{3}, a = b\sqrt{3}$
3.		Altitude of equilateral triangle Side of equilateral triangle	3. $h = \frac{1}{2}s\sqrt{3}$ $s = \frac{2}{3}h\sqrt{3}$

(contd.)

Right-Triangle Formulas

4. *(figure: square with diagonal d, side s, $45°$)*	Side of square Diagonal of square	4. $s = \frac{1}{2}d\sqrt{2}$ $d = s\sqrt{2}$
5. *(figure: triangle C, B, A with altitude h to D, segments a, b, p, q, c)*	Altitude to hypotenuse Leg of right triangle	5. $\frac{p}{h} = \frac{h}{q}$, $h^2 = pq$, $h = \sqrt{pq}$ $\frac{c}{a} = \frac{a}{p}$, $a^2 = pc$, $a = \sqrt{pc}$ $\frac{c}{b} = \frac{b}{q}$, $b^2 = qc$, $b = \sqrt{qc}$

Coordinate-Geometry Formulas

1. *(figure: line through $P_1(x_1, y_1)$, $M(x_M, y_M)$, $P_2(x_2, y_2)$)*	Midpoint M Distance P_1P_2 Slope of $\overrightarrow{P_1P_2}$	1. $x_M = \dfrac{x_1 + x_2}{2}$, $y_M = \dfrac{y_1 + y_2}{2}$ $d = \sqrt{(x_2 - x_1)^2 + (y_2 - y_1)^2}$ $m = \dfrac{y_2 - y_1}{x_2 - x_1}$, $m = \dfrac{\Delta y}{\Delta x}$, $m = \tan i$
2. *(figure: parallel lines L_1, L_2 and perpendicular L')*	Slopes of parallels, L_1 and L_2 Slopes of perpendiculars, L_1 and L'	2. Same slope, m $mm' = -1$ $m' = -\dfrac{1}{m}$, $m = -\dfrac{1}{m'}$
3. *(figure: lines L_1 parallel to x-axis, L_2 parallel to y-axis)*	Equation of L_1, parallel to x-axis Equation of L_2, parallel to y-axis	3. $y = k'$ $x = k$
4. *(figure: parallel lines L_1, L_2, L_3 through (x_1, y_1))*	Equation of L_1 with slope m and y-intercept b Equation of L_2 with slope m passing through the origin Equation of L_1 with x-intercept a and y-intercept b Equation of L_3 with slope m and passing through (x_1, y_1)	4. $y = mx + b$ $y = mx$ $\dfrac{x}{a} + \dfrac{y}{b} = 1$ $y - y_1 = m(x - x_1)$
5. *(figure: circle centered at origin with radius r)*	Equation of circle with center at origin and radius r	5. $x^2 + y^2 = r^2$

Introduction to the Graphing Calculator

In this text, a number of opportunities to utilize calculators have been provided to the reader. In virtually all these cases, a traditional hand-held calculator would be sufficient, and earlier in the text a generic version of such a calculator was presented. However, hand-held algebraic-entry graphing calculators have become quite prevalent, and although many of the applications of such calculators are well above the level of the mathematics in this book, it would be worthwhile to introduce these devices at this point. The goal here is to make you familiar with such calculators, and to make you at ease with them. In that way, when you engage in more advanced mathematics (such as college algebra, trigonometry, precalculus, and the calculus), it will be easier for you to use graphing calculators.

First, what does such a calculator look like? An example is indicated in Fig. A-1. This diagram illustrates the keyboard of the Hewlett-Packard HP 38G calculator. While there are other such calculators available to you on the market, the author of this text has found that this particular one offers the easiest retrieval of answers, and meshes particularly well with the needs of the mathematics student at both the precalculus levels and more advanced levels as well. In this Appendix you will be presented with a brief overview of the calculator, and of its use in solving equations and graphing simple functions.

The reader should turn the calculator on by pressing the $\boxed{\text{ON}}$ key. Note that the $\boxed{\text{ON}}$ key is also the $\boxed{\text{CANCEL}}$ key. Also, if you wish to lighten or darken your calculator's screen, simply hold down the $\boxed{\text{ON}}$ key while you press the $\boxed{-}$ or $\boxed{+}$ keys. (*Note*: Please refer to the *Reference Manual* provided with the HP 38G for a more detailed description of the many uses of the calculator.)

The HOME screen is the main area in which you will work. Press $\boxed{\text{HOME}}$ to find this area. Note that the HOME screen is divided into two main parts: the large rectangular area is the space in which entries and answers are indicated. The smaller rectangular area is the Editline. For example, if you press the keystrokes

$$\boxed{7}\ \boxed{5}\ \boxed{\text{ENTER}}$$

338

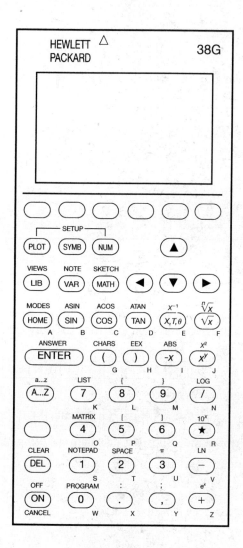

Fig. A-1
(Reprinted with the permission of the Hewlett-Packard Company.)

then you will find the following on the HOME screen:

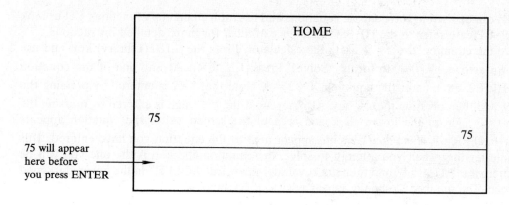

If you press the keystrokes

$$\boxed{7}\ \boxed{5}\ \boxed{/}\ \boxed{5}\ \boxed{\text{ENTER}}$$

then you will find the following on the HOME screen:

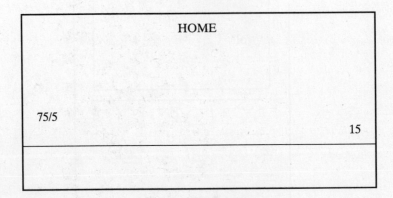

Notice that the HP 36G uses algebraic-entry notation. Thus, to perform a calculation, the operations required are first pressed, and then the $\boxed{\text{ENTER}}$ key is pressed. Again, if you wish to find $689 - 231$, press the keystrokes

$$\boxed{6}\ \boxed{8}\ \boxed{9}\ \boxed{-}\ \boxed{2}\ \boxed{3}\ \boxed{1}\ \boxed{\text{ENTER}}$$

To find 5 squared (5^2), press the keystrokes

$$\boxed{5}\ \boxed{x^y}\ \boxed{2}\ \boxed{\text{ENTER}}$$

The reader should perform the following calculations using the HP 36G or similar graphing calculator:

1. $483 + 286$.

2. $47 - 81$.

3. $843 * 35$.

4. $75/21$.

5. $45 x^y 2$, or $(45)^2$.

6. $\sqrt{15}$.

7. The reciprocal of 21.

8. The absolute value of -45.

The HP 38G can be used to solve simple equations and to graph elementary functions. Let us look at some examples. Please refer to the HP 38G *Reference Manual* for more detailed instructions.

Let us solve the equation $X - 2 = 9$ using the calculator. Press the $\boxed{\text{LIB}}$ (Library) key, and use the up and down arrows, $\boxed{\blacktriangledown}\ \boxed{\blacktriangle}$, to locate "Solve." Press $\boxed{\text{ENTER}}$. At any one of the equation lines (marked E1, E2, etc.), enter the equation $X - 2 = 9$. Note that "X" is entered by pressing the $\boxed{\text{"A} \dots \text{Z"}}$ key, and then pressing the $\boxed{\cdot}$ key. Also note that the "=" sign is entered by pressing the key under the "=" sign on the lower, darkened area of the screen. After the equation appears correctly, press $\boxed{\text{ENTER}}$. A check mark should appear next to the equation you have entered. This check mark indicates that when you attempt to solve, the equation checked is the one you will be solving. Next, press the $\boxed{\text{NUM}}$ key, and then the key under the word "SOLVE" in the lower, darkened area of the screen. The number 11 should appear next to the symbol X.

The reader should try to solve the equation $2X - 11 = 13$ using the HP 38G and the following series of steps: Press $\boxed{\text{LIB}}$, use the arrow keys to scroll to Solve, press $\boxed{\text{ENTER}}$, go to any of the "E" lines, and enter the equation $2X - 11 = 13$. Press $\boxed{\text{ENTER}}$. Press the key under the $\boxed{\checkmark\text{CHK}}$ box on the lower, darkened area of the screen. Press the $\boxed{\text{NUM}}$ key and then press the key under the word SOLVE in the lower, darkened area of the screen. You should see the answer 12 appear next to the symbol X.

The reader should solve each of the following using the HP 38G or similar graphing calculator:

1. $2X - 15 = 26$.

2. $3X = 4X + 4$.

3. $5Y = 6Y = 12.5$.

Let us now investigate the graph of the equation $Y = X - 4$. Press the $\boxed{\text{LIB}}$ key, and use the arrow keys to locate "Function." Press $\boxed{\text{ENTER}}$, and at any one of the "F" lines enter the equation $X - 4$. Make certain that you press $\boxed{\text{ENTER}}$. Note that you may enter the "X" by pressing the key under the symbol X in the lower, darkened area of the screen. Make certain that the equation you have entered is "checked" and then press the $\boxed{\text{PLOT}}$ key. The graph of the line will appear on a set of coordinate axes. Consult the *Reference Manual* for more details concerning the use of the HP 38G for graphing purposes.

The reader should try to graph the equation $Y = 2X - 11$ using the HP 38G and the following series of steps. Press $\boxed{\text{LIB}}$, use the arrow keys to scroll to Function, press $\boxed{\text{ENTER}}$, go to any of the "F" lines, and enter $2X - 11$. Press $\boxed{\text{ENTER}}$. Make certain that the expression is "checked," and that no other expressions are "checked." All expressions checked will be plotted when you press $\boxed{\text{PLOT}}$. Now press $\boxed{\text{PLOT}}$. The graph will appear on the coordinate axes on the calculator's screen.

The reader should graph each of the following using the HP 38G or similar graphing calculator:

1. $Y = 2X - 5$.

2. $Y = -5X = 13$.

3. $2Y = 3X = 7$. (*Hint*: Divide both sides of the equation by 2.)

Answers to Supplementary Problems

CHAPTER 1

1. (a) $3n + 8n = 11n$; (b) $10n - \frac{1}{2} = 9\frac{1}{2}n$; (c) $p = 3s$; (d) $A = ss$

2. (a) $p = 6s$; (b) $A = \frac{1}{2}bh$; (c) $p = s - c$; (d) $s = l + w$; (e) $V = lwh$

3. (a) $\{10, 15, 30, 50\}$; (b) $\{9, 11, 17, 25\}$; (c) $\{9, 8\frac{1}{2}, 7, 5\}$; (d) $\{\frac{3}{5}, \frac{5}{7}, \frac{11}{13}, \frac{19}{21}\}$; (e) $\{3, 4\frac{1}{2}, 9, 15\}$

4. (a) $d + 10$; (b) $x + y$; (c) $f + g + 15$; (d) $q + r + 30$; (e) $a + b + d + e$; (f) $p + s + 60$; (g) $j + m + 20$; (h) $t + v + w + 66$

5. (a) $8r$; (b) abc; (c) $7pq$; (d) $12tv$; (e) $35wxy$; (f) $13cdef$; (g) $100hk$; (h) $44mrs$; (i) $13abcd$

6. (a) $8mm$; (b) $5pq$; (c) $\frac{2}{3}c$; (d) $\frac{1}{2}bh$; (e) $f/7$; (f) $25/x$; (g) $\dfrac{uv}{9}$

7. (a) if $d = 0$; (b) if $t = 0$; (c) if $x = 0$; (d) if $a = 8$; (e) if $c = 7$; (f) if $x = 2$; (g) if $w = y$; (h) if $p = 0$ or $q = 0$; (i) if $x = 2y$

8. (a) $n + 25$ or $25 + n$; (b) $n + 30$ or $30 + n$; (c) $n + 35$ or $35 + n$; (d) $n + 40$ or $40 + n$; (e) $n + 45$ or $45 + n$; (f) $n + 50$ or $50 + n$; (g) $n - 30$; (h) $n - 35$; (i) $40 - n$; (j) $45 - n$; (k) $50 - n$; (l) $n - 55$

9. (a) $w - 15$; (b) $l - 50$; (c) $t - 60$; (d) $p + 100$; (e) $r - 20$; (f) $d + 30$; (g) $A + 30$; (h) $t - 40$; (i) $f + 8$; (j) $a - 5$

10. (a) $3x$; (b) $b/8$ or $\frac{1}{8}b$; (c) $12y$; (d) $\frac{3}{8}r$ or $3r/8$; (e) $10/y$; (f) $y/10$

342

11. (a) $b - \dfrac{c}{2}$; (b) $g/3 - 5$; (c) $4r/9$; (d) $\dfrac{m + 60}{2}$; (e) $\dfrac{3x}{4} - y$; (f) $2d - 25$; (g) $5x + 8$; (h) $4(r + 9)$;

(i) $\dfrac{m + p + q + 60}{4}$; (j) $b/3c$

12. (a) $3d - 25$; (b) $2w + 5$; (c) $5T + 8$; (d) $\dfrac{p}{2} + 50$; (e) $p/3 - 50$; (f) $100y + 2$

13. (a) $(7) + 10 = 2 + (15)$ and $17 = 17$
    (b) $(7) + (110) = (17) + 100$ and $117 = 117$
    (c) $(105) + (1002) = 5 + (1102)$ and $1107 = 1107$
    (d) $(1102) + (105) = (12) + (1100) + 5$ and $1117 = 1117$
    (e) $(10)10 = 2(50)$ and $100 = 100$
    (f) $(10)(1000) = (100)100$ and $10{,}000 = 10{,}000$

14. (a) 30; (b) 14; (c) 25; (d) 1; (e) 3; (f) 83; (g) 60; (h) $18\frac{1}{2}$

15. (a) 1; (b) 17; (c) 8; (d) $5\frac{1}{2}$; (e) 6; (f) 5; (g) 8; (h) 19; (i) 11; (j) 3; (k) 29; (l) 0; (m) 7; (n) 12; (o) 3

16. (a) 50; (b) 30; (c) 12; (d) 56; (e) 32; (f) 48; (g) 100; (h) 120; (i) 48; (j) 28; (k) 0; (l) 4

17. (a) 32; (b) 3; (c) 5; (d) 6; (e) $8\frac{1}{2}$; (f) 39; (g) 75; (h) 55; (i) 16; (j) 39; (k) 33; (l) 36

18. (a) 5; (b) 1; (c) 6; (d) 0; (e) 6; (f) 0; (g) valueless; (h) 0; (i) 6; (j) 3; (k) 0; (l) 0; (m) $1\frac{1}{2}$; (n) 3; (o) $\frac{2}{3}$

19. (a) $5xyz$; (b) 5, xyz; (c) 5, x, y, z; (d) $3a$, bc; (e) $3ab$, c; (f) $3a(b + c)$ is a single term.

20. (a) 7, 11; (b) 5, 5; (c) p, q; (d) $\frac{3}{4}$, x; (e) 1/10, w; (f) 8, $(x - 5)$; (g) $\frac{1}{4}$, $(y - 2)$

21.

	(a)	(b)	(c)	(d)	(e)	(f)	(g)
Numeral Coefficient	1	$\frac{1}{8}$	$\frac{1}{10}$	.03	$\frac{3}{10}$	$\frac{2}{3}$	$\frac{3}{5}$
Literal Coefficient	w	x	n	ab	y	a/b	$(a - b)$

22. (a) $7 \cdot 3^2$; (b) $7xy^3$; (c) $\dfrac{7x}{y^3}$; (d) $(7x)^2$; (e) $(a + 5)^2$; (f) $\dfrac{2r^2 w}{5s + r^2}$

23. (a) $4 \cdot 7 \cdot 7$; (b) $\frac{1}{2}yyyy$; (c) $\dfrac{5a}{bbbb}$; (d) $(ab)(ab)(ab)$; (e) $(x + 2)(x + 2)$; (f) $\dfrac{aa - bbb}{c + dd}$

24. (a) 18; (b) 12; (c) 36; (d) 7; (e) 25; (f) 13; (g) 54; (h) 216; (i) 19; (j) 1; (k) 72; (l) 108

CHAPTER 2

1. (a) $x = 6$; (b) $n = 11$; (c) $y = 12$

3. (a) $n - 8 = 13$; (b) $\frac{2}{3}n = 10$; (c) $3(n + 6) = 33$; (d) $n + 20 = 3n$; (e) $n + 5 = 2n - 4$

4. 1. (e); 2. (g); 3. (d); 4. (b); 5. (c); 6. (a); 7. (f)

5. (a) $n + 3n = 52$; (b) $n + n + 20 = 84$; (c) $n + 2n - 15 = 78$

6. (a) $a = 4$; (b) $b = 8$; (c) $c = 8$; (d) $d = 20$; (e) $x = 18$; (f) $y = 22$; (g) $h = 20$; (h) $k = 20$; (i) $m = 58$; (j) $n = 72$; (k) $x = 47$; (l) $y = 153$

7.   (*a*) $p = 12$; (*b*) $r = 16$; (*c*) $s = \frac{7}{5}$ or $1\frac{2}{5}$; (*d*) $n = \frac{1}{2}$; (*e*) $w = \frac{1}{3}$; (*f*) $x = \frac{7}{8}$; (*g*) $t = 30$; (*h*) $u = 65$; (*i*) $x = 60$; (*j*) $y = 18$; (*k*) $a = 4$; (*l*) $b = \frac{5}{2}$ or $2\frac{1}{2}$

8.   (*a*) $n = 16$; (*b*) $n = 32$; (*c*) $n = 3$; (*d*) $n = 192$; (*e*) $y = 12$; (*f*) $y = 18$; (*g*) $y = 5$; (*h*) $y = 45$; (*i*) $y = 36$; (*j*) $y = 320$; (*k*) $y = 320$; (*l*) $y = 4$; (*m*) $x = 8\frac{2}{3}$; (*n*) $x = 9\frac{1}{3}$; (*o*) $x = 27$; (*p*) $x = 3$

9.   (*a*) $x = 3$; (*b*) $y = 13$; (*c*) $a = 9$; (*d*) $b = 12$; (*e*) $z = 12$; (*f*) $h = 10\frac{1}{2}$; (*g*) $m = 55\frac{1}{3}$; (*h*) $n = 20$; (*i*) $x = 4.5$; (*j*) $y = 14$; (*k*) $r = 5$; (*l*) $s = 1$; (*m*) $t = \frac{1}{2}$; (*n*) $x = \frac{3}{8}$; (*o*) $y = 0$; (*p*) $l = 13$; (*q*) $n = 5.1$; (*r*) $h = 70$; (*s*) $t = 12$; (*t*) $x = 9$

10.   (*a*) division; (*b*) addition; (*c*) multiplication; (*d*) subtraction; (*e*) division; (*f*) multiplication

11.   (*a*) $x = 5$; (*b*) $y = \frac{1}{3}$; (*c*) $12 = z$; (*d*) $\frac{1}{12} = w$; (*e*) $r = \frac{3}{2}$; (*f*) $s = \frac{2}{3}$; (*g*) $\frac{5}{2} = t$; (*h*) $\frac{2}{5} = u$

12.   (*a*) {30}; (*b*) {50}; (*c*) {400}; (*d*) {2000}; (*e*) {1000}; (*f*) {1.6}; (*g*) {210}; (*h*) {500}

13.   (*a*) $s = 70$; (*b*) $t = 12$; (*c*) $600 = n$; (*d*) $20 = w$; (*e*) $m = 260$; (*f*) $x = 40$; (*g*) $23 = y$; (*h*) $4.1 = z$

14.   (*a*) $7 = x$; (*b*) $y = 5$; (*c*) $z = 9$; (*d*) $11 = w$; (*e*) $6 = x$; (*f*) $y = 3$; (*g*) $z = 56$; (*h*) $w = 8\frac{1}{2}$

15.   (*a*) \$9.63; (*b*) \$3.21; (*c*) \$57.78

16.   (*a*) \$1700; (*b*) \$15,000; (*c*) \$124

17.   (*a*) $x = 6$; (*b*) $y = 84$; (*c*) $80 = z$; (*d*) $150 = w$; (*e*) $a = 6$; (*f*) $b = 600$; (*g*) $6 = c$

18.   (*a*) {7}; (*b*) {12}; (*c*) {72}; (*d*) {.55}; (*e*) {88}; (*f*) {.03}

19.   (*a*) $x = 28$; (*b*) $y = 24$; (*c*) $x = 6$; (*d*) $y = 81$; (*e*) $z = 25$; (*f*) $c = 4$

20.   (*a*) $s = 40$; (*b*) $t = 100$; (*c*) $n = 84$; (*d*) $r = 10$; (*e*) $w = 42$

21.   (*a*) 42 km; (*b*) 50 km

22.   (*a*) \$400; (*b*) \$5000; (*c*) \$610

23.   (*a*) {45}; (*b*) {109}; (*c*) {15}; (*d*) {122}; (*e*) {624}; (*f*) {691}; (*g*) {22}; (*h*) {44}

24.   (*a*) $b = 7$; (*b*) $c = 7\frac{1}{4}$; (*c*) $12.2 = d$; (*d*) $6.8 = c$; (*e*) $f = 2\frac{7}{8}$; (*f*) $g = 2\frac{2}{3}$; (*g*) $6.49 = m$; (*h*) $9.46 = n$

25.   (*a*) \$1.00; (*b*) \$1.12

26.   (*a*) 4 ft 6 in; (*b*) 3 ft 7 in

27.   (*a*) {30}; (*b*) {30}; (*c*) {43}; (*d*) {43}; (*e*) {165}; (*f*) {241}; (*g*) {475}; (*h*) {1037}

28.   (*a*) $h = 9\frac{5}{8}$; (*b*) $j = 99\frac{1}{2}$; (*c*) $42.3 = m$; (*d*) $9.33 = n$; (*e*) $p = 3$; (*f*) $r = 20$; (*g*) $2.10 = s$; (*h*) $8.90 = t$

29.   (*a*) $87°$; (*b*) $26\frac{1}{2}°$; (*c*) $18\frac{1}{4}°$

30.   (*a*) 40; (*b*) 47; (*c*) 50; (*d*) 70; (*e*) 110

31.   (*a*) $x = 2$; (*b*) $x = 2\frac{1}{2}$; (*c*) $x = 4$; (*d*) $x = 1$; (*e*) $x = 7$; (*f*) $x = 8$; (*g*) $x = 8$; (*h*) $x = 4\frac{1}{2}$; (*i*) $x = 16$; (*j*) $x = 40$; (*k*) $x = 4$; (*l*) $x = 230$; (*m*) $x = 40$; (*n*) $x = 60$; (*o*) $x = 123$; (*p*) $x = 76$

32. (a) $n = 1$; (b) $m = 7$; (c) $n = 3$; (d) $p = 3$; (e) $n = 4\frac{1}{2}$; (f) $r = 4$; (g) $t = \frac{3}{2}$ or $1\frac{1}{2}$; (h) $s = 30$

33. (a) $\{9\}$; (b) $\{10\}$; (c) $\{5\frac{1}{2}\}$; (d) $\{3\}$; (e) $\{1\frac{1}{2}\}$; (f) $\{5\}$; (g) $\{15\}$; (h) $\{2\frac{1}{2}\}$

34. (a) $x = 8$; (b) $x = \frac{1}{8}$; (c) $y = 2$; (d) $y = \frac{1}{2}$; (e) $n = 4$; (f) $n = 1\frac{1}{2}$; (g) $w = 3\frac{3}{4}$; (h) $w = 6$

35. (a) $\{24\}$; (b) $\{52\}$; (c) $\{12\}$; (d) $\{10\}$; (e) $\{15\}$; (f) $\{12\}$; (g) $\{20\}$; (h) $\{35\}$; (i) $\{4\}$; (j) $\{9\}$; (k) $\{6\}$; (l) $\{3.6\}$

36. (a) $\{10\}$; (b) $\{90\}$; (c) $\{20\}$; (d) $\{7\}$; (e) $\{13\}$; (f) $\{12\}$; (g) $\{1.5\}$; (h) $\{.17\}$; (i) $\{12\}$; (j) $\{2\frac{2}{3}\}$; (k) $\{6\}$; (l) $\{5\}$; (m) $\{15\}$; (n) $\{\frac{1}{2}\}$; (o) $\{7\}$; (p) $\{.48\}$; (q) $\{.2\}$; (r) $\{70\}$; (s) $\{5\}$; (t) $\{2\}$

37. (a) 20 girls; (b) 6 girls

38. (a) \$4.50; (b) \$12.60

39. (a) \$180,000; (b) \$150,000

40. (a) $p = 10d$; (b) $d = \dfrac{p}{10}$; (c) $n = 20D$; (d) $h = \frac{1}{2}q$; (e) $q = \dfrac{2}{5}d$ or $\dfrac{2d}{5}$

41. (a) $c = 10d + 25q$; (b) $c = 5n + 100D$; (c) $n = 2d + 10h$; (d) $d = \dfrac{n}{2} + \dfrac{p}{10}$; (e) $q = 4D + \dfrac{n}{5}$

42. (a) $s = 3600h$; (b) $h = \dfrac{m}{60}$; (c) $h = 168w$; (d) $d = 30M$; (e) $d = 30M + 7w + 5$;

(f) $m = 60h + .5$ or $m = 60h + \frac{1}{2}$; (g) $d = 365y + 21$

43. (a) $i = 36y$; (b) $y = f/3$; (c) $m = c/100$

44. (a) $D = 5R$; (b) $D = R/2$; (c) $RT = 25$; (d) $RT = 300$; (e) $D = 20T$; (f) $D = 1200T$

45. (a) $\dfrac{d}{2} = r$; (b) $\dfrac{p}{5} = s$; (c) $\dfrac{D}{30} = T$; (d) $W = \dfrac{A}{25}$; (e) $\dfrac{c}{\pi} = d$; (f) $\pi = \dfrac{c}{d}$; (g) $N = \dfrac{c}{P}$; (h) $\dfrac{T}{P} = R$; (i) $\dfrac{v}{LW} = H$;

(j) $\dfrac{v}{2\pi r^2} = h$; (k) $C = \dfrac{5(F - 32)}{9}$; (l) $\dfrac{2A}{b + b'} = h$

46. (a) $p = 10s$; (b) $15R = D$; (c) $8W = A$; (d) $7w = d$; (e) $2\pi r = c$; (f) $M = FD$; (g) $2PF = A$; (h) $T = 5RQ$;

(i) $V = 3LWH$; (j) $T = 7RS$; (k) $L = KAV^2$; (l) $\dfrac{\pi r^2 h}{3}$

47. (a) $a = 60 - b$; (b) $g = 85 - 3c$; (c) $h = l + 10r$; (d) $3m - 4n = p$; (e) $10r + 5t = s$; (f) $h = j + 12 - 4g$;

(g) $b = c - d - 5a$; (h) $5a - 4c - 3e = f$; (i) $c = 100p + 10 - \dfrac{b}{2}$

48. (a) $P = \dfrac{3R - 40}{4}$; (b) $\dfrac{36 - 12f}{5} = i$; (c) $P = 25 - 2R$; (d) $\dfrac{2A}{h} = b$; (e) $\dfrac{3V}{\pi r^2} = h$; (f) $\dfrac{2A}{b + b'} = h$;

(g) $R = 8S + 2T$; (h) $5(8h - 12) = k$ or $40h - 60 = k$; (i) $\frac{3}{2}(20p - 8t) = q$ or $30p - 12t = q$

49. (a) $\dfrac{l - q}{n - l} = d$; (b) $l - \dfrac{2s}{n} - a$ or $\dfrac{2s - an}{n}$; (c) $\frac{5}{9}(F - 32) = C$

50. (a) 300; (b) 2; (c) 131; (d) -40; (e) 100; (f) 1024; (g) 16; (h) $22\frac{1}{2}$

51. (a) $7\frac{1}{2}$; (b) 10; (c) $4\frac{1}{2}$; (d) 13; (e) 15; (f) 14; (g) 120; (h) 119

52. (a) 57 kmph; (b) 0.6 m

53. (a) $13.85766x^{11}y$; (b) $.0755555q^3$; (c) $10,989y^{11}$

54. 466.20046

55. $A(3,0)$; $B(4,3)$; $C(3,4)$; $D(0,2)$; $E(-2,4)$; $F(-4,2)$; $G(-1,0)$; $H(-3\frac{1}{2},-2)$; $I(-2,-3)$; $J(0,-4)$; $K(1\frac{1}{2},-2\frac{1}{2})$; $L(4,-2\frac{1}{2})$

57. Perimeter of square formed is 20 units; its area is 25 square units.

58. Area of parallelogram = 30 square units. Area of $\triangle BCD$ = 15 square units.

59. (a) $(4,3)$; (b) $(2\frac{1}{2},3\frac{1}{2})$; (c) $(-4,6)$; (d) $(7,5)$; (e) $(-10,-2\frac{1}{2})$; (f) $(0,10)$; (g) $(4,-1)$; (h) $(-5,-2\frac{1}{2})$; (i) $(5,5)$; (j) $(-3,-10)$; (k) $(5,6)$; (l) $(0,-3)$

60. (a) $(4,0)$, $(0,3)$, $(4,3)$; (b) $(-3,0)$, $(0,5)$, $(-3,5)$; (c) $(6,-2)$, $(0,-2)$, $(6,0)$; (d) $(4,6)$, $(4,9)$, $(3,8)$; (e) $(2,-3)$, $(-2,2)$, $(0,5)$; (f) $(-\frac{1}{2},0)$, $(\frac{1}{2},\frac{1}{2})$, $(0,-1\frac{1}{2})$

61. (a) $(0,2)$, $(1,7)$, $(4,5)$, $(3,0)$; (b) $(-2,7)$, $(3,6)$, $(6,1)$, $(1,2)$; (c) $(-1,2)$, $(3,3)$, $(3,-4)$, $(-1,-5)$; (d) $(-2,-1)$, $(4,2\frac{1}{2})$, $(7,-4)$, $(1,-7\frac{1}{2})$

62. (a) $(2,6)$, $(4,3)$; (b) $(1,0)$, $(0,-2\frac{1}{2})$; (c) common midpoint, $(2,2)$

63. (a) $(-2,3)$; (b) $(-3,-6)$; (c) $(-1\frac{1}{2},-2)$; (d) (a,b); (e) $(2a,3b)$; (f) $(a,b+c)$

64. (a) $M(4,8)$; (b) $A(-1,0)$; (c) $B(6,-3)$

65. (a) $B(2,3\frac{1}{2})$; (b) $D(3,3)$; (c) $A(-2,9)$

66. (a) Prove that $ABCD$ is a parallelogram (since opposite sides are congruent) and has a rt. $\angle$.
       (b) The point $(3,2\frac{1}{2})$ is the midpoint of each diagonal.
       (c) Yes, since the midpoint of each diagonal is their common point.

67. (a) $D(3,2)$, $(1\frac{1}{2},1)$; (b) $E(0,2)$, $(3,1)$; (c) no, since the midpoint of each median is not a common point

68. (a) 5; (b) 6; (c) 10; (d) 12; (e) 5.4; (f) 7.5; (g) 9; (h) a

69. (a) 3, 3, 6; (b) 4, 14, 18; (c) 1, 3, 4; (d) a, $2a$, $3a$

70. (a) 13; (b) 5; (c) 15; (d) 5; (e) 10; (f) 15; (g) $3\sqrt{2}$; (h) $5\sqrt{2}$; (i) $\sqrt{10}$; (j) $2\sqrt{5}$; (k) 4; (l) $a\sqrt{2}$

72. (a) $\triangle ABC$; (b) $\triangle DEF$; (c) $\triangle GHJ$; (d) $\triangle KLM$ is *not* a rt. $\triangle$

73. (a) $5\sqrt{2}$; (b) $\sqrt{5}$; (c) $\sqrt{65}$

75. (a) 10; (b) 5; (c) $5\sqrt{2}$; (d) 13; (e) 4; (f) 3

76. (a) on; (b) on; (c) outside; (d) on; (e) inside; (f) inside; (g) on

77. (a) $\frac{9}{5}$; (b) $\frac{5}{6}$; (c) $\frac{5}{2}$; (d) 3; (e) 2; (f) 1; (g) 5; (h) -2; (i) -3; (j) $\frac{3}{2}$; (k) -1; (l) 1

78. (a) 3; (b) 4; (c) $-\frac{1}{2}$; (d) -7; (e) 5; (f) 0; (g) 3; (h) 5; (i) -4; (j) $-\frac{2}{3}$, (k) -1; (l) -2; (m) 5; (n) 6; (o) -4; (p) -8

79. (*a*) 72°; (*b*) 18°; (*c*) 68°; (*d*) 22°; (*e*) 45°; (*f*) 0°

80. (*a*) 0.0875; (*b*) 0.3057; (*c*) 0.3640; (*d*) 0.7002; (*e*) 1; (*f*) 3.2709; (*g*) 11.4301

81. (*a*) 0°; (*b*) 25°; (*c*) 45°; (*d*) 55°; (*e*) 7°; (*f*) 27°; (*g*) 37; (*h*) 53° (*i*) 66°

82. (*a*) $\overline{BC}, \overline{BD}, \overline{AD}, \overline{AE}$; (*b*) $\overline{BF}, \overline{CF}, \overline{DE}, \overline{DF}$; (*c*) $\overline{AF}, \overline{CD}$; (*d*) $\overline{AB}, \overline{EF}$

83. (*a*) 0; (*b*) no slope; (*c*) 5; (*d*) −5; (*e*) 0.5; (*f*) −0.0005

84. (*a*) 0; (*b*) no slope; (*c*) no slope; (*d*) 0; (*e*) 5; (*f*) −1; (*g*) 2

85. (*a*) $\frac{3}{2}$; (*b*) $\frac{7}{3}$; (*c*) −1; (*d*) 6

86. (*a*) −2; (*b*) −1; (*c*) −$\frac{1}{3}$; (*d*) −$\frac{2}{5}$; (*e*) −10; (*f*) 1; (*g*) $\frac{5}{4}$; (*h*) $\frac{4}{13}$; (*i*) no slope; (*j*) 0

87. (*a*) 0; (*b*) −2; (*c*) 3; (*d*) −1

88. (*a*) −$\frac{3}{2}$; (*b*) $\frac{2}{3}$; (*c*) −$\frac{3}{2}$

89. (*a*) −$\frac{1}{2}$; (*b*) 1; (*c*) 2; (*d*) −1

90. (*a*) −$\frac{3}{2}$; (*b*) $\frac{1}{4}$; (*c*) $\frac{5}{6}$

91. (*a*) and (*b*)

92. (*a*) 19; (*b*) 9; (*c*) 2

CHAPTER 3

1. (*a*) point; (*b*) line; (*c*) plane; (*d*) plane; (*e*) line; (*f*) point

2. (*a*) $\overline{AE}, \overline{DE}$; (*b*) $\overline{ED}, \overline{CD}, \overline{BD}, \overline{FD}$; (*c*) $\overline{AD}, \overline{BE}, \overline{CE}, \overline{EF}$; (*d*) F

3. (*a*) $AB = 16$; (*b*) $AE = 10\frac{1}{2}$

4. (*a*) 18; (*b*) 90°, (*c*) 50°; (*d*) 130°; (*e*) 230°

5. (*a*) ∠CBE; (*b*) ∠AEB; (*c*) ∠ABE; (*d*) ABC; ∠BCD; ∠BED; (*e*) ∠AED

6. (*a*) 130°; (*b*) 120°; (*c*) 75°; (*d*) 132°

7. (*a*) 75°; (*b*) 40°; (*c*) 10$\frac{1}{3}$° or 10°20′; (*d*) 9°11′

8. (*a*) 90°; (*b*) 120°; (*c*) 135°; (*d*) 270°; (*e*) 180°

9. (*a*) 90°; (*b*) 60°; (*c*) 15°; (*d*) 165°

10. (*a*) $\overline{AB} \perp \overline{BC}$ and $\overline{AC} \perp \overline{CD}$; (*b*) 129°; (*c*) 102°; (*d*) 51°; (*e*) 129°

11. (*a*)   △ABC, hypotenuse $\overline{AB}$, legs $\overline{AC}$ and $\overline{BC}$
    △ACD, hypotenuse $\overline{AC}$, legs $\overline{AD}$ and $\overline{CD}$
    △BCD, hypotenuse $\overline{BC}$, legs $\overline{BD}$ and $\overline{CD}$

    (b) $\triangle DAB$ and $\triangle ABC$
    (c) $\triangle AEB$, legs $\overline{AE}$ and $\overline{BE}$, base $\overline{AB}$, vertex angle $\angle AEB$
          $\triangle CED$, legs $\overline{DE}$ and $\overline{CE}$, base $\overline{CD}$, vertex angle $\angle CED$

12. (a) $\overline{AR} \cong \overline{BR}$ and $\angle PRA \cong \angle PRB$; (b) $\angle ABF \cong \angle CBF$; (c) $\angle CGA \cong \angle CGD$; (d) $\overline{AM} \cong \overline{MD}$

13. (a) vert. $\angle$s; (b) comp. adj. $\angle$s; (c) adj. $\angle$s; (d) supp. adj. $\angle$s; (e) comp. $\angle$s; (f) vert. $\angle$s

14. (a) 25°, 65°; (b) 18°, 72°; (c) 60°, 120°; (d) 61°, 119°; (e) 50°, 130°; (f) 56°, 84°; (g) 90°, 90°

15. (a) 48°, 27°; (b) 65°, 25°; (c) 148°, 32°

CHAPTER 4

1. (a) A is H; (b) P is D; (c) R is S; (d) E is K; (e) A is G; (f) triangles are geometric figures; (g) a rectangle is a quadrilateral

2. (a) $a = c = f$; (b) $g = 15$; (c) $f = a$; (d) $a = h$; (e) $b = e$

3. (a) 130; (b) 4; (c) yes; (d) $x = 8\frac{1}{2}$; (e) $y = 15$; (f) $x = 6$; (g) $x = \pm 6$

4. (a) $AC = 12$, $AE = 11$, $AF = 15$, $DF = 9$
    (b) $m\angle ADC = 92°$, $m\angle BAE = 68°$, $m\angle FAD = 86°$, $m\angle BAD = 128°$

5. (a) $AB = DF$; (b) $AB = AC$; (c) $\angle ECA \cong \angle DCB$; (d) $\angle BAD \cong \angle BCD$

6. (a) If equals are divided by equals, the quotients are equal.
    (b) Doubles of equals are equal.
    (c) If equals are multiplied by equals, the products are equal.
    (d) Halves of equals are equal.

7. (a) If equals are divided by equals, the quotients are equal.
    (b) If equals are multiplied by equals, the products are equal.
    (c) Doubles of equals are equal.
    (d) Halves of equals are equal.

8. (a) Their new rates of pay per hour will be the same.
    (b) Those stocks have the same value now.
    (c) The classes have the same registers now.
    (d) 100°C = 212°F
    (e) Their parts will be the same length.
    (f) He has a total of \$10,000 in Banks A, B, and C.
    (g) Their values are the same.

9. (a) Vertical angles are congruent.
    (b) All straight angles are congruent.
    (c) Supplements of congruent angles are congruent.
    (d) Perpendiculars form right angles and all right angles are congruent.
    (e) Complements of congruent angles are congruent.

10. In each answer, (H) indicates the hypothesis and (C) indicates the conclusion.
    (a) (H) Stars, (C) twinkle.
    (b) (H) Jet planes, (C) are the speediest.
    (c) (H) Water, (C) boils at 212° Fahrenheit.

(d) (H) If it is the American flag, (C) its colors are red, white and blue.
(e) (H) If you fail to do homework in the subject, (C) you cannot learn geometry.
(f) (H) If the umpire calls a fourth ball, (C) a batter goes to first base.
(g) (H) If A is B's brother and C is B's daughter, (C) then A is C's uncle.
(h) (H) An angle bisector, (C) divides the angle into two equal parts.
(i) (H) If it is divided into three equal parts, (C), a segment is trisected.
(j) (H) A pentagon, (C) has five sides and five angles.
(k) (H) Some rectangles, (C) are squares.
(l) (H) If their sides are made longer, (C) angles do not become larger.
(m) (H) If they are congruent and supplementary, (C) angles are right angles.
(n) (H) If one of its sides is not a straight line segment, (C) the figure cannot be a polygon.

11. (a) An acute angle is half a right angle. Not necessarily true.
    (b) A triangle having one obtuse angle is an obtuse triangle. True.
    (c) If the batter is out, then the umpire called a third strike. Not necessarily true.
    (d) If you are shorter than I, then I am taller than you. True.
    (e) If our weights are unequal, then I am heavier than you. Not necessarily true.

CHAPTER 5

1. (a) $\triangle I \cong \triangle II \cong \triangle III$, s.a.s. $\cong$ s.a.s.; (b) $\triangle I \cong \triangle III$, a.s.a. $\cong$ a.s.a.; (c) $\triangle I \cong \triangle II \cong \triangle III$, s.s.s. $\cong$ s.s.s.

2. (a) a.s.a. $\cong$ a.s.a.; (b) s.a.s. $\cong$ s.a.s.; (c) s.s.s. $\cong$ s.s.s.; (d) s.a.s. $\cong$ s.a.s.; (e) a.s.a. $\cong$ a.s.a.; (f) s.a.s. $\cong$ s.a.s.;
    (g) s.a.s. $\cong$ s.a.s.; (h) a.s.a. $\cong$ a.s.a.

3. (a) $\overline{AD} \cong \overline{DC}$; (b) $\angle ABD = \angle DBC$; (c) $\angle 1 \cong \angle 4$; (d) $\overline{BE} \cong \overline{ED}$; (e) $\overline{BD} \cong \overline{AC}$; (f) $\angle BAD \cong \angle CDA$

4. (a) $\angle 1 \cong \angle 3$, $\angle 2 \cong \angle 4$, $\overline{BD} \cong \overline{BE}$; (b) $\overline{AB} \cong \overline{AC}$, $\overline{BD} \cong \overline{DC}$, $\angle B \cong \angle C$;
    (c) $\angle E \cong \angle C$, $\angle A \cong \angle F$, $\angle EDF \cong \angle ABC$

5. (a) $x = 19$, $y = 8$; (b) $x = 4$, $y = 12$; (c) $x = 48$, $y = 12$

8. (a) $\angle b \cong \angle d$, $\angle E \cong \angle G$; (b) $\angle A \cong \angle 1 \cong \angle 4$, $\angle 2 \cong \angle C$; (c) $\angle 1 \cong \angle 5$, $\angle 4 \cong \angle 6$, $\angle EAD \cong \angle EDA$

9. (a) $\overline{BE} \cong \overline{EC}$; (b) $\overline{AB} \cong \overline{BD} \cong \overline{AD}$, $\overline{BC} \cong \overline{CD}$; (c) $\overline{BD} \cong \overline{DE}$, $\overline{EF} \cong \overline{FC}$, $\overline{AB} \cong \overline{AC}$

CHAPTER 6

1. (a) $x = 105°$, $y = 75°$; (b) $x = 60°$, $y = 40°$; (c) $x = 85°$, $y = 95°$; (d) $x = 50°$, $y = 50°$; (e) $x = 65°$, $y = 65°$;
    (f) $x = 40°$, $y = 30°$; (g) $x = 60°$, $y = 120°$; (h) $x = 90°$, $y = 35°$; (i) $x = 30°$, $y = 40°$; (j) $x = 80°$, $y = 10°$;
    (k) $x = 30°$, $y = 150°$; (l) $x = 85°$, $y = 95°$

2. (a) $x = 22°$, $y = 102°$; (b) $x = 40°$, $y = 100°$; (c) $x = 80°$, $y = 40°$

3. (a) Each angle measures 105°. (b) Each angle measures 70°. (c) Angles measure 72° and 108°.

7. (a) 25; (b) 9; (c) 20; (d) 8

8. (a) 8; (b) 10; (c) 2; (d) 14

10. (a) P is equidistant from B and C. P is on $\perp$ bisector of $\overline{BC}$.
       Q is equidistant from A and B. Q is on $\perp$ bisector of $\overline{AB}$.
       R is equidistant from A, C, and D. R is on $\perp$ bisectors of $\overline{AD}$ and $\overline{CD}$.

(b) P is equidistant from $\overline{AB}$ and $\overline{AD}$. P is on bisector of $\angle A$.
Q is equidistant from $\overline{AB}$ and $\overline{BC}$. Q is on bisector of $\angle B$.
R is equidistant from $\overline{BC}$, $\overline{CD}$, and $\overline{AD}$. R is on the bisectors of $\angle C$ and $\angle D$.

11. (a) P is equidistant from $\overline{AD}$, $\overline{AB}$, and $\overline{BC}$. Q is equidistant from $\overline{AD}$ and $\overline{AB}$ and equidistant from A and D. R is equidistant from $\overline{AB}$ and $\overline{BC}$ and equidistant from A and D.

(b) P is equidistant from $\overline{AD}$ and $\overline{CD}$ and equidistant from B and C. Q is equidistant from A, B, and C. R is equidistant from $\overline{AD}$ and $\overline{CD}$ and equidistant from A and B.

12. (a) $x = 50°$, $y = 110°$; (b) $x = 65°$, $y = 65°$; (c) $x = 30°$; $y = 100°$; (d) $x = 51°$, $y = 112°$; (e) $x = 52°$, $y = 40°$; (f) $x = 120°$, $y = 90°$

13. (a) $x = 55°$, $y = 125°$; (b) $x = 80°$, $y = 90°$; (c) $x = 56°$, $y = 68°$; (d) $x = 100°$, $y = 30°$; (e) $x = 30°$, $y = 120°$; (f) $x = 90°$, $y = 30°$

14. (a) $18°$, $54°$, $108°$; (b) $40°$, $50°$, $90°$; (c) $36°$, $36°$, $108°$; (d) $36°$, $72°$, $108°$, $144°$; (e) $50°$, $75°$; (f) $100°$, $60°$, and $20°$

16. (a) Since $x = 45$, each angle measures $60°$.
(b) Since $x = 25$, $x + 15 = 40$ and $3x - 35 = 40$; that is, two angles each measure $40°$.
(c) If $2x$, $3x$, and $5x$ represent the angles, $x = 18$ and $5x = 90$; that is, one of the angles measures $90°$.
(d) If x and $5x - 10$ represent the unknown angles, $x = 21$ and $5x - 10 = 95$; that is, one of the angles measures $96°$.

17. (a) 7 st. $\angle$s, 30 st. $\angle$s; (b) $1620°$, $5400°$, $180,000°$; (c) 30.12, 27, 202

18. (a) $20°$, $18°$, $9°$; (b) $160°$, $162°$, $171°$; (c) 3, 9, 20, 180; (d) 3, 12, 36, 72, 360

19. (a) $65°$, $90°$, $95°$, $110°$; (b) $140°$, $100°$, $60°$, $60°$

20. (a) $\triangle I \cong \triangle III$ by hy. leg $\cong$ hy. leg; (b) $\triangle I \cong \triangle III$ by s.a.a. $\cong$ s.a.a.

CHAPTER 7

1. (a) $x = 15$, $y = 25$; (b) $x = 20$, $y = 130$; (c) $x = 20$, $y = 140$

4. (a) $\square EFGH$; (b) $\square ABCD$ and $EBFD$; (c) $\square GHKJ$, $HILK$, $GILJ$; (d) $\square ACHB$, $CEFH$

5. (a) Two sides are congruent and $\parallel$. (b) Opposite sides are congruent. (c) Opposite angles are congruent. (d) $\overline{AD}$ and $\overline{BC}$ are congruent and parallel ($\overline{AD} \cong \overline{EF} \cong BC$).

6. (a) $x = 6$, $y = 12$; (b) $x = 5$, $y = 9$; (c) $x = 120$, $y = 30$; (d) $x = 15$, $y = 45$

7. (a) $x = 14$, $y = 6$; (b) $x = 18$, $y = 4\frac{1}{2}$; (c) $x = 8$, $y = 5$; (d) $x = 3$, $y = 9$

10. (a) $x = 5$, $y = 7$; (b) $x = 10$, $y = 35$; (c) $x = 2\frac{1}{2}$, $y = 17\frac{1}{2}$; (d) $x = 8$, $y = 4$; (e) $x = 25$, $y = 25$; (f) $x = 11$, $y = 118$

13. (a) $x = 6$, $y = 40$; (b) $x = 3$, $y = 5\frac{1}{2}$; (c) $x = 8\frac{1}{3}$, $y = 22$

14. (a) $x = 28$, $y = 25\frac{1}{2}$; (b) $x = 12$ (since y does not join midpoints, Pr. 3 does not apply); (c) $x = 19$, $y = 23\frac{1}{2}$

15. (a) $m = 19$; (b) $b' = 36$; (c) $b = 73$

16. (*a*) $x = 11$, $y = 33$; (*b*) $x = 32$, $y = 26$; (*c*) $x = 12$, $y = 36$

17. (*a*) $22\frac{1}{2}$; (*b*) 70

18. (*a*) 21; (*b*) 30; (*c*) 14; (*d*) 26

CHAPTER 8

5. (*a*) square; (*b*) isosceles triangle; (*c*) trapezoid; (*d*) right triangle

6. (*a*) 140°; (*b*) 60°; (*c*) 90°; (*d*) $(180 - x)°$; (*e*) $x°$; (*f*) $(90 + x)°$

7. (*a*) 100°; (*b*) 50°, 80°; (*c*) 54°, 27°; (*d*) 45°; (*e*) 35°; (*f*) 45°

8. (*a*) $x = 22$; (*b*) $y = 6$; (*c*) $AB + CD = 22$; (*d*) perimeter $= 44$; (*e*) $x = 21$; (*f*) $r = 14$

9. (*a*) 0; (*b*) 40; (*c*) 33; (*d*) 7

10. (*a*) tangent externally; (*b*) tangent internally; (*c*) the circles are 5 units apart; (*d*) overlapping

11. (*a*) concentric; (*b*) tangent internally; (*c*) tangent externally; (*d*) outside each other; (*e*) the smaller entirely inside the larger; (*f*) overlapping

13. (*a*) 40; (*b*) 90; (*c*) 170; (*d*) 180; (*e*) $2x$; (*f*) $180 - x$; (*g*) $2x - 2y$

14. (*a*) 20; (*b*) 45; (*c*) 85; (*d*) 90; (*e*) 130; (*f*) 174; (*g*) x; (*h*) $90 - \frac{1}{2}x$; (*i*) $x - y$

15. (*a*) 85; (*b*) 170; (*c*) c; (*d*) $2i$; (*e*) 60; (*f*) 30

16. (*a*) 60, 120, 180; (*b*) 80, 120, 160; (*c*) 100, 120, 14; (*d*) 36, 144, 180

17. (*a*) $m\angle x = 136°$; (*b*) $m\widehat{y} = 111°$; (*c*) $m\angle x = 130°$; (*d*) $m\angle y = 126°$; (*e*) $m\angle x = 110°$; $m\widehat{y} = 77°$

18. (*a*) 135°; (*b*) 90°; (*c*) $(180 - x)°$; (*d*) $(90 + x)°$; (*e*) 100°; (*f*) 80°; (*g*) 55°; (*h*) 72°

19. (*a*) 85°; (*b*) $y°$; (*c*) 110°; (*d*) 95°; (*e*) 72°; (*f*) 50°; (*g*) 145°; (*h*) 87°

20. (*a*) 50; (*b*) 60

21. (*a*) $m\widehat{x} = 65°$, $m\widehat{y} = 65°$; (*b*) $m\angle x = 90°$, $m\angle y = 55°$; (*c*) $m\angle x = 37°$, $m\angle y = 50°$

22. (*a*) 19; (*b*) 45; (*c*) 69; (*d*) 90; (*e*) 125; (*f*) 167; (*g*) $\frac{1}{2}x$; (*h*) $180 - \frac{1}{2}x$; (*i*) $x + y$

23. (*a*) 110; (*b*) 135; (*c*) 180; (*d*) 270; (*e*) $180 - 2x$; (*f*) $360 - 2x$; (*g*) $2x - 2y$; (*h*) $7x$

24. (*a*) 45°; (*b*) 60°; (*c*) 30°; (*d*) 18°

25. (*a*) $m\widehat{x} = 120°$, $m\angle y = 60°$; (*b*) $m\angle x = 62°$, $m\angle y = 28°$; (*c*) $m\angle x = 46°$, $m\angle y = 58°$

26. (*a*) 75°; (*b*) 75°; (*c*) 115°; (*d*) 100°; (*e*) 140°; (*f*) 230°; (*g*) 80°; (*h*) 48°

27. (*a*) 85°; (*b*) 103°; (*c*) 80°; (*d*) 72°; (*e*) 90°; (*f*) 110°; (*g*) 130°; (*h*) 110°

28. (*a*) $m\widehat{x} = 68°$, $m\angle y = 95°$; (*b*) $m\angle x = 90°$, $m\angle y = 120°$; (*c*) $m\widehat{x} = 34°$, $m\widehat{y} = 68°$

29. (*a*) 30°; (*b*) 37°; (*c*) 20°; (*d*) 36°; (*e*) 120°; (*f*) 130°; (*g*) 94°; (*h*) 25°

30. (*a*) 45°; (*b*) 75°; (*c*) 50°; (*d*) $36\frac{1}{2}°$; (*e*) 90°; (*f*) 140°; (*g*) 115°; (*h*) 45°; (*i*) 80°

31. (*a*) 20°; (*b*) 85°; (*c*) $(180 - x)°$; (*d*) $(90 + x)°$; (*e*) 90°; (*f*) 25°; (*g*) 42°; (*h*) 120°; (*i*) 72°; (*j*) 110°; (*k*) 145°; (*l*) $(180 - y)°$; (*m*) 240°; (*n*) $(180 + x)°$; (*o*) 270°

32. (*a*) $m\widehat{x} = 43°$, $m\angle y = 43°$; (*b*) $m\widehat{x} = 190°$, $m\angle y = 55°$; (*c*) $m\widehat{x} = 140°$, $m\angle y = 40°$

33. (*a*) 120°; (*b*) 150°; (*c*) 180°; (*d*) 50°; (*e*) $22\frac{1}{2}°$; (*f*) 45°

34. (*a*) $m\widehat{x} = 150°$, $m\widehat{y} = 40°$; (*b*) $m\widehat{x} = 190°$, $m\widehat{y} = 70°$; (*c*) $m\widehat{x} = 252°$, $m\widehat{y} = 108°$

35. (*a*) 25°; (*b*) 39°; (*c*) 50°; (*d*) 30°; (*e*) 40°; (*f*) 76°; (*g*) 45°; (*h*) 95°; (*i*) 75°; (*j*) 120°

36. (*a*) 74°; (*b*) 90°; (*c*) 55°; (*d*) 60°; (*e*) 40°; (*f*) 37°; (*g*) 84°; (*h*) 110°; (*i*) 66°; (*j*) 98°; (*k*) 75°; (*l*) 79°

37. (*a*) $m\angle x = 120°$, $m\angle y = 60°$; (*b*) $m\angle x = 45°$, $m\angle y = 22\frac{1}{2}°$; (*c*) $m\angle x = 36°$, $m\angle y = 72°$

38. (*a*) $m\widehat{x} = 40°$, $m\angle y = 80°$; (*b*) $m\widehat{x} = 45°$, $m\angle y = 67\frac{1}{2}°$; (*c*) $m\angle x = 78°$, $m\angle y = 103°$

CHAPTER 9

1. (*a*) 4; (*b*) $\frac{1}{3}$; (*c*) $\frac{6}{5}$; (*d*) $\frac{10}{7}$; (*e*) $\frac{9}{7}$; (*f*) 2; (*g*) $\frac{1}{5}$; (*h*) $\frac{3}{7}$; (*i*) $\frac{2}{3}$; (*j*) $\frac{7}{8}$; (*k*) 2; (*l*) $\frac{5}{7}$; (*m*) 20; (*n*) $\frac{1}{3}$; (*o*) 3

2. (*a*) 6; (*b*) $\frac{14}{5}$; (*c*) $\frac{1}{7}$; (*d*) $\frac{3}{2}$; (*e*) 3; (*f*) $\frac{7}{2}$; (*g*) 2; (*h*) 250; (*i*) $\frac{1}{20}$; (*j*) 8; (*k*) $\frac{5}{3}$; (*l*) $\frac{9}{2}$

3. (*a*) 2:3:10; (*b*) 12:6:1; (*c*) 5:2:1; (*d*) 1:4:7; (*e*) 4:3:1; (*f*) 8:2:1; (*g*) 50:5:1; (*h*) 6:2:1; (*i*) 8:2:1

4. (*a*) $\frac{6}{7}$; (*b*) 12; (*c*) $\frac{13}{3}$; (*d*) $\frac{1}{4}$; (*e*) 6; (*f*) $\frac{16}{9}$ (*g*) $\frac{1}{3}$; (*h*) $\frac{3}{2}$; (*i*) $\frac{2}{7}$; (*j*) 11; (*k*) $\frac{4}{5}$; (*l*) 60; (*m*) 3; (*n*) $\frac{3}{20}$; (*o*) $\frac{1}{2}$; (*p*) 14

5. (*a*) $\frac{1}{3}$; (*b*) $3c$; (*c*) $\frac{d}{2}$; (*d*) $\frac{2r}{D}$; (*e*) $\frac{b}{a}$; (*f*) $\frac{4}{S}$; (*g*) $\frac{S}{6}$; (*h*) $\frac{3r}{2t}$; (*i*) 1:4:10; (*j*) 3:2:1; (*k*) $x^2:x:1$; (*l*) 6:5:4:1

6. (*a*) $5x$ and $4x$, sum $= 9x$; (*b*) $9x$ and x, sum $= 10x$; (*c*) $2x$, $5x$, and $11x$, sum $= 18x$; (*d*) x, $2x$, $2x$, $3x$, and $7x$, sum $= 15x$

7. (*a*) $5x + 4x = 45$, $x = 5$, 25° and 20°; (*b*) $5x + 4x = 90$, $x = 10$, 50° and 40°; (*c*) $5x + 4x = 180$, $x = 20$, 100° and 80°; (*d*) $5x + 4x + x = 180$, $x = 18$, 90° and 72°.

8. (*a*) $7x + 6x = 91$, $x = 7$, 49°, 42° and 35°; (*b*) $7x + 5x = 180$, $x = 15$, 105°, 90° and 75°; (*c*) $7x + 3x = 90$, $x = 9$, 63°, 54° and 45°; (*d*) $7x + 6x + 5x = 180$, $x = 10$, 70°, 60° and 50°

9. (*a*) 16; (*b*) 16; (*c*) ±6; (*d*) $\pm2\sqrt{5}$; (*e*) ±5; (*f*) 2; (*g*) bc/a; (*h*) $\pm6y$

10. (*a*) 21; (*b*) $4\frac{2}{3}$; (*c*) ±6; (*d*) $\pm5\sqrt{3}$; (*e*) 8; (*f*) ±4; (*g*) 3; (*h*) $\pm\sqrt{ab}$

11. (*a*) 15; (*b*) 3; (*c*) 6; (*d*) $2\frac{2}{3}$; (*e*) $3\frac{1}{3}$; (*f*) 30; (*g*) 32; (*h*) $6a$

12. (*a*) 6; (*b*) 6; (*c*) 3; (*d*) $4b$; (*e*) $\sqrt{10}$; (*f*) $\sqrt{27}$ or $3\sqrt{3}$; (*g*) $\sqrt{pq}$; (*h*) $a\sqrt{b}$

13. (a) $\dfrac{c}{b} = \dfrac{d}{x}$; (b) $\dfrac{a}{p} = \dfrac{q}{x}$; (c) $\dfrac{h}{a} = \dfrac{a}{x}$; (d) $\dfrac{3}{7} = \dfrac{1}{x}$; (e) $\dfrac{c}{a} = \dfrac{b}{x}$

14. (a) $\dfrac{x}{y} = \dfrac{1}{2}$; (b) $\dfrac{x}{y} = \dfrac{3}{4}$; (c) $\dfrac{x}{y} = \dfrac{1}{2}$; (d) $\dfrac{x}{y} = \dfrac{h}{a}$; (e) $\dfrac{x}{y} = b$

15. Only (b) is not a proportion since $3(12) \neq 5(7)$; that is, $36 \neq 35$.

16. (a) $\dfrac{x}{2} = \dfrac{9}{3}$, $x = 6$; (b) $\dfrac{x}{1} = \dfrac{4}{5}$, $x = \dfrac{4}{5}$; (c) $\dfrac{x}{a} = \dfrac{b}{2}$, $x = \dfrac{ab}{2}$; (d) $\dfrac{x}{5} = \dfrac{1}{10}$, $x = \dfrac{1}{2}$; (e) $\dfrac{x}{20} = \dfrac{5}{4}$, $x = 25$

17. (a) d; (b) 35; (c) 5 (d) 4

18. (a) 21; (b) $\frac{3}{2}$; (c) 5

19. (a) 16; (b) $6\frac{2}{3}$; (c) 10

20. (a) yes, since $\frac{15}{10} = \frac{18}{12}$; (b) no, since $\frac{10}{13} \neq \frac{7}{9}$; (c) yes, since $\dfrac{3x}{5x} = \dfrac{36}{60}$

21. (a) 12; (b) 8; (c) 60

22. (a) 15; (b) 15; (c) $6\frac{1}{2}$

24. (a) $35°$; (b) $53°$

25. (a) $a = 16$; (b) $b = 15$; (c) $c = 126$

27. (a) $\angle ABE \cong \angle EDC$, $\angle BAE \cong \angle DCE$ (also vert. $\angle$s at E)
(b) $\angle BAF \cong \angle FEC$, $\angle B \cong \angle D$ (also $\angle EAD \cong \angle BFA$)
(c) $\angle A \cong \angle EDF$, $\angle F \cong \angle BCA$
(d) $\angle A \cong \angle A$, $\angle B \cong \angle C$
(e) $\angle C \cong \angle D$, $\angle CAB \cong \angle CAD$
(f) $\angle A \cong \angle A$, $\angle C \cong \angle DBA$

28. (a) $\angle D \cong \angle B$, $\angle AED \cong \angle FGB$; (b) $\angle ADB \cong \angle ABC$, $\angle A \cong \angle A$; (c) $\angle ABC \cong \angle AED$, $\angle BAE \cong \angle EDA$

29. (a) $\angle C \cong \angle F$, $\frac{14}{20} = \frac{21}{30}$; (b) $\angle A \cong \angle A$, $\frac{10}{25} = \frac{6}{15}$; (c) $\angle B \cong \angle B$, $\frac{16}{18} = \frac{20}{35}$

30. (a) $\frac{6}{18} = \frac{8}{24} = \frac{10}{30}$; (b) $\frac{24}{36} = \frac{28}{42} = \frac{30}{45}$; (c) $\frac{12}{18} = \frac{16}{24} = \frac{18}{27}$

32. (a) $q = 20$; (b) $p = 8$; (c) $b = 7$; (d) $a = 12$; (e) $AB = 35$; (f) $d = 2\frac{1}{4}$

33. (a) 8; (b) 6; (c) $26\frac{2}{3}$

34. (a) 42 ft; (b) 66 ft

37. (a) 8:5; (b) 3:5; (c) halved (in each case)

38. (a) 15; (b) 60; (c) 25, 35, 40; (d) 4; (e) 6, 3

39. (a) 3:7; (b) 7:2; (c) quadrupled; (d) 7

43. (a) 5; (b) 14; (c) 6; (d) 5; (e) 12; (f) 13; (g) 48; (h) 2

44. 30, 18

45. (*a*) 8; (*b*) 6; (*c*) 12; (*d*) 5; (*e*) 7; (*f*) 12; (*g*) 30; (*h*) $7\frac{1}{2}$; (*i*) 5; (*j*) 8

46. (*a*) 8; (*b*) 13; (*c*) 21; (*d*) 6; (*e*) 9; (*f*) 14; (*g*) 3; (*h*) 8

47. (*a*) $a = 4, h = \sqrt{12}$ or $2\sqrt{3}$; (*b*) $c = 9, h = \sqrt{20}$ or $2\sqrt{5}$; (*c*) $q = 4$ and $b = \sqrt{80}$ or $4\sqrt{5}$; (*d*) $p = 18, h = \sqrt{108} = 6\sqrt{3}$

48. (*a*) 25; (*b*) 39; (*c*) $\sqrt{41}$; (*d*) 10; (*e*) $7\sqrt{2}$

49. (*a*) $b = 16$; (*b*) $a = 2\sqrt{7}$; (*c*) $a = 8$; (*d*) $b = 2\sqrt{3}$; (*e*) $b = 5\sqrt{2}$; $b = \sqrt{3}$

50. (*a*) 9, 12; (*b*) 10, 24; (*c*) 80, 150°; (*d*) $2\sqrt{5}, 4\sqrt{5}$

51. (*a*) 41; (*b*) $5\sqrt{5}$

52. (*a*) 12; (*b*) $10\sqrt{2}$; (*c*) $5\sqrt{5}$

53. All except (*h*)

54. (*a*) yes; (*b*) no, since $(2x)^2 + (3x)^2 \neq (4x)^2$

55. (*a*) 8; (*b*) 6; (*c*) $\sqrt{19}$; (*d*) $5\sqrt{3}$

56. (*a*) 15; (*b*) $2\sqrt{5}$; (*c*) 6

57. (*a*) 16; (*b*) 30; (*c*) $4\sqrt{3}$; (*d*) 10

58. (*a*) 10; (*b*) 12; (*c*) 28; (*d*) 15

59. (*a*) 5; (*b*) 20; (*c*) 15; (*d*) 25

60. (*a*) 12; (*b*) 24

61. 12

62. 30

63. (*a*) 10 and $10\sqrt{3}$; (*b*) $7\sqrt{3}$ and 14; (*c*) 5 and 10

64. (*a*) $11\sqrt{3}$; (*b*) $a\sqrt{3}$; (*c*) 48; (*d*) $16\sqrt{3}$

65. (*a*) 25 and $25\sqrt{3}$; (*b*) 35 and $35\sqrt{3}$

66. (*a*) 28, $8\sqrt{3}$; (*b*) 17, $14\sqrt{3}$

67. (*a*) $17\sqrt{2}$; (*b*) $a\sqrt{2}$; (*c*) $34\sqrt{2}$; (*d*) 30

68. (*a*) $20\sqrt{2}$; (*b*) $40\sqrt{2}$

69. (*a*) 45, $13\sqrt{2}$; (*b*) 11, $27\sqrt{2}$; (*c*) $15\sqrt{2}$, 55

70. $6\sqrt{2}, 5\sqrt{2}$

CHAPTER 10

1. (a) 99 in^2; (b) 3 ft^2 or 432 in^2; (c) 500; (d) 120; (e) $36\sqrt{3}$; (f) $100\sqrt{3}$; (g) 300; (h) 150

2. (a) 48; (b) 432; (c) $25\sqrt{3}$; (d) 240

3. (a) 7 and 4; (b) 12 and 6; (c) 9 and 6; (d) 6 and 2; (e) 10 and 7; (f) 20 and 8

4. (a) 1296 in^2; (b) $30\frac{1}{4}$ yd^2; (c) 100 square decimeters (100 dm^2)

5. (a) 225; (b) $12\frac{1}{4}$; (c) 3.24; (d) $64a^2$; (e) 121; (f) $6\frac{1}{4}$; (g) $9b^2$; (h) 32; (i) $40\frac{1}{2}$; (j) 64

6. (a) 128; (b) 72; (c) 100; (d) 49; (e) 400

7. (a) 1600; (b) 400; (c) 100

8. (a) 9; (b) 36; (c) $9\sqrt{2}$; (d) $4\frac{1}{2}$; (e) $\frac{9}{2}\sqrt{2}$

9. (a) $2\frac{1}{2}$; (b) 52; (c) 10; (d) $5\sqrt{2}$; (e) 6; (f) 4

10. (a) 16 ft^2; (b) 6 ft^2 or 864 in^2; (c) 70; (d) 1.62 m^2

11. (a) $3x^2$; (b) $x^2 + 3x$; (c) $x^2 - 25$; (d) $12x^2 + 11x + 2$

12. (a) 36; (b) 15; (c) 16

13. (a) $2\frac{2}{3}$; (b) 20; (c) 9; (d) 3; (e) 15; (f) 12; (g) 8; (h) 7

14. (a) 11 in^2; (b) 3 ft^2 or 1 yd^2; (c) $4x - 28$; (d) $10x^2$; (e) $2x^2 + 18x$; (f) $\frac{1}{2}(x^2 - 16)$; (g) $x^2 - 9$

15. (a) 84; (b) 48; (c) 30; (d) 120; (e) 148; (f) 423; (g) $8\sqrt{3}$; (h) 9

16. (a) 24; (b) 2; (c) 4

17. (a) 8; (b) 10; (c) 8; (d) 18; (e) $9\frac{3}{5}$; (f) $12\frac{1}{2}$; (g) 12; (h) 18

18. (a) $25\sqrt{3}$; (b) $36\sqrt{3}$; (c) $12\sqrt{3}$; (d) $25\sqrt{3}$; (e) $b^2\sqrt{3}$; (f) $4x^2\sqrt{3}$; (g) $3r^2\sqrt{3}$

19. (a) $2\sqrt{3}$; (b) $\frac{49}{2}\sqrt{3}$; (c) $24\sqrt{3}$; (d) $18\sqrt{3}$

20. (a) $24\sqrt{3}$; (b) $54\sqrt{3}$; (c) $150\sqrt{3}$

21. (a) 15; (b) 8; (c) 12; (d) 5

22. (a) 140; (b) 69; (c) 225; (d) $60\sqrt{2}$; (e) 94

23. (a) 150; (b) 204; (c) 39; (d) $64\sqrt{3}$; (e) 160

24. (a) 4; (b) 7; (c) 18 and 9; (d) 9 and 6; (e) 10 and 5

25. (a) 17 and 9; (b) 23 and 13; (c) 17 and 11; (d) 5; (e) 13

26. (a) 36; (b) $38\frac{1}{2}$; (c) $12\sqrt{3}$; (d) $12x^2$; (e) 120; (f) 96; (g) 18; (h) $\frac{49}{2}\sqrt{2}$; (i) $32\sqrt{3}$; $98\sqrt{3}$

27. (*a*) 737; (*b*) 14; (*c*) 77

28. (*a*) 10; (*b*) 12 and 9; (*c*) 20 and 10; (*d*) 5; (*e*) $\sqrt{10}$

29. 12

34. (*a*) 1:49; (*b*) 49:4; (*c*) 1:3; (*d*) 1:25; (*e*) $81:x^2$; (*f*) 9:*x*; (*g*) 1:2

35. (*a*) 49:100; (*b*) 4:9; (*c*) 25:36; (*d*) 1:9; (*e*) 9:4; (*f*) 1:2

36. (*a*) 10:1; (*b*) 1:7; (*c*) 20:9; (*d*) 5:11; (*e*) 2:*y*; (*f*) 3*x*:1; (*g*) $\sqrt{3}:2$; (*h*) $1:\sqrt{2}$; (*i*) $x:\sqrt{5}$; (*j*) $\sqrt{x}:4$

37. (*a*) 6:5; (*b*) 3:7; (*c*) $\sqrt{3}:1$; (*d*) $\sqrt{5}:2$; (*e*) $\sqrt{3}:3$ or $1:\sqrt{3}$

38. (*a*) 100; (*b*) $12\frac{1}{2}$; (*c*) 12; (*d*) 100; (*e*) 105; (*f*) 18; (*g*) $20\sqrt{3}$

39. (*a*) 12; (*b*) 63; (*c*) 48; (*d*) $2\frac{1}{2}$; (*e*) 45

40. (*a*) 10; (*b*) 12; (*c*) 22

41. (*a*) 5; (*b*) 13; (*c*) $7\frac{1}{2}$

42. (*a*) 6; (*b*) 10; (*c*) 1.2

CHAPTER 11

1. (*a*) 200; (*b*) 24.5; (*c*) 112; (*d*) 13; (*e*) 9; (*f*) $3\frac{1}{3}$; (*g*) 4.5

2. (*a*) $12\frac{1}{2}$; (*b*) 23.47; (*c*) $7\sqrt{3}$; (*d*) 18.5; (*e*) $3\sqrt{2}$

3. (*a*) 24°; (*b*) 24°; (*c*) 156°

4. (*a*) 40°; (*b*) 9; (*c*) 140°

5. (*a*) 15°; (*b*) 15°; (*c*) 24

6. (*a*) 5°; (*b*) 72°; (*c*) 175°

7. (*a*) Regular octagon; (*b*) regular hexagon; (*c*) equilateral triangle; (*d*) regular decagon; (*e*) square; (*f*) regular dodecagon (12 sides)

9. (*a*) 9; (*b*) 30; (*c*) $6\sqrt{3}$; (*d*) 6; (*e*) $13\sqrt{3}$; (*f*) 6; (*g*) $20\sqrt{3}$; (*h*) 60

10. (*a*) $18\sqrt{2}$; (*b*) $7\sqrt{2}$; (*c*) 40; (*d*) $8\sqrt{2}$; (*e*) 3.4; (*f*) 28; (*g*) $5\sqrt{2}$; (*h*) $2\sqrt{2}$

11. (*a*) $30\sqrt{3}$; (*b*) 14; (*c*) 27; (*d*) 18; (*e*) $8\sqrt{3}$; (*f*) $4\sqrt{3}$; (*g*) $48\sqrt{3}$; (*h*) 42; (*i*) 6; (*j*) 10; (*k*) $\frac{5}{2}\sqrt{3}$; (*l*) $3\sqrt{3}$

12. (*a*) 817; (*b*) 3078

13. (*a*) $54\sqrt{3}$; (*b*) $96\sqrt{3}$; (*c*) $600\sqrt{3}$

14. (*a*) 576; (*b*) 324; (*c*) 100

15. (a) $36\sqrt{3}$; (b) $27\sqrt{3}$; (c) $\frac{16}{3}\sqrt{3}$; (d) $144\sqrt{3}$; (e) $3\sqrt{3}$; (f) $48\sqrt{3}$

16. (a) 10; (b) 10; (c) $5\sqrt{3}$

17. (a) 18; (b) $9\sqrt{3}$; (c) $6\sqrt{3}$; (d) $3\sqrt{3}$

18. (a) 1:8; (b) 4:9; (c) 9:10; (d) 8:11; (e) 3:1; (f) 2:5; (g) $4\sqrt{2}$:3; (h) 5:2

19. (a) 5:2; (b) 1:5; (c) 1:3; (d) 3:4; (e) 5:1

20. (a) 5:1; (b) 4:7; (c) x:2; (d) $\sqrt{2}$:1; (e) $\sqrt{3}$:y; (f) $\sqrt{x}$:$3\sqrt{2}$ or $\sqrt{2x}$: 6

21. (a) 1:4; (b) 1:25; (e) 36:1; (d) 9:100; (e) 49:25

22. (a) 12π; (b) 14π; (c) 10π; (d) $2\pi\sqrt{3}$

23. (a) 9π; (b) 25π; (c) 64π; (d) $\frac{1}{4}\pi$; (e) 18π

24. (a) $C = 10\pi$, $A = 25\pi$; (b) $r = 8$, $A = 64\pi$; (c) $r = 4$, $C = 8\pi$

25. (a) 12π; (b) 4π; (c) 7π; (d) 26π; (e) $8\pi\sqrt{3}$; (f) 3π

26. (a) 98π; (b) 18π; (c) 32π; (d) 25π; (e) 72π; (f) 100π

27. (a) (1) $C = 8\pi$, $A = 16\pi$, (2) $C = 4\sqrt{3}\pi$, $A = 12\pi$
 (b) (1) $C = 16\pi$, $A = 64\pi$, (2) $C = 8\sqrt{3}\pi$, $A = 48\pi$
 (c) (1) $C = 12\pi$, $A = 36\pi$, (2) $C = 6\pi$, $A = 9\pi$
 (d) (1) $C = 16\pi$, $A = 64\pi$, (2) $C = 8\pi$, $A = 16\pi$
 (e) (1) $C = 20\sqrt{2}\pi$, $A = 200\pi$, (2) $C = 20\pi$, $A = 100\pi$
 (f) (1) $C = 6\sqrt{2}\pi$, $A = 18\pi$, (2) $C = 6\pi$, $A = 9\pi$

28. (a) 10 ft; (b) 17 ft; (c) $3\sqrt{5}$ ft or 6.7 ft

29. (a) 2π; (b) 10π; (c) 8; (d) 11π; (e) 6π; (f) 10π

30. (a) 3π; (b) $12\frac{1}{2}$; (c) 5π; (d) 2π; (e) π; (f) 4π

31. (a) 6π; (b) $\pi/6$; (c) $25\pi/6$; (d) 25π; (e) $4\frac{1}{2}$; (f) 13; (g) 24π; (h) $8\pi/3$

32. (a) 6π; (b) 20; (c) 3π; (d) 16π

33. (a) 120°; (b) 240°; (c) 36°; (d) 180°; (e) 135°; (f) $(180/\pi)°$ or 57.3° to nearest tenth

34. (a) 72°; (b) 270°; (c) 40°; (d) 150°; (e) 320°

35. (a) 90°; (b) 270°; (c) 45°; (d) 36°

36. (a) 12; (b) 9; (c) 10; (d) 6; (e) 5; (f) $3\sqrt{2}$

37. (a) 4; (b) 10; (c) 10 cm; (d) 9

38. (a) $6\pi - 9\sqrt{3}$; (b) $24\pi - 36\sqrt{3}$; (c) $\frac{3}{2}\pi - \frac{9}{4}\sqrt{3}$; (d) $\dfrac{\pi r^2}{6} - \dfrac{r^2\sqrt{3}}{4}$; (e) $\dfrac{2\pi r^2}{3} - r^2\sqrt{3}$

39. (a) $4\pi - 8$; (b) $150\pi - 225\sqrt{3}$; (c) $24\pi - 36\sqrt{3}$; (d) $16\pi - 32$; (e) $50\pi - 100$

40. (a) $\dfrac{64\pi}{3} - 16\sqrt{3}$; (b) $24\pi - 16\sqrt{2}$; (c) $\dfrac{80\pi}{3} - 16$

41. (a) $\dfrac{16\pi}{3} - 4\sqrt{3}$; (b) $\dfrac{8\pi}{3} - 4\sqrt{3}$; (c) $4\pi - 8$

42. (a) $12\pi - 9\sqrt{3}$; (b) $\frac{3}{2}\pi - \frac{9}{4}\sqrt{3}$; (c) $9\pi - 18$

43. (a) $200 - 25\pi/2$; (b) $48 + 26\pi$; (c) $25\sqrt{3} - 25\pi/2$; (d) $100\pi - 96$; (e) $128 - 32\pi$; (f) $300\pi + 400$; (g) 39π; (h) 100

44. (a) 36π; (b) $36\sqrt{3} + 18\pi$; (c) 14π

CHAPTER 12

1. The description of each locus is left for the reader.

(a)   (b)   (c) 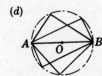 (d)

(e) 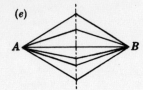 (f) 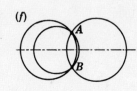 (g) 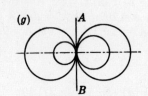 (h)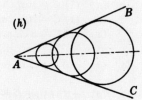

2. The diagrams are left for the reader.
  (a) The line parallel to the banks and midway between them
  (b) The perpendicular bisector of the segment joining the two floats
  (c) The bisector of the angle between the roads
  (d) The pair of bisectors of the angles between the roads

3. The diagrams are left for the reader.
  (a) A circle having the sun as its center and the fixed distance as its radius
  (b) A circle concentric to the coast, outside it, and at the fixed distance from it
  (c) A pair of parallel lines on either side of the row and 20 ft from it
  (d) A circle having the center of the clock as its center and the length of the clock hand as its radius.

4. (a) $\overline{EF}$; (b) $\overline{GH}$; (c) $\overline{EF}$; (d) $\overline{GH}$; (e) $\overline{EF}$; (f) $\overline{GH}$; (g) $\overline{AB}$; (h) a 90° arc from A to G with B as center

5. (a) $\overline{AC}$; (b) $\overline{BD}$; (c) $\overline{BD}$; (d) $\overline{AC}$; (e) E

6. In each case, the letter refers to the circumference of the circle. (a) A; (b) C; (c) B; (d) A; (e) C; (f) A and C; (g) B

7. The description of each locus is left for the reader.

(a) 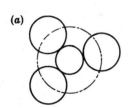 (b) 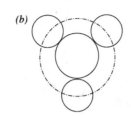 (c) 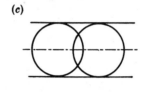 (d)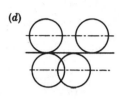

8. (a) $\overline{EF}$; (b) $\overline{GH}$; (c) line parallel to $\overline{AD}$ and $\overline{EF}$ midway between them; (d) $\overline{EF}$; (e) $\overline{BC}$; (f) $\overline{GH}$

(a) 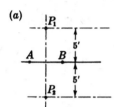 (b) 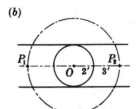 (c)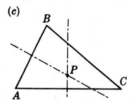

(d) 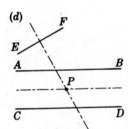 (e) 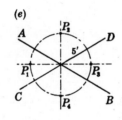 (f)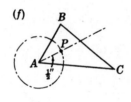

9. The explanation is left for the reader.

10. (a) The intersection of two of the angle bisectors
   (b) The intersection of two of the ⊥ bisectors of the sides
   (c) The intersection of the ⊥ bisector of $\overline{AB}$ and the bisector of $\angle B$
   (d) The intersection of the bisector of $\angle C$ and a circle with C as center and 5 as radius
   (e) The intersections of two circles, one with B as center and 5 as radius and the other with A as center and 10 as radius

11. (a) 1; (b) 1; (c) 4; (d) 2; (e) 2; (f) 1

13. (a) $x = -5$; (b) $y = 3\frac{1}{2}$; (c) $y = 3$ and $y = -3$; (d) $y = -5$; (e) $x = 4$ and $x = -4$; (f) $x = 5$ and $x = -1$;
   (g) $y = 4$; (h) $x = 1$; (i) $x = 9$

14. (a) $x = 6$; (b) $y = 5$; (c) $x = 6$; (d) $x = 5$; (e) $x = 6$; (f) $y = 3$

15. (a) $x = y$; (b) $y = x + 5$; (c) $x = y - 4$; (d) $y - x = 10$; (e) $x + y = 12$; (f) $x - y = 2$ or $y - x = 2$;
   (g) $x = y$ and $x = -y$; (h) $x + y = 5$

16. (a) line having y-intercept 5, slope 2; (b) line passing through (2, 3), slope 4;
   (c) line passing through (−2, −3), slope $\frac{5}{4}$; (d) line passing through origin, slope $\frac{1}{2}$;
   (e) line having y-intercept 7, slope −1; (f) line passing through origin, slope $\frac{1}{3}$

17. (a) $y = 4x$; (b) $y = -2x$; (c) $y = \frac{3}{2}x$ or $2y = 3x$; (d) $y = -\frac{2}{3}x$ or $5y = -2x$; (e) $y = 0$

18. (a) $y = 4x + 5$; (b) $y = -3x + 2$; (c) $y = \frac{1}{3}x - 1$ or $3y = x - 3$; (d) $y = 3x + 8$; (e) $y = -4x - 3$;
(f) $y = 2x$ or $y - 2x = 0$

19. (a) $\dfrac{y-4}{x-1} = 2$ or $y = 2x + 2$; (b) $\dfrac{y-3}{x+2} = 2$ or $y = 2x + 7$; (c) $\dfrac{y}{x+4} = 2$ or $y = 2x + 8$;

(d) $\dfrac{y+7}{x} = 2$ or $y = 2x - 7$

20. (a) $y = 4x$; (b) $y = \frac{1}{2}x + 3$; (c) $\dfrac{y-2}{x-1} = 3$; (d) $\dfrac{y+2}{x+1} = \dfrac{1}{3}$; (e) $y = 2x$

21. (a) circle with center at origin and radius 7; (b) $x^2 + y^2 = 16$; (c) $x^2 + y^2 = 64$ and $x^2 + y^2 = 4$

22. (a) $x^2 + y^2 = 25$; (b) $x^2 + y^2 = 81$; (c) $x^2 + y^2 = 4$ or $x^2 + y^2 = 144$

23. (a) 3; (b) $\frac{4}{3}$; (c) 2, (d) $\sqrt{3}$

24. (a) $x^2 + y^2 = 16$; (b) $x^2 + y^2 = 121$; (c) $x^2 + y^2 = \frac{4}{9}$ or $9x^2 + 9y^2 = 4$; (d) $x^2 + y^2 = \frac{9}{4}$ or $4x^2 + 4y^2 = 9$;
(e) $x^2 + y^2 = 5$; (f) $x^2 + y^2 = \frac{3}{4}$ or $4x^2 + 4y^2 = 3$

CHAPTER 13

1. (a) <; (b) >; (c) >; (d) >; (e) >; (f) <

2. (a) >; (b) >; (c) <; (d) >

3. (a) >; (b) <; (c) <; (d) >

4. (a) more; (b) less

5. (a) >; (b) >; (c) <; (d) >; (e) <; (f) <

6. (c), (d), and (e)

7. (a) 5 to 7; (b) 6 to 10; (c) 4 to 10; (d) 3 to 9; (e) 2 to 8; (f) 1 to 13

8. (a) $\angle B$, $\angle A$, $\angle C$; (b) $\overline{DF}$, $\overline{EF}$, $\overline{DE}$; (c) $\angle 3$, $\angle 2$, $\angle 1$

9. (a) $m\angle BAC > m\angle ACD$; (b) $AB > BC$

10. (a) $\overline{BC}$, $\overline{AB}$, $\overline{AC}$; (b) $\angle BOC$, $\angle AOB$, $\angle AOC$; (c) $\overline{AD}$, $\overline{AB} \cong \overline{CD}$, $\overline{BC}$; (d) $\overline{OG}$, $\overline{OH}$, $\overline{OJ}$

CHAPTER 14

1. (a) Ornament, jewelry, ring, wedding ring; (b) vehicle, automobile, commercial automobile, truck;
(c) polygon, quadrilateral, parallelogram, rhombus;
(d) angle, obtuse angle, obtuse triangle, isosceles obtuse triangle.

2. (a) A regular polygon is an equilateral and an equiangular polygon.
(b) An isosceles triangle is a triangle having at least two congruent sides.
(c) A pentagon is a polygon having five sides.
(d) A rectangle is a parallelogram having one right angle.

(e)   An inscribed angle is an angle formed by two chords and having its vertex on the circumference of the circle.

(f)   A parallelogram is a quadrilateral whose opposite sides are parallel.

(g)   An obtuse angle is an angle larger than a right angle and less than a straight angle.

3.   (a) $x + 2 \neq 4$; (b) $3y = 15$; (c) She does not love you. (d) His mark was not more than 65.
     (e) Joe is not heavier than Dick. (f) $a + b = c$

4.   (a)   A nonsquare does not have congruent diagonals. False (for example, when applied to a rectangle or a regular pentagon).

     (b)   A non-equiangular triangle is not equilateral. True.

     (c)   A person who is not a bachelor is a married person. This inverse is false when applied to an unmarried female.

     (d)   A number that is not zero is a positive number. This inverse is false when applied to negative numbers.

5.   (a)   Converse true, inverse true, contrapositive true
     (b)   Converse false, inverse false, contrapositive true
     (c)   Converse true, inverse true, contrapositive true
     (d)   Converse false, inverse false, contrapositive true

6.   (a)   Partial converses: interchange (2) and (3) or (1) and (3).
           Partial inverses: negate (1) and (3) or (2) and (3).
     (b)   Partial converses: interchange (1) and (4) or (2) and (4) or (3) and (4).
           Partial inverses: negate (1) and (4) or (2) and (4) or (3) and (4).

7.   (a) Necessary and sufficient; (b) Necessary but not sufficient; (c) Neither necessary nor sufficient;
     (d) Sufficient but not necessary; (e) Necessary and sufficient; (f) Sufficient but not necessary;
     (g) Necessary but not sufficient

CHAPTER 17

1.   (a) C; (b) D; (c) B; (d) $\overline{AD}$

2.   Rectangle $A'C'B'D'$, where $A' = A$, $C' = D$, $B' = B$, and $D' = C$

3.   True

4. $\overline{AB}$, $\overline{CD}$, $\overline{EF}$ (where E = midpoint of $\overline{AD}$, F = midpoint of $\overline{CB}$), and $\overline{GH}$ (where G = midpoint of $\overline{AC}$, H = midpoint of $\overline{DB}$)

6.   Each has an axis of symmetry.

7.   (a) $B' = (-2, 0)$; (b) $A' = (-2, -2)$; (c) $(0, 0)$; (d) $\triangle A'OB'$, where A' and B' are $(-2, -2)$ and $(-2, 0)$

8.   (a) E; (b) C; (c) $\overline{DC}$; (d) $\overline{B'C'}$, where $B' = (-1, 1)$ and $C' = (1, -1)$;
     (e) $A'B'C'D'$, where $A' = D$, $B' = C$, $C' = B$, $D' = A$

9.   (a) $(3, 11)$; (b) $(7, 8)$; (c) $(1, 3)$; (d) $(2, 4)$

10. $A' = (2, 6)$, $B' = (2, 7)$, $C' = (4, 7)$, $D' = (4, 6)$

11. $h = -3$, $k = -4$; $T(-8, -6) = (-11, -10)$

12.   (a) $(4, -3)$; (b) $(3+h, 7+k)$; (c) $(e+h, f+k)$; (d) $(5, 2)$

13.   (a) $\overline{EF}$; (b) $E' = (1, 1)$, $F' = (2, 0)$; (c) $E' = (0, 2)$, $F' = (1, 1)$; (d) $O' = O$, $E' = E$, $F' = F$

14.   (a) $(-1, 4)$; (b) $(-3, 0)$; (c) $\overline{y'x'}$, where $y' = (-3, 0)$ and $x' = (-1, 4)$

15. $O' = O$, $E' = (0, -1)$, $F' = (-1, 0)$; $\triangle E'O'F'$ is the image

16.   (a) $(0, 1)$ and $(-1, 1)$; (b) $(0, -1)$ and $(1, 0)$;
       (c) $O'A'B'C'$, where $O' = O$, $A' = (0, 1)$, $B' = (-1, 1)$, and $C' = (-1, 0)$

17.   (a) $\left(-\frac{1}{3}, 1\right)$; (b) $\left(\frac{5}{2}, -\frac{3}{2}\right)$; (c) $(0, 0)$; (d) $(5, 30)$

18. $n = \frac{5}{3}$; $\left(0, -\frac{35}{3}\right)$

19.   (a) $O' = O$, $A' = \left(0, \frac{1}{2}\right)$, $B' = \left(\frac{1}{2}, 0\right)$; images is $\triangle OA'B'$; (b) midpoint $= \left(\frac{1}{2}, \frac{1}{2}\right)$; image is $\left(\frac{3}{2}, \frac{3}{2}\right)$

INDEX

Abscissa, 46
Acute angles, 69, 127
Acute triangle, 74
Addends, 3, 4
Addition:
 Associative Law of, 9
 Commutative Law of, 3
 expressing algebraically, 6
Addition rule, 25, 33, 34
Additive identity, 2
Adjacent angles, 76, 77, 78
Algebra:
 applying to trapezoids, 144
 and arithmetic, relating fundamentals, 1–3
Algebraic equations, 2
Algebraic expression, 6, 8
 containing parentheses, 11, 12
 evaluating, 11
 not containing parentheses, 10
Algebraic postulates, 85–86
Alternate interior angles, 115
Altitude, 203
Analytic geometry:
 areas in, 234
 locus in, 265
 and reflections, 322
Angle bisector, 74
Angle-measure sum:
 applying ratios, 128–129
 equilateral triangle, 129
 isosceles triangle, 129
 numerical applications, 128
 polygons, applying algebra to, 133–134
 principles, 126–128
 using algebra to prove, 130
 See also Sum of measures of angles
Angle measurement formulas, 133
Angles, 68–72
 acute, 69, 127
 adding and subtracting, 71
 base, 143, 213
 basic theorems, 90–92
 central, 67, 161, 171, 242
 combining, 293
 complementary, 76, 77
 congruent, 70, 91, 128, 172, 196, 292, 308, 313–314
 consecutive, 146
 constructing, 298

Angles (Cont.):
 duplicating, 292–293
 exterior, 114, 126, 127
 finding, 72
 finding parts, 71
 formulas, 335
 inscribed, 172, 173, 310–311
 interior, 114, 115, 126, 127, 131
 measuring, 69, 178–179
  in a circle, 171
  principles, 172–174
 naming, 68, 71
 obtuse, 127
 reflex, 70
 right, 69, 91, 173
 straight, 91
 sum of measures in a triangle, 307–308
 supplementary, 78, 173
 of triangle, sum of measures, 125–130, 307–308
 types, 69
 vertical, 76, 77, 91
 See also Pairs of angles
Apothem, 242, 317
Arcs, 67, 68, 161, 177, 309
 intercept, 161
 length of, 249
 measuring, 171, 178
 minor, 161
Area, 225–241
 in analytic geometry, 234
 of circle, 247–248
 of closed plane figure, 225
 of combination figure, 251
 equal-areas problem, 232
 of equilateral triangle, 228–229
 formulas, 336
 of parallelogram, 227, 231, 315–316
 of polygon, 231
 of quadrilateral, 234
 of rectangle, 225–226
 of regular polygon, 246, 317
 of rhombus, 230–231
 of sector, 249–250
 of segment, 249–251
 of similar polygon, 233
 of square, 225–226
 of trapezoid, 229–230, 235, 316–317
 of triangle, 227–229, 231–232, 234, 316
Arithmetic and algebra, relating fundamentals, 1–3

Associative Law:
  of Addition, 9
  for four numbers, 10
  of Multiplication, 9
  for three numbers, 10
Assumptions, 284
Axioms, 272
Axis of symmetry, 320

Base, 14, 15
Base angles, 143, 213
Bisecting, 70
Bisectors:
  constructions, 293–295
  perpendicular, 70, 74, 164, 259

Calculator, 44–45
  graphing, 338–341
Center of circle, locus of, 261–262
Center of regular polygon, 242
Central angle, 67, 161, 171
  of regular polygon, 242
Chords, 67, 161, 163, 174, 309, 311
  congruent, 164
  intersecting, 205
Circles, 67–68, 161–188
  applying principles, 165, 175–177
  area of, 247–248
  circumference of, 67, 161, 247–248
  circumscribed, 248
  concentric, 162
  congruent, 67, 162–164
  constructions, 299–301
  distance formula, 50
  equal, 162
  inequality theorems, 274–275, 276–277
  inscribed, 162, 248
  intersection formulas, 336
  line of centers, 170–171
  matching test of vocabulary, 164–165
  outside each other, 168–169
  overlapping, 168
  principles, 163–164
  problem, 166
  Pythagorean theorem, 210–211
  ratios of segments and areas in, 248
  relationships, 161–166
  segments intersecting inside and outside,
    205–207
  tangent externally, 168
  tangent internally, 168
  in varying relative positions, 168
Circumference of circle, 67, 161, 247–248
Circumscribed circle, 248
Circumscribed polygon, 162
Circumscribing regular polygons, 301–302

Coefficient, 13, 14
Coin formulas, 39
Collinear points, 52
Combination figure, area of, 251
Commutative Law:
  of Addition, 3
  of Multiplication, 4
Complementary angles, 76, 77
Concentric circle, 162
Conclusion, 92–94
Congruency:
  principles, 134
  problem, 104–105
  proving, 136–137
Congruent angles, 70, 91, 128, 172, 196, 292, 308,
    313–314
Congruent chords, 164
Congruent circles, 67, 162–164
Congruent figures, 100
Congruent polygons, 231
Congruent segments, 66
Congruent triangles, 100–113, 134, 307
  applying algebra to, 104
  basic principles, 100–101
  determining reason for congruency, 135
  finding parts needed to prove, 102–103
  methods of proving, 101
  selecting, 101–102, 135
  selecting corresponding parts, 103
Consecutive angles, 146
Constant, 1
Constructions, 291–305
  angle congruent to given angle, 292
  angles, 298
  bisectors, 293–295
  circles, 299–301
  line segment congruent to given line segment,
    292
  parallel lines, 298–299
  parallelogram, 299
  perpendiculars, 293–295
  similar triangles, 302–303
  tangents, 299–301
Continued ratio, 189, 190
Contrapositive of statement, 284–286
Converses:
  forming, 94
  partial, of theorem, 286–287
  of statement, 92–93, 284–286
Coordinate geometry, formulas, 337
Coordinates, 46, 47
Cubes, 14

Decimal coefficients, 27
Decimal divisors, 30
Deductive reasoning in geometry, 283–284

Defined terms, 282–283
Definitions, 282–283
  faulty, 283
  requirements, 282
  sequence, 283
Diameter, 67, 162
Dilations, 330
  of squares, 331
  of triangles, 330
Distance formula, 49–50
  circles, 50
  parallelograms, 50
  triangle, 49
Distances, 121–125
  applying principles, 124–125
  between two geometric figures, 121
  between two points, 49
  finding, 123
  principles of, 121–122
Division, 5, 7
  expressing algebraically, 7
  by zero, 5
Division rule, 25, 28
Domain, 1
Duplicating segments and angles, 292–293

Equal polygons, 231
Equal-products problem, 205
Equality, properties, 21
Equality rules, 25, 26
Equations, 20
  checking or verifying, 21
  matching sentences, 22
  solving, 20, 23, 25–31, 33, 36
  translating verbal problems into, 22
Equilateral triangles, 73, 105–106, 127, 129, 212
  area of, 228–229
  line relationships in, 245
Equivalent expressions, 20
Equivalent proportion, 191
Evaluation, 1
Experimentation, 84
Exponents, 14, 15
Expression, 13
Exterior angles, 114, 126, 127
Extreme of a proportion, 191

Factors, 4, 13
Figures, images of, 327
Fixed points, 318
Formulas, 1, 3
  deriving, 38–40
  reference, 335–337
  transforming, 40–42
Fourth proportionals, 190, 192
Fractional coefficients, 29, 30, 36

Geometric postulates, 86–87, 90
Geometry:
  analytic (see Analytic geometry)
  deductive reasoning in, 283–284
  historical background, 64
  transformational, 318–334
Graphing calculator, 338–341
Graphs, 46–47
Group relationships, 84

Hexagon, 302
Hypotenuse, 207, 208, 212, 314, 315
Hypothesis, 92–94

Identity, 20
If-then form, 92, 93
Images:
  from images, 326
  of figures, 327
  of points, 319–320, 325
  of triangles, 320, 326, 328–329
  under reflections, 321–324
  under translation, 326
Inclination of a line, 53
Indirect reasoning, 278
Inequalities, 272–281
Inequality axioms, 272–273, 275–276
Inequality postulate, 273
Inequality problem, 277
Inequality symbols, 275
Inequality theorems:
  circles, 274–277
  triangles, 273, 276
Inscribed angle, 172, 173, 310–311
Inscribed circle, 162, 248
Inscribed polygon, 162
Inscribed regular polygon, 251, 301–302
Interior angles, 114, 115, 126, 127, 131
Inverse:
  partial, of theorem, 286–287
  of statement, 284–286
Inverse operations, 23, 24, 34, 35, 40
Isosceles right triangle, 127, 213
Isosceles trapezoid, 143, 213
Isosceles triangles, 73, 105–108, 129
  Pythagorean theorem, 209

Length formulas, 40
Line ratios:
  similar polygons, 204
  similar triangles, 204
Line relationships:
  in equilateral triangle, 245
  in regular hexagon, 245
  in square, 245

Line segments, 65–67
    combining, 292–293
    finding lengths and points of, 67
    naming, 66
Line symmetry, 320–321
Lines, 64–65
    of centers of two circles, 168
    parallel (*see* Parallel lines)
Literal addends, 3, 4
Literal coefficient, 13, 14
Literal factors, 4
Locating a point satisfying given conditions,
        123–124
Locus, 259–271
    in analytic geometry, 265
    applying principles, 266–267
    of center of circle, 261–262
    constructing, 262
    determining, 259, 261
    fundamental theorems, 259–260
    of points, 259–261, 263
    principles of, 259–260
    proving, 263–264
Logically equivalent statements, 285

Mean of a proportion, 191
Mean proportionals, 192
    in right triangle, 207–208
Measurement, 84 (*see also* specific applications)
Median, 74, 143, 153–156, 203, 231
Midpoint, 153–154, 156
    formula, 48
    of segment, 48
Minor arc, 161
Multiplication, 5, 7
    Associative Law of, 9
    Commutative Law of, 4
    expressing algebraically, 7
Multiplication rule, 25, 29, 31
Multiplicative identity, 2
Multiplicative property:
    of one, 2
    of zero, 2

Natural number coefficients, 27
Natural number divisors, 30
Natural numbers, 31, 33
Necessary conditions, 287–288
Negative of statement, 284, 286
Negative slope, 51
Number scale, 46
Numerical addends, 3
Numerical coefficient, 13, 14
Numerical expressions:
    containing parentheses, 12
    evaluating, 11

Numerical expressions (*Cont.*):
    not containing parentheses, 10
Numerical factors, 4

Observation, 84
Obtuse angle, 69, 127
Obtuse triangle, 74, 75
Octagon, 302
Ordinate, 46
Origin, 46

Pairs of angles, 76–79
    finding, 79
    formed by two lines cut by transversal, 114
    naming, 78
    principles, 77
    using two unknowns, 79
Parallel lines, 114–120
    algebraic applications, 119–120
    applying principles and their converses, 119
    constructing, 298–299
    numerical applications, 118
    principles of, 116–118
    problems, 120
    properties of, 117–118
Parallelograms, 145–151
    applying principles, 147–148
    applying properties, 147
    area of, 227, 231, 315–316
    constructing, 299
    determining, 147
    distance formula, 50
    principles involving properties of, 145–146
    problems, 148
    quadrilaterals as, 146
    special, 148–154
Parallels, three or more, 152–156
Parentheses, 8
    algebraic expression containing, 11, 12
    algebraic expression not containing, 10
    numerical expression containing, 12
    numerical expression not containing, 10
Partial converse of theorem, 286–287
Partial inverse of theorem, 286–287
Patterns in reflections, 324
Pentagon, 72
Per cents as coefficients, 28, 31
Perimeter, 203, 317
Perpendicular, 70
    constructions, 293–295
Perpendicular bisector, 70, 74, 164, 259
pi (π), 247
Plane, 65
Point symmetry, 321, 322
Points, 64
    images of, 319–320, 325

Points (*Cont.*):
   locus of, 259–261, 263
   naming, 66
   rotations of, 328
Polygons, 72
   angle principles, 132–133
   applying angle-measure formulas, 133
   area of, 231
   circumscribed, 162
   congruent, 231
   equal, 231
   inscribed, 162
   interior and exterior angles, 126
   names, 131
   of same size or shape, 231
   similar (*see* Similar polygons)
   sum of measures
      of angles, 130–134
      of exterior angles, 132
      of interior angles, 131
Positive slope, 51
Postulates, 85–90
   algebraic, 85–86
   geometric, 86–87, 90
Powers, 14
   reading, 14
   of variable bases, 15
Problem solving, 28, 31, 32, 34
   using two operations, 37
Proof:
   by deductive reasoning, 83–84
   methods of, 83–91
   theorems, 94–95, 306–317
Proportional, mean, 192
Proportional segments, 193–196
   applying principles, 195–196
   principles, 194
   problem, 196
Proportions, 190–193
   changing equal products into, 192
   changing into new proportions, 193
   combining numerators and denominators, 193
   eight arrangements, 194
   equivalent, 191
   extreme, 191
   mean, 191
   principles, 191
   problem, 202
   similar polygons, 201–203, 233
   similar triangles, 201, 203
   unknowns in, 192
Pythagorean Theorem, 208
   circle, 210–211
   isosceles triangle, 209
   rhombus, 210
   trapezoid, 210

Quadrant, 46
Quadrilaterals:
   area of, 234
   graphing, 47
   as parallelograms, 146
   sum of angles, 126, 133–134
   tangent sides, 169

Radius:
   of circle, 67, 161
   of regular polygon, 242
Ratios, 189–190
   algebraic, 190
   angle problems, 190
   continued, 189, 190
   of lines and areas of regular polygons, 247
   numerical, 190
   right triangle, 209
   of segments and areas of circles, 248
   of segments and areas of regular polygons, 246–247
   of similitude, 203
   three angles having fixed ratio, 190
   two quantities with different units, 189
   two quantities with same unit, 189
Reasoning:
   deductive (*see* Deductive reasoning)
   improvement of, 282–290
   indirect, 278
Rectangle, 149
   area of, 225–226
Reflections, 318–319
   and analytic geometry, 322
   images under, 321–324
   patterns in, 324
Reflex angle, 70
Reflexive property, 21
Regular hexagons, line relationships in, 245
Regular polygons, 132, 242–247
   applying angle-measure formulas, 133
   area of, 246, 317
   circumscribing, 301–302
   inscribing, 251, 301–302
   measures of lines and angles, 243–244
   principles, 242–243
   problem, 244
   ratios of lines and areas of, 247
   ratios of segments and areas of, 246–247
   segments of, 245, 246
Replacement set, 1, 3
Rhombus, 149
   applying algebra to, 151
   area of, 230–231
   Pythagorean Theorem, 210
Right angles, 69, 91, 173
Right triangles, 73, 127, 308–309, 314, 315

Right triangles (*Cont.*):
  formulas, 336–337
  isosceles, 127, 213
  mean proportionals in, 207–208
  principles of 30°-60°-90° triangle, 211
  principles of 45°-45°-90° triangle, 212
  ratios, 209
  special, 211–213
Root, 21
Rotation, 71, 327
  of points, 328
  symmetry of, 329

Scalene triangle, 73
Secants, 162, 174, 312
Sector, area of, 249–250
Segments, 203
  area of, 249–251
  duplicating, 292–293
  intersecting inside and outside a circle, 205–207
  midpoint of, 48
  proportional, 193–196
  proving equal products of lengths of, 205
  of regular polygons, 245, 246
Semicircle, 67
Similar polygons, 196, 231
  line ratios, 204
  proportions, 203, 233
Similar triangles, 196–202, 313–314
  applying principles, 199–201
  constructing, 302–303
  finding heights using ground shadows, 202
  line ratios, 204
  principles, 197–198
  problem, 202
  proportion, 201–203, 233
  proving a proportion, 197
  ratio of areas and segments of, 233
Similarity, 189–224
Slope:
  of a line, 51–54
  of parallel and perpendicular lines, 51, 53
Solution set, 20
Square unit, 225
Squares, 14, 149
  area of, 225–226
  dilations of, 331
  line relationships in, 245
Statement:
  contrapositive of, 284–285, 286
  converse of, 92–93, 284–286
  forms, 92
  inverse of, 284–286
  logically equivalent, 285
  negative of, 284, 286
Straight angles, 70, 91

Subject-predicate form, 92, 93
Substitution rule, 20
Subtraction, expressing algebraically, 6
Subtraction rule, 25, 31, 32
Sufficient conditions, 287–288
Sum of measures of angles:
  exterior angles of polygons, 132
  of polygon, 130–134
  of quadrilateral, 126
  of triangle, principles, 126
  of triangles, 125–130, 307–308
  See also Angle-measure sum
Supplementary angles, 78, 173
Syllogism, 85
Symbols, 5
Symmetry:
  axis of, 320
  line, 320–321
  point, 321, 322
  property, 21
  rotational, 329

Tangents, 162, 167–169, 312, 313
  applying principles, 169–170
  constructions, 299–301
  principles, 167
  problem, 171
Term, 13
Theorems, 90, 284
  partial converse of, 286–287
  partial inverse of, 286–287
  proving, 94–95, 306–317
Time formulas, 40
Transformational geometry, 318–334
Transformations, 318
  properties of, 331
  requiring addition or subtraction, 41
  requiring division, 41
  requiring multiplication, 41
  requiring two operations, 41
Transitive property, 21
Translations, 324–325
  images under, 326
Transversal, 114
Trapezoids, 143–144, 153–155
  applying algebra to, 144
  area of, 229–230, 235, 316–317
  isosceles, 213
  principles, 144
  proof of principles, 144
  Pythagorean Theorem, 210
Triangles, 72–76, 153–156
  acute, 74
  altitude, 74–75
  area of, 227–229, 231–232, 234, 316
  classifying, 73–74

Triangles (*Cont.*):
  congruent (*see* Congruent triangles)
  constructing, 295–298
  dilations of, 330
  distance formula, 49
  equilateral (*see* Equilateral triangles)
  graphing, 47
  images of, 320, 326, 328–329
  inequality theorems, 273, 276
  isosceles (*see* Isosceles triangles)
  naming, 75
  obtuse, 74, 75
  right (*see* Right triangles)
  scalene, 73
  similar (*see* Similar triangles)
  special lines, 74, 75–76
  sum of angles, 125–130, 307–308
  tangent sides, 169
  tests for, 208
Truth, determining, 94
Truth set, 20

Unconditional equation, 20

Undefined terms, 64–65, 283
Unknowns, 44
  finding, 331
  in proportions, 192
  representing, 22

Values, 1
Variable expression, 1, 3
Variables, 1, 20, 42–43
Vertex, 73
Vertical angles, 76, 77, 91

x-axis, 46
x-coordinate, 46

y-axis, 46
y-coordinate, 46

Zero:
  division by, 5
  multiplicative property of, 2
  one variable is, 12